TOUCHSTONES

Literature and the Writing Process

Essays / Poems / Short Stories

Asalean Springfield and Gloria C. Johnson

SIMON & SCHUSTER CUSTOM PUBLISHING

Cover art by Dr. Brenda Chappell.

Printed in the United States of America

10 9 8 7 6 5 4 3 2 1

Please visit our Web site at www.sscp.com

ISBN 0-536-01208-3

BA 98129

SIMON & SCHUSTER CUSTOM PUBLISHING
160 Gould Street/Needham Heights, MA 02194
Simon & Schuster Education Group

Copyright Acknowledgments

Grateful acknowledgment is made to the following sources for permission to reprint material copyrighted or controlled by them:

Selections from *Writing Themes About Literature,* by Edgar V. Roberts, 1964, Prentice-Hall, Inc.

"Ways of Reading," by David Bartholomae and Anthony Petrosky, reprinted from *Ways of Reading: An Anthology for Writers*, 1988, Bedford/ St. Martin's Press. Copyright © 1929 by Harcourt Brace Jovanovich, Inc., and renewed 1957 by Leonard Woolf.

"Computers and the Writing Process," by Derek Owens, reprinted by permission from *Discovering Literature*, 1995, Prentice-Hall, Inc. Copyright © 1995 by Derek Owens.

"The Myth of the Cave," by Plato, translated by Benjamin Jowett, 1984.

"Letter To His Father"," by Franz Kafka, translated by Ernest Kaiser and Eithene Wilkins, 1954, Schocken Books, a division of Random House.

"A WASP Stings Back," by Robert Claiborne, reprinted from *Newsweek*, 1974.

"The Postcolonial and The Postmodern," by Kwame Anthony Appiah, reprinted from *In My Father's House*, 1992, by permission of Oxford University Press. Copyright © 1992 by Kwame Anthony Appiah.

"Africana Womanism: An Historical, Global Perspective for Women of African Descent," by Clenora Hudson-Weems, reprinted from *Call and Response*, 1998, Houghton Mifflin Co.

"The Classroom as Curriculum," by American Association of University Women, reprinted by permission from *The AAUW Report: How Schools Shortchange Girls*, 1992. Copyright © 1992 by the American Association of University Women Educational Foundation, Washington DC.

"How It Feels to be Colored Me," by Zora Neale Hurston, reprinted by permission from *I Love Myself When I Am Laughing . . . And Then Again When I Am Looking Mean and Impressive*, 1979. Copyright © 1979 by the Estate of Zora Neale Hurston.

"Black Youths' Rate of Suicide Rising Sharply," by Pam Belluck, reprinted by permission from *The New York Times*, March 20, 1998. Copyright © 1998 by the New York Times Co.

"A Black Experience," by Leonard C. Archer, reprinted from *Black Images in the American Theatre*, 1973, Pagent Poseidon, LTD.

"Address: Democratic National Convention, San Francisco, July 17, 1984," by Jesse Jackson, reprinted from *Proceedings of the 1984 Democratic National Convention*, 1984, Democratic National Committee.

"Imagination and Truth: An Interview with James Alan McPherson," by Nell Beram, reprinted by permission from the *Hungry Mind Review*, Spring 1998. Copyright © 1998 by Nell Beram.

"One Art," by Elizabeth Bishop, reprinted from *The Complete Poems, 1927-1979*, Farrar, Straus & Giroux.

"The Mother," by Gwendolyn Brooks, reprinted from *Blacks,* 1991, Third World Press.

"Beware: Do Not Read This Poem," by Ishmael Reed, reprinted from *New and Collected Poems,* 1976, Random House.

"The Self-Hatred of Don L. Lee," by Don L. Lee, reprinted from *The Black Poets,* Bantam Books (Doubleday).

"Big Momma," by Don L. Lee, reprinted from *Think Black,* 1969, Broadside Press.

"Serpent Knowledge," by Robert Pinsky, reprinted by permission from *An Explanation of America.* Copyright © 1979 by Princeton University Press.

"What's Going On," by Marvin Gaye, A. Cleveland, and R. Benson, 1970, Jobete Music Co., Inc.

"The Revolution Will Not Be Televised," by Gil Scott-Heron, reprinted from *So Far, So Good,* 1970, Third World Press.

"Reflection on the Vietnam War Memorial," by Jeffrey Harrison, reprinted from *Yellow House on the Corner,* 1980.

"Digging," by Seamus Heaney, reprinted from *Death of a Naturalist,* 1966, Faber & Faber, Ltd. (UK).

"Sunlight," by Seamus Heaney, reprinted from *Wintering Out,* 1980, Faber & Faber, Ltd. (UK).

"Black Woman," Ed. R. Barksdale, reprinted by *Black Writers of America,* 1972, Prentice-Hall, Inc.

"A Rose for Emily," by William Faulkner, 1931, Random House.

"Rock, Church," by Langston Hughes, 1996. Copyright © 1996 by Ramona Bass and Arnold Rampersad, executors of the Estate of Langston Hughes.

"No Place to Make Love," by Langston Hughes, 1996. Copyright © 1996 by Ramona Bass and Arnold Rampersad, executors of the Estate of Langston Hughes.

"On The Road," by Langston Hughes, 1996. Copyright © 1996 by Ramona Bass and Arnold Rampersad, executors of the Estate of Langston Hughes.

"Sonny's Blues," by James Baldwin, reprinted from *Going to Meet the Man,* 1958, Doubleday, a division of Bantam Doubleday Dell Publishing Group, Inc.

"The Gilded Six-Bits," by Zora Neale Hurston, reprinted from *Story,* 1933, *Scholastic Magazine.*

"Barbados," by Paule Marshall, reprinted from *Soul Clap Hands and Sing,* 1961.

"A Worn Path," by Eudora Welty, reprinted from *A Curtain of Green and Other Stories,* 1941, by permission of Harcourt Brace and Company. Copyright © 1941, renewed 1969 by Eudora Welty.

"The Red Convertible," by Louise Erdrich, reprinted by permission from *Love Medicine,* 1993, Henry Holt and Co. Inc. Copyright © 1984, 1993 by Louise Erdrich.

"Something for the Time Being," by Nadine Gordimer, reprinted by permission from *Selected Stories,* 1975, Viking Penguin, a division of Penguin Putnam, Inc. Copyright © 1960, renewed 1988 by Nadine Gordimer.

"Samuel," by Grace Paley, reprinted from *Enormous Changes at the Last Minute,* 1974, Farrar, Straus & Giroux.

"The Yellow Wallpaper," by Charlotte Perkins Gilman, reprinted from *New England Magazine,* 1892.

"Death of a Son," by Njabulo S. Ndebele, reprinted from *Triquarterly,* Northwestern University Press.

"The Wild Man of the Green Swamp," by Maxine Hong Kingston, reprinted by permission from *China Men,* 1980, Alfred A. Knopf, Inc. Copyright © 1980 by Maxine Hong Kingston.

"Kubota," by Garrett Hongo, reprinted from *Ploughshares.*

"The Conversion of the Jews," by Phillip Roth, reprinted by permission from *Goodbye Columbus,* 1959, Houghton Mifflin, Co. Copyright © 1959, renewed 1987 by Philip Roth. All rights reserved.

"Two Kinds," by Amy Tan, reprinted from *The Joy Luck Club,* 1989, Putnam Publishing Group.

"1920," by Toni Morrison, reprinted from *Sula,* 1973, Random House.

"Sister Monroe," by Maya Angelou, reprinted from *I Know Why the Caged Bird Sings,* 1969, Random House.

"Swaddling Clothes," by Yukio Mishima, reprinted from *Death and Midsummer and Other Stories,* Translated by Ivan Morris, 1966, New Directions Publishing Corporation.

Dedication

Dr. Crawford B. Lindsay, Chairperson
Department of English
Tennessee State University
in the department from 1953-1973

and

Mrs. Alma Dunn Jones
Coordinator of English, Lower Division
Tennessee State University
in the department from 1932-1972

Foreword

Touchstones: Literature and The Writing Process includes selections from three basic types of literature: the essay, poetry, and short stories. These literary works create a confluence of world cultures as an intricate American experience. It recognizes the historical influence of Ancient Western literature, continental cultural forces, and the more direct linkage of British to American literature by means of far reaching thematic threads consistent with the shared history of the two nations. Post-colonial developing world and Eastern nations are approached indirectly through literature written by American internationals. The text has a unique historical arrangement; each of the three divisions has somewhat parallel chronologies, which give way, however, to a cluster of related themes within relevant historical periods.

Non-fiction essays are included to provide means for constructive reading, inspired writing, and the use of the computer to facilitate research for compositions. The familiar essays focus attention on contemporary cultural diversity, whereas the somewhat more classical poems and short stories do more to provide texts for both analytical and theoretical discourses; thus modern critical approaches to writing about literature are introduced.

Touchstones is a customized reader for the student of eclectic interests. To address global demands unique to every literate society by means of fundamental ideas institutionalized therein is the social function of the combined literary selections. The text provides rich materials for informing oral and written discussions and critical analysis required of the university student.

Acknowledgments

Touchstones was compiled by faculty members of the Department of Languages, Literature, and Philosophy at Tennessee State University for writing about literature that surveys the contemporary integrated cultural milieu, thus fostering a universal perspective among a diverse student population. Contributors to the text are Dr. Helen Houston, Professor of English; Dr. Katherine Bryant, Coordinator of Writing Across the Curriculum; Dr. Louise Watkins, Professor of English; Dr. Mike Darby, Associate Professor of English; Dr. Carolyn Moran, Assistant Professor of English; Dr. Jocelyn Irby, Assistant Professor of English; Mr. Alonzo Moody, Assistant Professor of Linguistics; Ms. Lynn Lewis, Assistant Professor of English; Mr. James Byrdsong, Retired Professor of Literature. Thank you for your cooperation and sincere commitment. Expressions of gratitude are also extended to Dr. Brenda Chappell, Professor of Art and free lance artist, who created the art for the cover of the text, and Ms. Tonya McCorkle the typist, in the Department of Languages, Literature, and Philosophy. Most assuredly, we appreciate the technical assistance of Antony C. Springfield, a graduate student at Meharry Medical College.

Table of Contents

III. Short Stories

Glossary of Literary Terms

Abstract terms: words that refer to ideas or concepts rather than to material objects.

Aesthetics: the study of the beautiful; a branch of philosophy which attempts to define the nature of art and the criteria by which it may be judged.

Allegory: a type of narrative that attempts to reinforce its thesis by making its characters represent specific abstract ideas or qualities.

Alliteration: the repeated use of a consonant, especially at the beginning of stressed words or syllables.

Allusion: a reference to a person, place, thing, or event that allows a writer to compress a great deal of meaning into a very few words.

Analogy: a correlation of two essentially unlike things which share one or more common features.

Concrete terms: words that refer to material objects rather than to ideas or concepts.

Connotation: the acquired meaning of a word or the nuances of meaning in a word or image.

Deconstruction: a critical movement that holds that there is an inherently unstable relationship between words and meaning. The critic examines elements of a literary text for its internal signification in order to demonstrate that every text finally generates innumerable, contradictory, and indeterminate meanings.

Denotation: the meaning of a word, without overtones; the dictionary meaning of a word.

Diction: the choice of words; the artistic arrangement of words.

Didactic: intended to be instructive.

Double Entendre: a form of pun in which one of the two meanings is risqué.

Dramatic Monologue: a poem that is given over entirely to the speech of a single character, at some specific critical moment, thereby unintentionally revealing his or her essential temperament and personality.

Elegy: a subjective poem, usually meditating on death of a specific person or a group.

Epic: a long narrative poem, elevated and dignified in style and dealing with the lofty deeds of heroes and their people.

Genre: a form, class, or type of literary work, for example, essay, poem, short story or play.

Imagery: the use of words that appeal to the senses.

Irony: refers to some contrast or discrepancy between appearance and reality. Irony takes three specific forms: in verbal irony there is a contrast between what is literally said and what is actually meant; in dramatic irony the state of affairs known to the reader is the reverse of what its participants suppose it to be; in situational irony a set of circumstances turns out to be the reverse of what is expected or is appropriate.

Lyric: a subjective poem that expresses personal thought and emotion.

Metaphor: figure of speech which compares two normally unrelated things that share common features.

Metonymy: a figure of speech in which one object or idea takes the place of another with which it is closely associated.

Milieu: the environment or surroundings in which an author lives or in which a work is written.

Narrative: a poem that tells a story.

New Historicism: an approach to literary history as part of a larger cultural history; historicism was the offspring of Hegelian idealism, and later the evolutionary naturalism of Herbert Spencer. Several major historicists studied literature in the context of social, political, and cultural history. They saw a nation's literary history as an expression of its evolving spirit.

Paradox: a statement that is apparently contradictory but at another level is in fact not so; a self-contradictory statement that turns out to be valid.

Persona: the fictitious speaker or "I," who is to be distinguished from the poet himself or herself.

Personification: a figure of speech in which an idea or thing is given human attributes.

Poetry: metrical writing; intensified language of experience.

Postmodernism: a term that involves diverse attempts to break away from modernist forms, which had become conventional, as well as to overthrow the elitism of "high" art by recourse to the models of mass culture.

Psychological Criticism: the theory treats a work primarily as an expression, in fictional form, of the state of mind and the structure of personality of the author.

Reader Response Criticism: a critical approach that begins by assuming the texts are meant to be read; that a text does not really exist until it is in fact read; and that to an extent, the meaning of the text is created or produced by the reader.

Satire: a literary work that ridicules or rebukes human vice or folly.

Structure: systems of internal relationships of parts in a poem; the overall design of a literary work.

Symbol: something that stands for itself and perceptibly stands for an abstract idea to which it is related.

Tone: the attitude of the poet or of the persona toward the subject of the poem or toward the audience.

Essays

What Is a Theme?

Edgar V. Roberts

As you begin any of the assignments in this book, you should bear in mind the two basic requirements of themes. The first is the need for a *central idea* or *point.* A theme, like an essay, should be a short, accurate, and forceful presentation of ideas or descriptions, well contrived as a totality or unity. A theme should not ramble in any way, but should be clearly united around a dominating central thought. A theme is a brief "mind's full" on any particular subject; that is, it presents and considers the subject in several of its various aspects. The theme cannot cover all aspects, as might a book or a long essay.

The second basic need is that a theme must have a *clearly discernible organization.* Let's examine these two essentials in detail.

The Central Idea or Point

Themes are so named because throughout such a composition there runs a basic or central idea—a theme—that unifies the paper into a logical whole. On every subject you encounter there should be some dominating idea or mood that will suggest itself to you or one that you will derive from your own intensive concentration. For example, when you look at a room, you might feel that it is cheerful; when you listen to the latest news, you might decide that it is depressing. Were you to write a theme describing the room or another discussing the news, you would have to keep your central idea foremost in your reader's mind *throughout* your theme, or else you would not have a theme.

One of your first objectives should be to decide on a central idea for your theme. Usually you are working on an assignment by your instructor. With this problem or topic in front of you, your task will be to formulate a central idea that will in effect be your solution to the problem raised by the assignment. As always, study the work carefully and take notes describing the major actions and statements. If the author has stated an idea that he develops in the work, make a note of that. Then, consider your notes with care and try to work up a central idea on the basis of your observations and conclusions.

Let us assume, for the moment, that your assignment has been a "specific problem" theme, the question being: "In Katherine Mansfield's story 'Miss Brill', is Miss Brill worth your sympathy?" The following material represents a collection of notes and observations that are typical of what you yourself might write when reading the story for this assignment. Notice that page numbers are given, so that you could easily go back to the story to refresh your memory on any particular details when you start sketching out your theme.

p. 14 Miss Brill is introduced. She loves her fox fur, and lives alone. She is observant. Goes to the park often. Is familiar with the park and the band there.

p. 15 She eavesdrops, but not with a bad intention. She lives in others, not herself. She recalls a dull conversation of the week before.

p. 16 Many people are in the park, some seeming to be like Miss Brill. She is quiet and harmless, with no more aggression than a bag of popcorn. She is observant. Life seems to have passed her by, and therefore she enjoys looking at the people in the park. She seems to have some sympathy for the lady in the "ermine toque."

pp. 16–17 She discovers a way to identify herself with life in the park: she is an actress, with a part to play. She almost expects that everyone will join the band in singing. She would be a part of the choir.

p. 17 A boy and a girl come. They insult Miss Brill so that she can hear them.

pp. 17–18 She goes home to her small, dark room. She cries. Her emotions have gone from calm, to excitement, to being terribly hurt, to being just about heartbroken. She is like a hurt bird, a helpless victim. She is a nobody, but a nobody I feel sorry for. Her real sorrow results from the fact that the insults make her aware of just how insignificant she is. The only joy she knew is taken away from her. What more would she have? Not much.

In response to the assignment question, this set of notes leads naturally to the following central idea: "Miss Brill is worth my sympathy." The conclusion is not startling in any way, and a reader might just as easily have arrived at it without the system of notes. The virtue of the notes, however, is that they will provide a factual basis for developing the theme.

Once you have established the central idea, you should concentrate on the use you can make of it. All the ideas you now bring out about the work you have read will be related to the central idea. When you finally write your theme you should state the idea in your introduction, for you should leave your reader in no doubt about what you wish to assert. Also, throughout your theme, you should think about reminding your reader that references to the story or poem, and all conclusions you are drawing, are relevant to the central idea. Anything not related to this point will not belong in your theme.

The need for a central idea will also make you aware of the need for paragraph transitions, because you are proving or showing *one* central idea, not a *number* of ideas. Transitions form bridges to connect one part of the theme with another; having a central idea always in mind makes continuity between paragraphs both essential and natural.

Organizing Your Theme around the Central Idea

The next job is to establish a clearly ascertainable organization for your theme. You should look carefully again at your notes and at any ideas they suggest. It may be that some ideas have already come to you, for quite often a person decides even before he has finished reading the assigned work that he would like to write on such or such a point. Get these ideas down on paper, for when you can see them in front of you, you can work with them. If you try to keep them only in your mind, you may forget them. Here are some raw thoughts on the subject of the story we have been examining. They represent a stage beyond the first stage of note taking. Each one of the topics could make separate parts of a possible theme.

1. Miss Brill has little human contact except what she imagines.
2. She is lonely and vulnerable.
3. She doesn't have much insight into her life until the end of the story, or else she is self-indulgent.
4. She is no threat to anyone in the park. Therefore the young people who insult her must be at least somewhat unkind if not vicious.
5. The author views Miss Brill with apparent understanding and kindness, but she underplays the ending.
6. If Miss Brill can keep people at a distance, she seems to function fairly well. When they get close, they hurt her.

These are six possible thoughts on the story. Working with these, together with others that may occur as you go along, you should create topics that will serve as the basis for developing the entire theme. Let us say that you choose number 4 as a topic, then number 2, and then number 5. These topics may be entitled: (4) harmlessness, (2) loneliness and vulnerability, and (5) artistic treatment. If you arrange these topics in 1, 2, 3 order, you now have the material for a potential thesis sentence and an outline.

The Thesis Sentence

A thesis sentence is like an itinerary of a journey, a plan of action. Just as few people would ever think of taking a trip before planning a route, so you should never start to write before having a clear notion of the topics you will consider. The thesis sentence does just that; it connects the central idea and the plan of topics. The first thing to do, then, is to put the central idea together with the topics, in order to plan the form of the thesis sentence. Again, let us use the materials that have been developed thus far:

Central Idea	Topics
Miss Brill is worth my sympathy.	1. harmlessness
	2. loneliness and vulnerability
	3. artistic treatment

From this arrangement we can now write the following thesis sentence:

> My feeling of her worthiness results from her harmlessness, from her loneliness and vulnerability, and from Katherine Mansfield's skillful treatment of her plight.

The Topic Sentence

Just as the entire theme is to be organized around the thesis sentence, each of the paragraphs should be organized around a topic sentence. The topic sentences are derived, grammatically speaking, from the topics in the predicate of the thesis sentence. Thus, the first topic will be Miss Brill's harmlessness. But something more must be done than just announcing the topic; the topic must be shown to have a bearing on the central idea. Let us suppose that the original raw thought can be brought back into consideration here, so that we can create the following topic sentence:

> Her harmlessness [topic] makes the hurt done to her seem unjustified and unnecessarily cruel [connecting topic to central idea].

Notice that the words "unjustified" and "unnecessarily" indicate a judgment by you the writer which is related to your conclusion that Miss Brill is worthy of your sympathy. With such a judgment you could then write a paragraph in which you might develop an argument like this one: "Unjustified and unnecessary hurt done to anyone makes me defensive and sympathetic to the person receiving the hurt." In such a way the topic sentence eventually enables you to expand upon the central idea, and at the same time it channels your thoughts into a development related to the central idea.

You should follow the same process in forming the other topic sentences, so that when you finish them you can put them together in the form of an outline. The particular type of outline that we have been developing here is the "analytical sentence outline." This type requires (1) that the central idea be modified so that it can be used in the grammatical subject of the thesis sentence, (2) that the topics in the predicate of the thesis sentence become the subjects of the topic sentences, and (3) that the predicates of the topic sentences have a clear bearing on the central idea. Such a plan will ensure that you are always thinking within a clear, definite pattern of organization.

As an optional final part of the outline, there may be a concluding or summarizing sentence, which should generally govern the conclusion of your paper. Because this sentence is not derived from the predicate of the thesis sentence, it is technically independent of the material to be included in the body of your theme. But it is a part of the thematic organization, and hence should bear a close relationship to the central idea. It may represent a summary of some of the leading ideas; it may suggest an evaluation or a criticism of some of these ideas; it may also suggest further avenues of exploration that you did not examine in the body of the theme.

As you plan your outline and start to write, bear in mind that writing is in many ways a process of discovery. New ideas often come as you write. If these ideas do not conform to your original plan, change the plan to accommodate them.

When completed, the analytical sentence outline should have the following appearance:

Theme:	**Miss Brill as a Sympathetic Character**
Paragraph 1	INTRODUCTION containing CENTRAL IDEA and THESIS SENTENCE
	CENTRAL IDEA: Miss Brill is worth my sympathy.
	THESIS SENTENCE: My feeling of her worthiness results from her harmlessness, from her loneliness and vulnerability, and from Katherine Mansfield's skillful treatment of her plight.
	BODY containing three TOPIC SENTENCES
Paragraph 2	TOPIC SENTENCE: Her harmlessness makes the hurt done to her seem unjustified and unnecessarily cruel.
Paragraph 3	TOPIC SENTENCE: Her loneliness and vulnerability make her pitiable.
Paragraph 4	TOPIC SENTENCE: Katherine Mansfield's skillful treatment of Miss Brill's plight encourages the right proportion of sympathy.
	CONCLUSION
Paragraph 5	TOPIC SENTENCE: Sympathy for Miss Brill is the dominant effect of the story.

By the time you have created an outline like this one, you will have been thinking and organizing for a considerable time, and you should be well prepared to write your theme. The outline should now be put into operation. In your introduction you should include both the central idea and the thesis sentence. The various topic sentences will belong at the beginnings of the paragraphs in the body of your theme. You may, of course, use the sentences just as they are, or you may, as you become more experienced as a writer, wish to modify them to make them seem less obvious. "It is the purpose of art," said one very wise person, "to hide art." If the machinery of your themes is creaky at first, your experience and development may eventually oil it and allow it to run smoothly and noiselessly.

The Sample Theme

So that you may see more clearly the relationship of the outline and the theme, the following theme is based on the material we have been discussing. It is drawn mainly from the ideas contained in the notes, and is wholly dependent for its structure on the analytical sentence outline.

Miss Brill as a Sympathetic Character

[1]To raise the question of whether Miss Brill, of Katherine Mansfield's story "Miss Brill," is sympathetic is to imply that there are reasons for which a reader may be put off by her. There are a number

of reasons. Objectively, Miss Brill is an odd person, one of the many scarecrows sitting on the park benches. She is not communicative; she does not lead an exciting or interesting life; she is simple and almost verges on stupidity; and finally, she is self-indulgent to the point of being almost unconnected with reality. Despite these bad qualities, *Miss Brill is worthy of sympathy.* * *My feeling results from her harmless character, from her loneliness and vulnerability, and from Miss Mansfield's skillful treatment of her plight.*†

*Central idea
†Thesis sentence

[2]*Miss Brill's harmlessness makes the hurt done to her by the young couple seem unjustified and unnecessarily cruel.* Her manner, her thoughts, and her activities all make a person feel that she is to be let alone and tolerated, but never to be harmed. If she eavesdrops on people in the park, the simple solution would be for the offended people to move to another bench, perhaps with a glare on their faces but with no other expression of anger toward her. She does not mistreat anyone, or lure any small children to their destruction. Nor could she do so. In view of these facts, the hurt done her by the young couple is far beyond anything she deserves. It seems almost calculated to hurt her in the worst possible way. Even if one did not like Miss Brill, his heart would have to go out to her as a result of the insult. Mine does.

Topic sentence

[3]*She is, then, no threat to anyone, and she is made to seem even pitiable because of her loneliness and vulnerability.* The thought of this lone creature imagining herself part of the park scene, and almost believing that everyone in the park will join the band in singing, makes one realize just how solitary she is. She has almost nothing. The best one could claim for her, even by stretching the imagination, is that she is a nobody. She has no one, and no one cares. All of her joy, pathetic as it is, hangs from the very thin thread of her imagination. With only that, she is defenseless and alone, like a hurt bird. There can be no delight, but only sorrow, when the defenseless are destroyed.

Topic sentence

[4]*This is not to say that the story is a sentimental one, for Katherine Mansfield's skillful treatment of Miss Brill's plight encourages exactly the right proportion of sympathy.* Particularly crucial in this regard is the treatment of Miss Brill when she returns home after the insults. She does not buy her honeycake, but rushes home. As she puts away her fur she is crying, but Katherine Mansfield does not overdo her tears. Instead she makes the reader figure out what is happening. This brief moment, when the reader ponders just whom Miss Brill hears crying, is enough to keep everything from going overboard. Also, the story ends abruptly at this point, with the reader being left to imagine the life that Miss Brill will have afterward, with no more of the joys she had experienced before the insults. There is no indulgence, but there is great sympathy here.

[5]Sympathy, rather than deep sorrow, is the major effect of the story. Miss Brill is shown realistically as a person of solitude and harmlessness. Her faults are at worst those of an eccentric, and her happiness is marginal and fleeting. If a person has many things and

> loses one of them, the loss causes little sympathy or grief. But if he has only one thing, and that is taken from him, as joy is stolen from Miss Brill, then that person has nothing. Is Miss Brill worthy of sympathy? She most certainly is.

Main Problem In Writing

Once you have understood and applied the principles of thematic development and thesis-sentence organization, you will still be faced with the problem of how to write well. There is little difficulty in recognizing superior examples of student writing when you see them, but there is usually much difficulty in understanding precisely what constitutes the superiority. For this reason the most difficult and perplexing questions you will ask as you write are these: (1) "How can I improve my writing?" or (2) "If I got a C on my last paper, why wasn't the grade a B or an A? How can I improve my grades?" These questions are really both the same, with a different emphasis. Your concern is with improvement.

As an undergraduate, you should not be offended if you are told that you probably have not yet acquired a great deal of knowledge and understanding of literature. Your mind is growing, and it still has many facts to assimilate and digest. As you accumulate these facts and develop your understanding, you will find that ease of expression will also develop. But at the moment your thoughts about literature might be expressed thus: "When I first read a work, I have a hard time following it. Yet when my instructor explains it, my understanding is greatly increased. I would like to develop the ability to understand the work without my instructor's help. How can I succeed in this aim? How can I become an independent reader?"

In answer, you started trying to overcome the problem the day you enrolled in college. This action testifies to your desire to improve. Bear in mind also that education is a process, and that what baffled you as a freshman may seem child's play when you are a junior or senior. But in the meantime you want to know how to assist growth. There is no magic answer, no shortcut to knowledge. You must work constantly to develop the habits of the good reader described earlier.

The second major obstacle to writing well is inexperience. As a result, when you start you may be tempted simply to retell a story or do no more than list the ideas in an argument. Even when you begin with specific approaches, you may wind up retelling the story. This is really the path of least resistance, but it is inadequate for most of your themes.

Your education is aimed first at acquiring knowledge and second at digesting and using knowledge. Retelling a story indicates only that you have read the material. To show your understanding of it, you must also show that you can put what you have read into a meaningful pattern.

There are a number of ways in which you may set up patterns of development that can assist you in showing your understanding. One way is to make a deliberate point of referring to events or statements in the work in a reverse order. Talk about the conclusion or the middle of the work first. But rarely should you refer first to the opening of the work. Beginning your discussion with references to parts of the work other than the opening will almost force you to discuss your understanding of the work rather than to retell events. If you look back at paragraph 2 of the sample theme on "Miss Brill," you will see that this technique has been used. One of the last events in the story, Miss Brill's being insulted by a young couple, is the main reference material of that paragraph, yet the paragraph is the first in the body of the theme. You can see that this technique permits you to impose your own organization on your theme and frees you from the organization of the work being analyzed.

Another important method is to consider the reader for whom you are writing your theme. Imagine that you are writing to another student like yourself, a student who has read the assigned work but who has not thought very much about it. You can immediately see what you would write for such a mythical reader. He knows the events or has followed the thread of the argument. He knows who says what and when it is said. As a result, you do not need to tell this reader about everything that goes on, but should regard your role as that of an explainer or interpreter. Tell him what things mean, but *do not tell him the things that happen.*

To look at the situation in still another way, you may have read stories and novels about Sherlock Holmes and Dr. Watson. Holmes always points out to Watson that all the facts are available to both of them, but that, though Watson sees, he does not observe. Your role is like that of Holmes, explaining and interpreting facts, and drawing conclusions that Dr. Watson has been unable to draw for himself. Once again, if you look back at the sample theme on "Miss Brill," you will notice that everywhere *the assumption has been made that the reader has read the story already.* References to the story are thus made primarily to remind the reader of something he or she already knows, but *the principal emphasis of the theme is to draw conclusions and develop arguments.*

From *Ways of Reading*

David Bartholomae

Before explaining how we organized this book, we would like to say more about the purpose and place of the kind of strong, aggressive, labor-intensive reading we've been referring to.

Readers face many kinds of experiences, and certain texts are written with specific situations in mind and invite specific ways of reading. Some texts, for instance, serve very practical purposes—they give directions or information. Others, like the short descriptive essays often used in English textbooks and anthologies, celebrate common ways of seeing and thinking and ask primarily to be admired. These texts seem self-contained; they announce their own meanings with little effort and ask little from the reader, making it clear how they want to be read and what they have to say. They ask only for a nod of the head or for the reader to take notes and give a sigh of admiration ("yes, that was very well said"). They are clear and direct. It is as though the authors could anticipate all the questions their own essays might raise and solve all the problems a reader might imagine. There is not much work for a reader to do, in other words, except, perhaps, to take notes and, in the case of textbooks, to work step-by-step, trying to remember as much as possible.

This is how assigned readings are often presented in university classrooms. Introductory textbooks (in biology or business, for instance) are good examples of books that ask little of readers outside of note-taking and memorization. In these texts the writers are experts and your job, as novice, is to digest what they have to say. And, appropriately, the task set before you is to summarize—so you can speak again what the author said, so you can better remember what you read. Essay tests are an example of the writing tasks that often follow this kind of reading. You might, for instance, study the human nervous system through textbook readings and lectures and then be asked to write a summary of what you know from both sources. Or a teacher might ask you during a class discussion to paraphrase a paragraph from a textbook describing chemical cell communication to see if you understand what you've read.

Another typical classroom form of reading is reading for main ideas. With this kind of reading you are expected to figure out what most people (or most people within a certain specialized group of readers) would take as the main idea of a selection. There are good reasons to read for main ideas. For one, it is a way to learn how to imagine and anticipate the values and habits of a particular group—test-makers or, if you're studying business, Keynesian economists, perhaps. If you are studying business, to continue this example, you must learn to notice what Keynesian economists notice—for instance, when they analyze the problems of growing government debt—to share key terms, to know the theoretical positions they take, and to adopt for yourself their common examples and interpretations, their jargon, and their established findings.

There is certainly nothing wrong with reading for information or reading to learn what experts have to say about their fields of inquiry. These are not, however, the only ways to read, although they are the ones most often taught. Perhaps because we think of ourselves as writing teachers, we are concerned with presenting other ways of reading in the college and university curriculum.

A danger arises in assuming that reading is only a search for information or main ideas. There are ways of thinking through problems and working with written texts which are essential to academic life, but which are not represented by summary and paraphrase or by note-taking and essay exams.

Student readers, for example, can take responsibility for determining the meaning of the text. They can work as though they were doing something other than finding ideas already there on the page and they can be guided by their own impressions or questions as they read. We are not, now, talking about finding hidden meanings. If such things as hidden meanings can be said to exist, they are hidden by readers' habits and prejudices (by readers' assumptions that what they read should tell them what they already know), or by readers' timidity and passivity (by their unwillingness to take the responsibility to speak their minds and say what they notice).

Reading to locate meaning in the text places a premium on memory, yet a strong reader is not necessarily a person with a good memory. This point may seem minor, but we have seen too many students haunted because they could not remember everything they read or retain a complete essay in their minds. A reader could set herself the task of remembering as much as she could from Walker Percy's "The Loss of the Creature," an essay filled with stories about tourists at the Grand Canyon and students in a biology class, but a reader could also do other things with that essay; a reader might figure out, for example, how both students and tourists might be said to have a common problem seeing what they want to see. Students who read Percy's essay as a memory test end up worrying about bits and pieces (bits and pieces they could go back and find if they had to) and turn their attention away from the more pressing problem of how to make sense of a difficult and often ambiguous essay.

A reader who needs to have access to something in the essay can use simple memory aids. A reader can go back and scan, for one thing, to find passages or examples that might be worth reconsidering. Or a reader can construct a personal index, making marks in the margin or underlining passages that seem interesting or mysterious or difficult. A mark is a way of saying, 'This is something I might want to work on later." If you mark the selections in this book as you read them, you will give yourself a working record of what, at the first moment of reading, you felt might be worth a second reading.

If Percy's essay presents problems for a reader, they are problems of a different order from summary and recall altogether. The essay is not the sort that tells you what it says. You would have difficulty finding one sentence that sums up or announces, in a loud and clear voice, what he is talking about. At the point you think Percy is about to summarize, he turns to one more example that complicates the picture, as though what he is discussing defies his attempts to sum things up. Percy is talking about tourists and students, about such things as individual "sovereignty" and our media culture's "symbolic packages," but if he has a point to make, it cannot be stated in a sentence or two.

In fact, Percy's essay is challenging reading in part because it does not have a single, easily identifiable main idea. A reader could infer that it has several points to make, none of which can be said easily and some of which, perhaps, are contradictory. To search for information, or to ignore the rough edges in search of a single, paraphrasable idea, is to divert attention from the task at hand, which is not to remember what Percy says but to speak about the essay and what it means to you, the reader. In this sense, the Percy essay is not the sum of its individual parts; it is, more accurately, what its readers make of it.

A reader could go to an expert on Percy to solve the problem of what to make of the essay—perhaps to a teacher, perhaps to a book in the library. And if the reader pays attention, he could remember what the expert said or she could put down notes on paper. But in doing either, the reader only rehearses what he or she has been told, abandoning the responsibility to make the essay meaningful. There are ways of reading, in other words, in which Percy's essay, "The Loss of the Creature," is not what it means to the experts but what it means to you as a reader willing to take the chance to construct a reading. You can be the authority on Percy; you don't have to turn to others. The meaning of the essay, then, is something you develop as you go along, something for which you must take final responsibility. The meaning is forged from reading the essay, to be sure, but it is determined by what you do with the essay, by the connections you can make and your explanation of why those connections are important, and by your account of what Percy might mean when he talks about "symbolic packages" or a "loss of sovereignty" (phrases Percy uses as key terms in the essay). This version of Percy's essay will finally be yours; it will not be exactly what Percy said. (Only his words in the order he wrote them would say exactly what he said.) You will choose the path to take through his essay and support it as you can with arguments, explanations, examples, and commentary.

If an essay or story is not the sum of its parts but something you as a reader create by putting together those parts that seem to matter personally, then the way to begin, once you have read a selection in this collection, is by reviewing what you recall, by going back to those places that stick in your memory—or, perhaps, to those sections you marked with checks or notes in the margins. You begin by seeing what you can make of these memories and notes. You should realize that with essays as long and complex as those we've included in this book, you will never feel, after a single reading, as though you have command of everything you read. This is not a problem. After four or five readings (should you give any single essay that much attention), you may still feel that there are parts you missed or don't understand. This sense of incompleteness is part of the experience of reading, at least the experience of reading serious work. And it is part of the experience of a strong reader. No reader could retain one of these essays in her mind, no matter how proficient her memory or how experienced she might be. No reader, at least no reader we would trust, would admit that he understood everything that Michel Foucault or Adrienne Rich or Patricia J. Williams had to say. What strong readers know is that they have to begin, and they have to begin regardless of their doubts or hesitations. What you have after your first reading of an essay is a starting place, and you begin with your marked passages or examples or notes, with questions to answer, or with problems to solve. Strong readings, in other words, put a premium on individual acts of attention and composition.

Strong Readers, Strong Texts

We chose essays and stories for this book that invite strong readings. Our selections require more attention (or a different form of attention) than a written summary, a reduction to gist, or a recitation of main ideas. They are not "easy" reading. The challenges they present, however, do not make them inaccessible to college students. The essays are not specialized studies; they have interested, pleased, or piqued general and specialist audiences alike. To say that they are challenging is to say, then, that they leave some work for a reader to do. They are designed to teach a reader new ways to read (or to step outside habitual ways of reading), and they anticipate readers willing to take the time to learn. These readers need not be experts on the subject matter. Perhaps the most difficult problem for students is to believe that this is true.

You do not need experts to explain these stories and essays, although you could probably go to the library and find an expert guide to most of the selections we've included. Let's take, for example, Adrienne Rich's, "When We Dead Awaken: Writing as Re-Vision." This essay looks at the history of women's writing (and at Rich's development as a poet). It argues that women have been trapped within a patriarchal culture—speaking in men's voices and telling stories prepared by men—and, as a consequence, according to Rich, "We need to know the writing of the past, and know it differently than we have ever known it; not to pass on a tradition but to break its hold over us."

You could go to the library to find out how Rich is regarded by experts, by literary critics or feminist scholars, for example; you could learn how her work fits into an established body of work on women's writing and the representation of women in modern culture. You could see what others have said about the writers she cites: Virginia Woolf, Jane Austen, and Elizabeth Bishop. You could see how others have read and made use of Rich's essay. You could see how others have interpreted the poems she includes as part of her argument. You could look for standard definitions of key terms, like "patriarchy" or "formalism."

Though it is often important to seek out other texts and to know what other people are saying or have said, it is often necessary and even desirable to begin on your own. Rich can also be read outside any official system of interpretation. She is talking, after all, about our daily experience. And when she addresses the reader, she addresses a person—not a term-paper writer. When she says, "We need to know the writing of the past, and know it differently than we have ever known it," she means us and what we know and how we know what we know. (Actually the "we" of her essay refers most accurately to women readers, leading men to feel the kind of exclusion women must feel when the reader is always "he." But it is us, the men who are in the act of reading this essay, who feel and respond to this pressure.)

The question, then, is not what Rich's words might mean to a literary critic, or generally to those who study contemporary American culture. The question is what you, the reader, can make of those words and Rich's use of them in the essay, given your own experience, your goals, and the work you do with what she has written. In this sense, "When We Dead Awaken: Writing as Re-Vision" is not what it means to others (those who have already decided what it means) but what it means to you, and this meaning is something you compose when you write about the essay; it is your account of what Rich says and how what she says might be said to make sense.

A teacher, poet, and critic we admire, I. A. Richards, once said, "Read as though it made sense and perhaps it will." To take command of complex material like the essays and stories in this book, you need not subordinate yourself to experts; you can assume the authority to provide such a reading on your own. This means you must allow yourself a certain tentativeness and recognize your limits. You should not assume that it is your job to solve the problems between men and women. You can speak with authority while still acknowledging that complex issues *are* complex.

There is a paradox here. On the one hand, the essays are rich, magnificent, too big for anyone to completely grasp all at once, and before them, as before inspiring spectacles, it seems appropriate to stand humbly, admiringly. And yet, on the other hand, a reader must speak with authority.

In "The American Scholar," Ralph Waldo Emerson says, "Meek young men grow up in libraries, believing it their duty to accept the views, which Cicero, which Locke, which Bacon, have given, forgetful that Cicero, Locke, and Bacon were only young men in libraries when they wrote these books." What Emerson offers here is, not a fact but an attitude. There is creative reading, he says, as well as creative writing. It is up to you to treat authors as your equals, as people who will allow you to speak too. At the same time, you must respect the difficulty and complexity of their texts and of the issues and questions they examine. Little is to be gained, in other words, by turning Rich's essay into a message that would fit on a poster in a dorm room: "Be Yourself" or "Stand on Your Own Two Feet."

Reading With and Against the Grain

From this pushing and shoving with and against texts, we come then to a difficult mix of authority and humility. A reader takes charge of a text; a reader gives generous attention to someone else's (a writer's) key terms and methods, commits his time to her examples, tries to think in her language, imagines that this strange work is important, compelling, at least for the moment.

Most of the questions in *Ways of Reading* will have you moving back and forth in these two modes, reading with and against the grain of a text, reproducing an author's methods, questioning his or her direction and authority. With the essay "When We Dead Awaken," for example, we have asked students to give a more complete and detailed reading of Rich's poems (the poems included in the essay) than she does, to put her terms to work, to extend her essay by extending the discussion of her examples. We have asked students to give themselves over to her essay—recognizing that this is not necessarily an easy thing to do. Or, again in Rich's name, we have asked students to tell a story of their own experience, a story similar to the one she tells, one that can be used as an example of the ways a person is positioned by a dominant culture. Here we are saying, in effect, read your world in Rich's terms. Notice what she would notice. Ask the questions she would ask. Try out her conclusions.

To read generously, to work inside someone else's system, to see your world in someone else's terms—we call this "reading with the grain." It is a way of working *with* a writer's ideas, in conjunction with someone else's text. As a way of reading, it can take different forms. In the reading and writing assignments that follow the selections in this book, you will sometimes be asked to summarize and paraphrase, to put others' ideas into your terms, to provide your account of what they are saying. This is a way of getting a tentative or provisional hold on a text, its examples and ideas; it allows you a place to begin to work. And sometimes you will be asked to extend a writer's project—to add your examples to someone else's argument, to read your experience through the frame of another's text, to try out the key terms and interpretive schemes in another writer's work. In the assignments that follow the Rich essay, for example, students are asked both to reproduce her argument and to extend her terms to examples from their own experience.

We have also asked students to read against the grain, to read critically, to turn back, for example, *against* Rich's project, to ask questions they believe might come as a surprise, to look for the limits of her vision, to provide alternate readings of her examples, to find examples that challenge her argument, to engage her, in other words, in dialogue. How might her poems be read to counter what she wants to say about them? If her essay argues for a new language for women, how is this language represented in the final poem or the final paragraphs, when the poem seems unreadable and the final paragraph sounds familiarly like the usual political rhetoric? If Rich is arguing for a collective movement, a "we" represented by the "we" of her essay, who is included and who is excluded by the terms and strategies of her writing? To what degree might you say that this is a conscious or necessary strategy?

Many of the essays in this book provide examples of writers working against the grain of the common readings or everyday language. This is true of John Berger and Susan Willis, for example, who define the material of our culture (exercise programs for Willis and art museums for Berger) against what they take to be the usual misreadings of them. It is true of John Edgar Wideman, who reads against his own text while he writes it—asking questions that disturb the story as it emerges on the page. It is true of Harriet Jacobs, Patricia J. Williams, and Virginia Woolf, whose writings show the signs of their efforts to work against the grain of the standard essay, habitual ways of representing what it means to know something, to be somebody, to speak before others.

This, we've found, is the most difficult work for students to do, this work against the grain. For good reasons and bad, students typically define their skill by reproducing rather than questioning or revising the work of their teachers (or the work of those their teachers ask them to read). It is important to read generously and carefully and to learn to submit to projects that others have begun. But it is also important to know what you are doing—to understand where this work comes from, whose interests it serves, how and where it is kept together by will rather than desire, and what it might have to do with you. To fail to ask the fundamental questions—where am I in this? how can I make my mark? whose interests are represented? what can I learn by reading with or against the grain?—to fail to ask these questions is to mistake skill for understanding, and it is to misunderstand the goals of a liberal education. All of the essays in this book, we would argue, ask to be read, not simply reproduced; they ask to be read and to be read with a difference. Our goal is to make that difference possible.

Computers and the Writing Process

Derek Owens

Preparing to Write

Once you've familiarized yourself with the basics of your word-processing software, you're ready to use the computer for a specific assignment. But before you begin, you need to focus. As you sit down, ask yourself these key questions: What is the assignment asking me to do? What will the audience expect? What do I know about this topic? What do I want to say to my audience?

A good way to begin is to jot down notes in a preliminary file. Whether you consider this "prewriting" or "freewriting," the task is the same: to clear your mind by putting down any thoughts, questions, insights, and ideas that come to you. Don't worry about spelling or grammar at this point; the goal here is to gather raw material that might later lead you down more specific paths. If at this time you wander off on tangents, fine; remember, this is not a first draft but an assemblage of early impressions. It's okay if things are sloppy; at this point you are generating a quantity of information. Make sure you save this file, perhaps under a name like "Rawnotes.Ex1."

Organizing Your Notes

You've now spent one or more sessions at the computer, compiling preliminary thoughts and ideas. Now it's time to put this information into some sort of order. Reread your jumble of ideas and seek patterns within the entries. Ask yourself questions: What do some of these entries have in common? Where do they differ? Once you've located several themes or threads within this information, use the block move or cut/paste commands to rearrange this material into sections with a subtitle for each one. (For example, different thoughts might be gathered under subheadings such as "Introductory Comments," "Ideas for Counterarguments," "Historical Background," "Research Data," "Descriptive Anecdotes," "Concluding Ideas," "Misc. Fragments," and so forth.)

Once you've grouped every thought or question under some heading, go back and review these sections. Which material do you think might belong at the beginning of your paper? Which at the end? How might you arrange the material in the middle? Again, using the cut/paste command, rearrange these entire sections so that they proceed in what looks like a sensible order.

Keep in mind that at this point nothing is set in stone. There's no reason to worry if what you've put in one section really belongs there or not. Your paper will go through a variety of revisions, and there will be time to make changes later. For now, your main goal is to inject some semblance of organization into your notes. The sooner you can see things in sections and categories rather than as a big lump of words, the better.

The advantage to organizing your thoughts in this manner is that the computer can help free you from any inclination to think that only one plan for a paper exists. A paper can be structured in numerous ways. The more flexible you are in experimenting with different outlines, the more likely it is that you'll create a design indicating a strong, creative synthesis of possible approaches.

Using Computers in Your Research

If you are writing a research paper, take a printout of your initial jottings and questions with you to the library as you begin the general search process. If your computer is hooked into a network, you might not even need to travel to the library to check its holdings since you might have access to the library's on-line catalog via phone lines. Some colleges will even have general references on a mainframe computer, and you will be able to peruse these references and do your initial reading while seated at your own computer.

If you belong to certain related discussion lists or "listservs" on the network, you can send general inquiries across that network, soliciting information from any subscribers to that list. You might introduce readers to the nature of your research and ask them if they know any related books or articles that would be worth exploring.

You can take notes at the computer, too, by creating files that divide your research topic into important categories. If categories get too large or ill defined, you can split them into separate files. And if you have a laptop computer, you can take your computer with you to the library, where you can research and write all in one session.

Writing a First Draft

At this point you should have a detailed, working outline full of thoughts and impressions that could feasibly be built on and turned into entire paragraphs. Print out this document and reread it, annotating it if you wish. You're now in a good position to begin writing a first draft.

Some people need the physical connection of hand to pen to paper when writing. If this is the case for you, go ahead and write your first draft by hand and write successive drafts at the keyboard. But keep in mind that writing out a draft longhand has two disadvantages compared with composing at the computer: even if you're a poor typist, a computer keyboard encourages speed, something that can come in handy if you're a student with a busy schedule. And secondly, later on when you work from a rough handwritten copy at the computer, you may be tempted to simply retype that same hard copy, rather than composing and revising it anew, a process that not only is inefficient but does not teach you how to dramatically revise your work. Yes, it does take time to sensitize yourself to the medium and cultivate an approach in which you're typing rough drafts directly into the computer, but in the long run this may not only save you valuable time, it may enhance your ability to revise.

Revising

Whether you're going to write only one first draft or (ideally) a series of intermittent drafts, there's one thing you ought to do after each one: print out a hard copy, read it, and make changes on that copy. Once you're satisfied with your changes, transfer them into your latest draft along with any other major and minor revisions.

There are several reasons to periodically print out and read hard copies of your work. For one, your eye will catch mistakes on paper that it will miss on the screen. (Incidentally, it's a great idea to read your work out loud as well, so that your ear will catch errors your eye might not see.) Also, there's a definite advantage to seeing the entire text on a piece of paper as opposed to visualizing your paper only through the keyhole of the computer screen. After all, chances are your final text will be presented to an audience in hard copy format, so it would be wise to familiarize yourself with what such a printout looks like at various phases throughout the revising process.

Fine-Tuning Your Final Draft

During the final stages of your revision, when you have assured yourself that the content of your paper is solid and organized to your liking, it is important to address smaller but equally vital concerns—namely, the style of your prose.

Throughout your revisions you will have stopped to revise particular sentences and clauses, but you will not have devoted systematic, document-wide attention to the construction of your sentences. A word processor encourages you to be a wordsmith—to quibble with word choices and to experiment with alternate phrasings. Ultimately the goal is to achieve a consistent style that allows you to communicate your main ideas as clearly, succinctly, and artfully as you can.

Your word processor provides an opportunity to rework your grammar, usage, and punctuation. At this late stage in your writing you ought not take anything for granted; that is, let no mark on your paper go unexamined. Are there passages where your language is markedly different from other sections? Will such unexpected rhetorical transitions enhance your prose or distract the reader? Have you overused certain words or phrases? Does every comma belong where you've put it? Might your text benefit from the inclusion of more (or fewer) semicolons, dashes, and colons? Are your paragraphs relatively consistent in length (or, conversely, could your text benefit from having a greater variety of paragraph lengths)?

Finally, run your final draft through the spell-checker. After all, you've worked too hard on your paper to let trivial errors of spelling, doubled words, or inadvertently used homonyms mar the end product. And even after you've run it through the spell-checker, it's essential to proofread the final draft one last time before turning it in, to catch the types of glitches the spell-checker cannot find.

This may seem like a lot of revisions, and it is. But good writers take pride in their work and are careful not to turn in anything unless they are sure it meets their own standard of excellence.

Printing

When you have proofread your document, print two final copies: one for you and one for your instructor. Make sure the format—in the line spacing, font, and margins—meets the requirements of your audience. It may be necessary to print out several copies before all of these final details are in order.

You should save at least some of the various drafts you've worked on prior to this point; in some cases an instructor might wish to see them to gauge your progress as a writer. While it's not always necessary to print your final paper out on a laser printer, which in some computer labs might require a fee, at least make sure that your text is dark and readable. Dot matrix printouts are sometimes hard to read, which is why some instructors don't permit them. Finally, remember to recycle any unnecessary copies you may have printed out.

Looking to the Future

So far we have discussed just one very important, albeit small feature of computers: the word-processing software. But for a writer, technological advancements in computer science and artificial intelligence offer exciting tools for the construction of ideas, and with them new ways of thought. "Hypertext" offers us alternative ways of creating prose in a nonlinear linked format where the reader chooses the order in which the prose is to be read. Soon, when hypermedia becomes more accessible to general users, written texts will be accompanied by sound, three-dimensional graphics, and video on a regular basis. As more and more writers join networked discussion lists and international listservs, the possibilities for collaborative writing will increase considerably. And in the case of virtual reality, where users put on goggles, gloves, and body suits and "enter" virtual worlds, the possibilities for writers will undergo profound changes.

While it may be some time before such innovative technologies replace the traditional two-dimensional college research paper, such advancements are nevertheless available right now, and every day become a more vivid part of our surroundings. Whether you are excited or skeptical about such shifts, we have reached a point in history where a writer can be at a significant disadvantage if he or she remains computer illiterate, especially now that very young children are learning how to use computers in their classrooms. Learning how to write with a word processor will not only help you create better writing, but it might offer you a limited initiation into the technological environments rapidly altering our daily landscape.

The Myth of the Cave[1]

Translated by Benjamin Jowett

Plato

And now, I said, let me show in a figure how far our nature is enlightened or unenlightened:—Behold! human beings living in an underground den, which has a mouth open towards the light and reaching all along the den; here they have been from their childhood, and have their legs and necks chained so that they cannot move, and can only see before them, being prevented by the chains from turning round their heads. Above and behind them a fire is blazing at a distance, and between the fire and the prisoners there is a raised way; and you will see, if you look, a low wall built along the way, like the screen which marionette players have in front of them, over which they show the puppets.

I see.

And do you see, I said, men passing along the wall carrying all sorts of vessels, and statues and figures of animals made of wood and stone and various materials, which appear over the wall? Some of them are talking, others silent.

You have shown me a strange image, and they are strange prisoners.

Like ourselves, I replied; and they see only their own shadows, or the shadows of one another, which the fire throws on the opposite wall of the cave?

True, he said; how could they see anything but the shadows if they were never allowed to move their heads?

And of the objects which are being carried in like manner they would only see the shadows?

Yes, he said.

And if they were able to converse with one another, would they not suppose that they were naming what was actually before them?

Very true.

And suppose further that the prison had an echo which came from the other side, would they not be sure to fancy when one of the passers-by spoke that the voice which they heard came from the passing shadow?

No question, he replied.

To them, I said, the truth would be literally nothing but the shadows of the images.

That is certain.

And now look again, and see what will naturally follow if the prisoners are released and disabused of their error. At first, when any of them is liberated and compelled suddenly to stand up and turn his neck round and walk and look toward the light, he will suffer sharp pains; the glare will distress him, and he will be unable to see the realities of which in his former state he had seen the shadows; and then conceive some one saying to him, that what he saw before was an illusion, but that now, when he is approaching nearer to being and his eye is turned towards more real existence, he has a clearer vision—what will be his reply? And you may further imaging that his instructor is pointing to the objects as they pass and requiring him to name them—will he not be perplexed? Will he not fancy that the shadows which he formerly saw are truer than the objects which are now shown to him?

Far truer.

And if he is compelled to look straight at the light, will he not have a pain in his eyes which will make him turn away to take refuge in the objects of vision which he can see, and which he will conceive to be in reality clearer than the things which are now being shown to him?

True, he said.

And suppose once more, that he is reluctantly dragged up a steep and rugged ascent, and held fast until he is forced into the presence of the sun himself, is he not likely to be pained and irritated? When he approaches the light his eyes will be dazzled, and he will not be able to see anything at all of what are now called realities.

Not all in a moment, he said.

He will require to grow accustomed to the sight of the upper world. And first he will see the shadows best, next the reflections of men and other objects in the water, and then the objects themselves; then he will gaze upon the light of the moon and the stars and the spangled heaven; and he will see the sky and the stars by night better than the sun or the light of the sun by day?

Certainly.

Last of all he will be able to see the sun, and not mere reflections of him in the water, but he will see him in his own proper place, and not in another; and he will contemplate him as he is.

Certainly.

He will then proceed to argue that this is he who gives the season and the years, and is the guardian of all that is in the visible world, and in a certain way the cause of all things which he and his fellows have been accustomed to behold?

Clearly, he said, he would first see the sun and then reason about him.

And when he remembered his old habitation, and the wisdom of the den and his fellow-prisoners, do you not suppose that he would felicitate himself on the change, and pity them?

Certainly, he would.

And if they were in the habit of conferring honors among themselves on those who were quickest to observe the passing shadows and to remark which of them went before, and which followed after, and which were together; and who were therefore best able to draw conclusions as to the future, do you think that he would care for such honors and glories, or envy the possessors of them? Would he not say with Homer,

> Better to be the poor servant of a poor master,

and to endure anything, rather than think as they do and live after their manner?

Yes, he said, I think that he would rather suffer anything than entertain these false notions and live in this miserable manner.

Imagine once more, I said, such an one coming suddenly out of the sun to be replaced in his old situation; would he not be certain to have his eyes full of darkness?

To be sure, he said.

And if there were a contest, and he had to compete in measuring the shadows with the prisoners who had never moved out of the den, while his sight was still weak, and before his eyes had become steady (and the time which would be needed to acquire this new habit of sight might be very considerable), would he not be ridiculous? Men would say of him that up he went and down he came without his eyes; and that it was better not even to think of ascending; and if any one tried to loose another and lead him up to the light, let them only catch the offender, and they would put him to death.

No question, he said.

The entire allegory, I said, you may now append, dear Glaucon, to the previous argument; the prison house is the world of sight, the light of the fire is the sun, and you will not misapprehend me if you interpret the journey upwards to be the ascent of the soul into the intellectual world according to my poor belief, which, at your desire, I have expressed—whether rightly or wrongly God knows. But, whether true or false, my opinion is that in the world of knowledge the idea of good appears last of all, and is seen only with an effort; and, when seen, is also inferred to be the universal author of all things beautiful and right, parent of light and of the lord of light in this visible world, and the immediate source of reason and truth in the intellectual; and that this is the power upon which he who would act rationally either in public or private life must have his eye fixed.

I agree, he said, as far as I am able to understand you.

Moreover, I said, you must not wonder that those who attain to this beatific vision are unwilling to descend to human affairs; for their souls are ever hastening into the upper world where they desire to dwell; which desire of theirs is very natural, if our allegory may be trusted.

Yes, very natural.

And is there anything surprising in one who passes from divine contemplations to the evil state of man, misbehaving himself in a ridiculous manner; if, while his eyes are blinking and before he has become accustomed to the surrounding darkness, he is compelled to fight in courts of law, or in other places, about the images or the shadows of images of justice, and is endeavouring to meet the conceptions of those who have never yet seen absolute justice?

Anything but surprising, he replied.

Anyone who has common sense will remember that the bewilderments of the eyes are of two kinds, and arise from two causes, either from coming out of the light or from going into the light, which is true of the mind's eye, quite as much as of the bodily eye; and he who remembers this when he sees anyone whose vision is perplexed and weak, will not be too ready to laugh; he will first ask whether that soul of man has come out of the brighter life, and is unable to see because unaccustomed to the dark, or having turned from darkness to the day is dazzled by excess of light. And he will count the one happy in his condition and state of being, and he will pity the other; or, if he have a mind to laugh at the soul which comes from below into the light, there will be more reason in this than in the laugh which greets him who returns from above out of the light into the den.

That, he said, is a very just distinction.

But then, if I am right, certain professors of education must be wrong when they say that they can put a knowledge into the soul which was not there before, like sight into blind eyes.

They undoubtedly say this, he replied.

Whereas, our argument shows that the power and capacity of learning exists in the soul already; and that just as the eye was unable to turn from darkness to light without the whole body, so too the instrument of knowledge can only by the movement of the whole soul be turned from the world of becoming into that of being, and learn by degrees to endure the sight of being, and of the brightest and best of being, or in other words, of the good.

Very true.

And must there not be some art which will effect conversion in the easiest and quickest manner; not implanting the faculty of sight, for that exists already, but has been turned in the wrong direction, and is looking away from the truth?

Yes, he said, such an art may be presumed.

And whereas the other so-called virtues of the soul seem to be akin to bodily qualities, for even when they are not originally innate they can be implanted later by habit and exercise, the virtue of wisdom more than anything else contains a divine element which always remains, and by this conversion is rendered useful and profitable; or, on the other hand, hurtful and useless. Did you never observe the narrow intelligence flashing from the keen eye of a clever rogue—how eager he is, how clearly his paltry soul sees the way to his end; he is the reverse of blind, but his keen eyesight is forced into the service of evil, and he is mischievous in proportion to his cleverness?

Very true, he said.

But what if there had been a circumcision of such natures in the days of their youth; and they had been severed from those sensual pleasures, such as eating and drinking, which, like leaden weights, were attached to them at their birth, and which drag them down and turn the vision of their souls upon the things that are below—if, I say, they had been released from these impediments and turned in the opposite direction, the very same faculty in them would have seen the truth as keenly as they see what their eyes are turned to now.

Very likely.

Yes, I said; and there is another thing which is likely, or rather a necessary inference from what has preceded, that neither the uneducated and uninformed of the truth, nor yet those who never make an end of their education, will be able ministers of State; not the former, because they have no single aim of duty which is the rule of all their actions, private as well as public; nor the latter, because they will not act at all except upon compulsion, fancying that they are already dwelling apart in the islands of the blest.

Very true, he replied.

Then, I said, the business of us who are founders of the State will be to compel the best minds to attain that knowledge which we have already shown to be the greatest of all—they must continue to ascend until they arrive at the good; but when they have ascended and seen enough we must not allow them to do as they do now.

What do you mean?

I mean that they remain in the upper world: but this must not be allowed; they must be made to descend again among the prisoners in the den, and partake of their labors and honors, whether they are worth having or not.

But is not this unjust? he said; ought we to give them a worse life, when they have a better?

You have again forgotten, my friend, I said, the intention of the legislator, who did not aim at making any one class in the State happy above the rest; the happiness was to be in the whole State, and he held the citizens together by persuasion and necessity, making them benefactors of the State, and therefore benefactors of one another; to this end he created them, not to please themselves, but to be his instruments in binding up the State.

True, he said, I had forgotten.

Observe, Glaucon, that there will be no injustice in compelling our philosophers to have a care and providence of others; we shall explain to them that in other States, men of their class are not obliged to share in the toils of politics: and this is reasonable, for they grow up at their own sweet will, and the government would rather not have them. Being self-taught, they cannot be expected to show any gratitude for a culture which they have never received. But we have brought you into the world to be rulers of the hive, kings of yourselves and of the other citizens, and have educated you far better and more perfectly than they have been educated, and you are better able to share in the double duty. Wherefore each of you, when his turn comes, must go down to the general underground abode, and get the habit of seeing in the dark. When you have acquired the habit, you will see ten thousand times better than the inhabitants of the den, and you will know what the several images are, and what they represent, because you have seen the beautiful and just and good in their truth. And thus our State which is also yours will be a reality, and not a dream only, and will be administered in a spirit unlike that of other

States, in which men fight with one another about shadows only and are distracted in the struggle for power, which in their eyes is a great good. Whereas the truth is that the State in which the rulers are most reluctant to govern is always the best and most quietly governed, and the State in which they are most eager, the worst.

Quite true, he replied.

And will our pupils, when they hear this, refuse to take their turn at the toils of State, when they are allowed to spend the greater part of their time with one another in the heavenly light?

Impossible, he answered; for they are just men, and the commands which we impose upon them are just; there can be no doubt that every one of them will take office as a stern necessity, and not after the fashion of our present rulers of State.

Yes, my friend, I said; and there lies the point. You must contrive for your future rulers another and a better life than that of a ruler, and then you may have a well-ordered State; for only in the State which offers this, will they rule who are truly rich, not in silver and gold, but in virtue and wisdom, which are the true blessings of life. Whereas if they go to the administration of public affairs, poor and hungering after their own private advantage, thinking that hence they are to snatch the chief good, order there can never be; for they will be fighting about office, and the civil and domestic broils which thus arise will be the ruin of the rulers themselves and of the whole State.

Most true, he replied.

And the only life which looks down upon the life of political ambition is that of true philosophy. Do you know of any other?

Indeed, I do not, he said.

And those who govern ought not to be lovers of the task? For, if they are, there will be rival lovers, and they will fight.

No question.

Who then are those whom we shall compel to be guardians? Surely they will be the men who are wisest about affairs of State, and by whom the State is best administered, and who at the same time have other honors and another and a better life than that of politics?

They are the men, and I will choose them, he replied.

And now shall we consider in what way such guardians will be produced, and how they are to be brought from darkness to light—as some are said to have ascended from the world below to the gods?

By all means, he replied.

The process, I said, is not the turning over of an oyster-shell,[2] but the turning round of a soul passing from a day which is little better than night to the true day of being, that is, the ascent from below which we affirm to be true philosophy?

Quite so.

Notes

1. Plato, born in Athens, the son of an aristocratic family, wrote thirty dialogues in which Socrates is the chief speaker. Socrates, about twenty-five years older than Plato, was a philosopher who called himself a gadfly

to Athenians. For his efforts at stinging them into thought, the Athenians executed him in 399 B.C. "The Myth of the Cave" is the beginning of Book VII of Plato's dialogue entitled *The Republic.* Socrates is talking with Glaucon.

For Plato, true knowledge is philosophic insight or awareness of the Good, not mere opinion or the knack of getting along in this world by remembering how things have usually worked in the past. To illustrate his idea that awareness of the Good is different from the ability to recognize the things of this shabby world, Plato (through his spokesman Socrates) resorts to an allegory: men imprisoned in a cave see on a wall in front of them the shadows or images of objects that are really behind them, and they hear echoes, not real voices. (The shadows are caused by the light from a fire behind the objects, and the echoes by the cave's acoustical properties.) The prisoners, unable to perceive the real objects and the real voices, mistakenly think that the shadows and the echoes are real, and some of them grow highly adept at dealing with this illusory world. Were Plato writing today, he might have made the cave a movie theater: we see on the screen in front of us images caused by an object (film, passing in front of light) that is behind us. Moreover, the film itself is an illusory image, for it bears only the traces of a yet more real world—the world that was photographed—outside of the movie theater. And when we leave the theater to go into the real world, our eyes have become so accustomed to the illusory world that we at first blink with discomfort—just as Plato's freed prisoners do when they move out of the cave—at the real world of bright day, and we long for the familiar darkness. So too, Plato suggests, dwellers in ignorance may prefer the familiar shadows of their unenlightened world ("the world of becoming") to the bright world of the eternal Good ("the world of being") that education reveals.

2. An allusion to a game in which two parties fled or pursued according as an oyster-shell which was thrown into the air fell with the dark or light side uppermost. (Translator's note)

Letter to His Father

Translated by Ernest Kaiser and Eithene Wilkins

Franz Kafka

*Franz Kafka (1883–1924) was born in Prague, Czechoslovakia to Jewish parents with whom he lived most of his life. Kafka was plagued by emotional and physical infirmities, and he completed and published very few of his works. Before he died of tuberculosis at the age of forty-one, he instructed his friend Max Brod to destroy his remaining manuscripts, but fortunately Brod did not do so. Three unfinished novels—*The Trial *(1925),* The Castle *(1926), and* Amerika *(1927)—brought Kafka international recognition, albeit posthumously. His stories, translated from German into English, can be found in* The Great Wall of China *(1933),* The Penal Colony *(1948), and* The Complete Stories *(1976).*

In the long missive later published as "Letter to His Father," Kafka, at age thirty-six, reproaches his father for his cruel, intimidating, and boorish behavior.

Dearest Father:

You asked me recently why I maintain that I am afraid of you. As usual I was unable to think of any answer to your question, partly for the very reason that I am afraid of you, and partly because an explanation of the grounds for this fear would mean going into far more details than I could even approximately keep in mind while talking. And if I now try to give you an answer in writing, it will still be very incomplete, because even in writing this fear and its consequences hamper me in relation to you and because [anyway] the magnitude of the subject goes far beyond the scope of my memory and power of reasoning. . . .

Compare the two of us: I, to put it in a very much abbreviated form, a Löwy with a certain basis of Kafka, which, however, is not set in motion by the Kafka will to life, business, and conquest, but by a Löwyish spur that urges more secretly, more diffidently, and in another direction, and which often fails to work entirely. You, on the other hand, a true Kafka in strength, health, appetite, loudness of voice, eloquence, self-satisfaction, worldly dominance, endurance, presence of mind, knowledge of human nature, a certain way of doing things on a grand scale, of course with all the defects and weaknesses that go with all these advantages and into which your temperament and sometimes your hot temper drive you. . . .

However it was, we were so different and in our difference so dangerous to each other that, if anyone had tried to calculate in advance how I, the slowly developing child, and you, the full-grown man, would stand to each other, he could have assumed that you would simply trample me underfoot so that nothing was left of me. Well, that didn't happen. Nothing alive can be calculated. But perhaps something worse happened. And in saying this I would all the time beg of you not to forget that I never, and not even for a single moment, believe any guilt to be on your side. The effect you had on me was the effect you could not help having. But you should stop considering it some particular malice on my part that I succumbed to that effect.

I was a timid child. For all that, I am sure I was also obstinate, as children are. I am sure that Mother spoilt me too, but I cannot believe I was particularly difficult to manage; I cannot believe that a kindly word, a quiet taking of me by the hand, a friendly look, could not have got me to do anything that was wanted of me. Now you are after all at bottom a kindly and soft-hearted person (what follows will not be in contradiction to this, I am speaking only of the impression you made on the child), but not every child has the endurance and fearlessness to go on searching until it comes to the kindliness that lies beneath the surface. You can only treat a child in the way you yourself are constituted, with vigor, noise, and hot temper, and in this case this seemed to you, into the bargain, extremely suitable, because you wanted to bring me up to be a strong brave boy. . . .

There is only one episode in the early years of which I have a direct memory. You may remember it, too. Once in the night I kept on whimpering for water, not, I am certain, because I was thirsty, but partly to be annoying, partly to amuse myself. After several vigorous threats had failed to have any effect, you took me out of bed, carried me out onto the *pavlatche* and left me there alone for a while in my nightshirt, outside the shut door. I am not going to say that this was wrong—perhaps at that time there was really no other way of getting peace and quiet that night—but I mention it as typical of your methods of bringing up a child and their effect on me. I dare say I was quite obedient afterwards at that period, but it did me inner harm. What was for me a matter of course, that senseless asking for water, and the extraordinary terror of being carried outside were two things that I, my nature being what it was, could never properly connect with each other. Even years afterwards I suffered from the tormenting fancy that the huge man, my father, the ultimate authority, would come almost for no reason at all and take me out of bed at night and carry me out onto the *pavlatche*, and that therefore I was such a mere nothing for him.

That then was only a small beginning, but this sense of nothingness that often dominates me (a feeling that is in another respect, admittedly, also a noble and fruitful one) comes largely from your influence. What I would have needed was a little encouragement, a little friendliness, and a little keeping open of my road, instead of which you blocked it for me, though of course with the good intention of making me go another road. But I was not fit for that. You encouraged me, for instance, when I saluted and marched smartly, but I was no future soldier, or you encouraged me when I was able to eat heartily or even drink beer with my meals, or when I was able to repeat songs, singing what I had not understood, or prattle to you using your own favorite expressions, imitating you, but nothing of this had anything to do with my future. And it is characteristic that even today you really only encourage me in anything when you yourself are involved in it, when what is at stake is your sense of self-importance.

At that time, and at that time everywhere, I would have needed encouragement. I was, after all, depressed even by your mere physical presence. I remember, for instance, how we often undressed together in the same bathing hut. There was I, skinny, weakly, slight; you strong, tall, broad. Even inside the hut I felt myself a miserable specimen, and what's more, not only in your eyes but in the eyes of the whole world, for you were for me the measure of all things. But then when we went out of the bathing hut before the people, I with you holding my hand, a little skeleton, unsteady, barefoot on the boards, frightened of the water, incapable of copying your swimming strokes, which you, with the best of intentions, but actually to my profound humiliation, always kept on showing me, then I was frantic with desperation and all my bad experiences in all spheres at such moments fitted magnificently together. . . .

In keeping with that, furthermore, was your intellectual domination. You had worked your way up so far alone, by your own energies, and as a result you had unbounded confidence in your opinion. For me as a child that was not yet so dazzling as later for the boy growing up. From your armchair you ruled the world. Your opinion was correct, every other was mad, wild, *meshugge*, not normal. With all this your self-confidence was so great that you had no need to be consistent at all and yet never ceased to be in the right. It did sometimes happen that you had no opinion whatsoever about a matter and as a result all opinions that were at all possible with respect to the matter were necessarily wrong without exception. You were capable, for instance, of running down the Czechs, and then the Germans, and then the Jews, and what is more, not only selectively but in every respect, and finally nobody was left except yourself. For me you took on the enigmatic quality that all tyrants have whose rights are based on their person and not on reason. At least so it seemed to me.

Now where I was concerned you were in fact astonishingly often in the right, which was a matter of course in talk, for there was hardly ever any talk between us, but also in reality. Yet this too was nothing particularly incomprehensible; in all my thinking I was, after all, under the heavy pressure of your personality, even in that part of it—and particularly in that—which was not in accord with yours. All these thoughts, seemingly independent of you, were from the beginning loaded with the burden of your harsh and dogmatic judgments; it was almost impossible to endure this, and yet to work out one's thoughts with any measure of completeness and permanence. I am not here speaking of any sublime thoughts, but of every little enterprise in childhood. It was only necessary to be happy about something or other, to be filled with the thought of it, to come home and speak of it, and the answer was an ironical sigh, a shaking of the head, a tapping of the table with one finger: "Is that all you're so worked up about?" or "I wish I had your worries!" or "The things some people have time to think about!" or "What can you buy yourself with that?" or "What a song and dance about nothing!" Of course, you couldn't be expected to be enthusiastic about every childish triviality, toiling and moiling as you used to. But that wasn't the point. The point was, rather, that you could not help always and on principle causing the child such disappointments, by virtue of your antagonistic nature, and further that this antagonism was ceaselessly intensified through accumulation of its material, that it finally became a matter of established habit even when for once you were of the same opinion as myself, and that finally these disappointments of the child's were not disappointments in ordinary life but, since what it concerned was your person, which was the measure of all things, struck to the very core. Courage, resolution, confidence, delight in this and that, did not endure to the end when you were against whatever it was or even if your opposition was merely to be assumed; and it was to be assumed in almost everything I did. . . .

You have, I think, a gift for bringing up children: you could, I am sure, have been of use to a human being of your own kind with your methods; such a person would have seen the reasonableness of what you told him, would not have troubled about anything else, and would quietly have done things the way he was told. But for me a child everything you shouted at me was positively a heavenly commandment, I never forgot it, it remained for me the most important means of forming a judgment of the world, above all of forming a judgment of you yourself, and there you failed entirely. Since as a child I was together with you chiefly at meals, your teaching was to a large extent teaching about proper behavior at table. What was brought to the table had to be eaten up, there could be no discussion of the goodness of the food—but you yourself often found the food uneatable, called it "this swill," said "that brute" (the cook) had ruined it. Because in accordance with your strong appetite and your particular habit you ate everything fast, hot and in big mouthfuls, the child had to hurry, there was a somber silence at table, interrupted by admonitions: "Eat first, talk afterwards," or "faster, faster, faster," or "there you are, you see, I finished ages ago." Bones mustn't be cracked with the teeth, but you could. Vinegar must not be sipped noisily, but you could. The main thing was that the bread should be cut straight. But it didn't matter that you did it with a knife dripping with gravy. One had to take care that no scraps fell on the floor. In the end it was under your chair that there were most scraps. At table one wasn't allowed to do anything but eat, but you cleaned and cut your fingernails, sharpened pencils, cleaned your ears with the toothpick. Please, Father, understand me rightly: these would in themselves have been utterly insignificant details, they only became depressing for me because you, the man who was so tremendously the measure of all things for me, yourself, did not keep the commandments you imposed on me. Hence the world was for me divided into three parts: into one in which I, the slave, lived under laws that had been invented only for me and which I could, I did not know why, never completely comply with; then into a second world, which was infinitely remote from mine, in which you lived, concerned with government, with the issuing of orders and with annoyance about their not being obeyed; and finally into a third world where everybody else lived happily and free from orders and from having to obey. I was continually in disgrace, either I obeyed your orders, and that was a disgrace, for they applied, after all, only to me, or I was defiant, and that was a disgrace too, for how could I presume to defy you, or I could not obey because, for instance, I had not your strength, your appetite, your skill, in spite of which you expected it of me as a matter of course; this was the greatest disgrace of all. What moved in this way was not the child's reflections, but his feelings. . . .

It was true that Mother was illimitably good to me, but all that was for me in relation to you, that is to say, is no good relation. Mother unconsciously played the part of a beater during a hunt. Even if your method of upbringing might in some unlikely case have set me on my own feet by means of producing defiance, dislike, or even hate in me, Mother canceled that out again by kindness, by talking sensibly (in the maze and chaos of my childhood she was the very pattern of good sense and reasonableness), by pleading for me, and I was again driven back into your orbit, which I might perhaps otherwise have broken out of, to your advantage and to my own. Or it was so that no real reconciliation ever came about, that Mother merely shielded me from you in secret, secretly gave me something, or allowed me to do something, and then where you were concerned I was again the furtive creature, the cheat, the guilty one, who in his worthlessness could only pursue backstairs methods even to get the things he regarded as his right. Of course, I then became used to taking such courses also in quest of things to which, even in my own view, I had no right. This again meant an increase in the sense of guilt.

It is also true that you hardly ever really gave me a whipping. But the shouting, the way your face got red, the hasty undoing of the braces and the laying of them ready over the back of the chair, all that was almost worse for me. It is like when someone is going to be hanged. If he is really hanged, then he's dead and it's all over. But if he has to go through all the preliminaries to being hanged and only when the noose is dangling before his face is told of his reprieve, then he may suffer from it all his life long. Besides, from so many occasions when I had, as you clearly showed you thought, deserved to be beaten, when you were however gracious enough to let me off at the last moment, here again what accumulated was only a huge sense of guilt. On every side I was to blame, I was in debt to you.

You have always reproached me (and what is more either alone or in front of others, you having no feeling for the humiliation of this latter, your children's affairs always being public affairs) for living in peace and quiet, warmth, and abundance, lack for nothing, thanks to your hard work. I think here of remarks that must positively have worn grooves in my brain, like: "When I was only seven I had to push the barrow from village to village." "We all had to sleep in one room." "We were glad when we got potatoes." "For years I had open sores on my legs from not having enough clothes to wear in winter." "I was only a little boy when I was sent away to Pisek to go into business." "I got nothing from home, not even when I was in the army, even then I was sending money home." "But for all that, for all that—Father was always Father to me. Ah, nobody knows what that means these days! What do these children know of things? Nobody's been through that! Is there any child that understands such things today?" Under other conditions such stories might have been very educational, they might have been a way of encouraging one and strengthening one to endure similar torments and deprivations to those one's father had undergone. But that wasn't what you wanted at all; the situation had, after all, become quite different as a result of all your efforts, and there was no opportunity to distinguish oneself in the world as you had done. Such an opportunity would first of all have had to be created by violence and revolution, it would have meant breaking away from home (assuming one had had the resolution and strength to do so and that Mother wouldn't have worked against it, for her part, with other means). But all that was not what you wanted at all, that you termed ingratitude, extravagance, disobedience, treachery, madness. And so, while on the one hand you tempted me to it by means of example, story, and humiliation, on the other hand you forbade it with the utmost severity. . . .

(Up to this point there is in this letter relatively little I have intentionally passed over in silence, but now and later I shall have to be silent on certain matters that it is still too hard for me to confess—to you and to myself. I say this in order that, if the picture as a whole should be somewhat blurred here and there, you should not believe that what is to blame is any lack of evidence; on the contrary, there is evidence that might well make the picture unbearably stark. It is not easy to strike a median position.) Here, it's enough to remind you of early days. I had lost my self-confidence where you were concerned, and in its place had developed a boundless sense of guilt. (In recollection of this boundlessness I once wrote of someone accurately: "He is afraid the shame will outlive him, even.") I could not suddenly undergo a transformation when I came into the company of other people; on the contrary, with them I came to feel an even deeper sense of guilt, for, as I have already said, in their case I had to make good the wrongs done them by you in the business, wrongs in which I too had my share of responsibility. Besides, you always, of course, had some objection to make, frankly or covertly, to everyone I associated with, and for this too I had to beg his pardon. The mistrust that you tried to instill into me, at business and at home, towards most people (tell me of any single person who was of importance to me in my childhood whom you didn't at least once tear to shreds with your criticism), this mistrust, which oddly enough was no particular burden to you (the fact was that you were strong enough to bear it, and besides it was in reality perhaps only a token of the autocrat), this mistrust, which for me as a little boy was nowhere confirmed in my own eyes, since I everywhere saw only people excellent beyond all hope of emulation, in me turned into mistrust of myself and into perpetual anxiety in relation to everything else. There, then, I was in general certain of not being able to escape from you.

A WASP Stings Back

Robert Claiborne

Contexts for Reading: *Although Robert Claiborne was born in England in 1919, he was educated in the United States and received a B.A. from New York University in 1942. Since the 1950s, he has been an editor and writer.* Our Marvelous Native Tongue: The Life and Times of the English Language *(1983) and* Saying What You Mean: A Commonsense Guide to American Usage *(1986) are two of his most recent works. "A WASP Stings Back," which first appeared in* Newsweek *in 1974, argues that White Anglo-Saxon Protestants are themselves the targets of stereotyping and prejudice.*

As You Read: *Word choice contributes to the style of a writer by setting the tone of the work. Notice how the choice of phrases such as "media gurus," "anti-Wasp chic," and "emotional uptightness" sets the tone for Claiborne's defense of "English-speaking peoples."*

Over the past few years, American pop culture has acquired a new folk antihero: the Wasp. One slick magazine tells us that the White Anglo-Saxon Protestants rule New York City, while other media gurus credit (or discredit) them with ruling the country—and, by inference, ruining it. A Polish-American declares in a leading newspaper that Wasps have "no sense of honor." *Newsweek* patronizingly describes Chautauqua as a citadel of "Wasp values," while other folklorists characterize these values more explicitly as a compulsive commitment to the work ethic, emotional uptightness and sexual inhibition. The Wasps, in fact, are rapidly becoming the one minority that every other ethic group—blacks, Italians, chicanos, Jews, Poles and all the rest—feels absolutely free to dump on. I have not yet had a friend greet me with "Did you hear the one about the two Wasps who . . . ?"—but any day now!

I come of a long line of Wasps; if you disregard my French great-great-grandmother and a couple of putatively Irish ancestors of the same vintage, a rather pure line. My mother has long been one of the Colonial Dames, an organization some of whose members consider the Daughters of the American Revolution rather parvenu. My umptyumpth Wasp great-grandfather, William Claiborne, founded the first European settlement in what is now Maryland (his farm and trading post were later ripped off by the Catholic Lord Baltimore, Maryland politics being much the same then as now).

The Stereotype

As a Wasp, the mildest thing I can say about the stereotype emerging from the current wave of anti-Wasp chic is that I don't recognize myself. As regards emotional uptightness and sexual inhibition, modesty forbids comment—though I dare say various friends and lovers of mine could testify on these points if they cared to. I will admit to enjoying work—because I am lucky enough to be able to work at what I enjoy—but not, I think, to the point of compulsiveness. And so far as ruling America, or even New York, is concerned, I can say flatly that (a) it's a damn lie because (b) if I *did* rule them, both would be in better shape than they are. Indeed I and all my Wasp relatives, taken in a lump, have far less clout with the powers that run this country than any one of the Buckleys or Kennedys (Irish Catholic), the Sulzbergers or Guggenheims (Jewish), or the late A. P. Giannini (Italian) of the Bank of America.

Admittedly, both corporate and (to a lesser extent) political America are dominated by Wasps—just as (let us say) the garment industry is dominated by Jews, and organized crime by Italians. But to conclude from this that The Wasps are the American elite is as silly as to say that The Jews are cloak-and-suiters or The Italians are gangsters. Wasp, like other ethnics, come in all varieties, including criminals—political, corporate and otherwise.

The Values

More seriously, I would like to say a word for the maligned "Wasp values," one of them in particular. As a matter of historical fact, it was we Wasps—by which I mean here the English-speaking peoples—who invented the idea of *limited governments:* that there are some things that no king, President or other official is allowed to do. It began more than seven centuries ago, with Magna Carta, and continued (to cite only the high spots) through the wrangles between Parliament and the Stuart kings, the Puritan Revolution of 1640, the English Bill of Rights of 1688, the American Revolution and our own Bill of Rights and Constitution.

The Wasp principle of limited government emerged through protracted struggle with the much older principle of unlimited government. This latter was never more cogently expressed than at the trial of Charles I, when the hapless monarch informed his judges that, as an anointed king, he was not accountable to any court in the land. A not dissimilar position was taken more recently by another Wasp head of state—and with no more success: Executive privilege went over no better in 1974 than divine right did in 1649. The notion that a king, a President or any other official can do as he damn well pleases has never played in Peoria—or Liverpool or Glasgow, Melbourne or Toronto. For more than 300 years, no Wasp nation has endured an absolute monarchy, dictatorship or any other form of unlimited government—which is something no Frenchman, Italian, German, Pole, Russian or Hispanic can say.

It is perfectly true, of course, that we Wasps have on occasion imposed unlimited governments on other (usually darker) people. We have, that is, acted in much the same way as have most other nations that possessed the requisite power and opportunity—including many Third World nations whose leaders delight in lecturing us on political morality (for recent information on this point, consult the files on Biafra, Bangladesh and Brazil, Indian tribes of). Yet even here, Wasp values have played an honorable part. When you start with the idea that Englishmen are entitled to self-government, you end by conceding the same right to Africans and Indians. If you begin by declaring that all (white) men are created equal, you must sooner or later face up to the fact that blacks are also men—and conform your conduct, however reluctantly, to your values.

The Faith

Keeping the Wasp faith hasn't always been easy. We Wasps, like other people, don't always live up to our own principles, and those of us who don't, if occupying positions of power, can pose formidable problems to the rest of us. Time after time, in the name of anti-Communism, peace with honor or some other slippery shibboleth, we have been conned or bullied into tolerating government interference with our liberties and privacy in all sorts of covert—and sometimes overt—ways; time after time we have had to relearn the lesson that eternal vigilance is the price of liberty.

It was a Wasp who uttered that last thought. And it was a congress of Wasps who, about the same time, denounced the executive privileges of George III and committed to the cause of liberty their lives, their fortunes and—*pace,* my Polish-American compatriot—their sacred honor.

The Postcolonial and the Postmodern

Kwame Anthony Appiah

You were called Bimbircokak
And all was well that way
You have become Victor-Emile-Louis-Henri-Joseph
Which
So far as I recall
Does not reflect your kinship with
Rockefeller.

Yambo Ouologuem

In 1987 the Center for African Art in New York organized a show entitled *Perspectives: Angles on African Art.* The curator, Susan Vogel, had worked with a number of "cocurators," whom I list in order of their appearance in the table of contents: Ekpo Eyo, quondam director of the Department of Antiquities of the National Museum of Nigeria; William Rubin, director of painting and sculpture at the Museum of Modern Art and organizer of its controversial Primitivism exhibit; Romare Bearden, African-American painter; Ivan Karp, curator of African ethnology at the Smithsonian; Nancy Graves, European-American painter, sculptor, and filmmaker; James Baldwin, who surely needs no qualifying glosses; David Rockefeller, art collector and friend of the mighty; Lela Kouakou, Baule artist and diviner, from Ivory Coast (this a delicious juxtaposition, richest and poorest, side by side); Iba N'Diaye, Senegalese sculptor; and Robert Farris Thompson, Yale professor and African and African-American art historian. Vogel describes the process of selection in her introductory essay. The one woman and nine men were each offered a hundred-odd photographs of "African Art as varied in type and origin, and as high in quality, as we could manage" and asked to select ten for the show. Or, I should say more exactly, that this is what was offered to eight of the men. For Vogel adds, "In the case of the Baule artist, a man familiar only with the art of his own people, only Baule objects were placed in the pool of photographs." At this point we are directed to a footnote to the essay, which reads:

> Showing him the same assortment of photos the others saw would have been interesting, but confusing in terms of the reactions we sought here. Field aesthetic studies, my own and others, have shown that African informants will criticize sculptures from other ethnic groups in terms of their own traditional criteria, often assuming that such works are simply inept carvings of their own aesthetic tradition.

I shall return to this irresistible footnote in a moment. But let me pause to quote further, this time from the words of David Rockefeller, who would surely never "criticize sculptures from other ethnic groups in terms of [his] own traditional criteria," discussing what the catalog calls a "Fante female figure":

> I own somewhat similar things to this and I have always liked them. This is a rather more sophisticated version than the ones that I've seen, and I thought it was quite beautiful . . . the total composition has a very contemporary, very Western look to it. It's the kind of thing that goes very well with contemporary Western things. It would look good in a modern apartment or house.

We may suppose that David Rockefeller was delighted to discover that his final judgment was consistent with the intentions of the sculpture's creators. For a footnote to the earlier "Checklist" reveals that the Baltimore Museum of Art desires to "make public the fact that the authenticity of the Fante figure in its collection has been challenged." Indeed, work by Doran Ross suggests this object is almost certainly a modern piece introduced in my hometown of Kumasi by the workshop of a certain Francis Akwasi, which "specializes in carvings for the international market in the style of traditional sculpture. Many of its works are now in museums throughout the West, and were published as authentic by Cole and Ross" (yes, the same Doran Ross) in their classic catalog *The Arts of Ghana.*

But then it is hard to be *sure* what would please a man who gives as his reason for picking another piece (this time a Senufo helmet mask), "I have to say I picked this because I own it. It was given to me by President Houphouet Boigny of Ivory Coast." Or one who remarks, "concerning the market in African art":

> The best pieces are going for very high prices. Generally speaking, the less good pieces in terms of quality are not going up in price. And that's a fine reason for picking the good ones rather than the bad. They have a way of becoming more valuable.
>
> I like African art as objects I find would be appealing to use in a home or an office. . . . I don't think it goes with everything, necessarily—although the very best perhaps does. But I think it goes well with contemporary architecture.

There is something breathtakingly unpretentious in Mr. Rockefeller's easy movement between considerations of finance, of aesthetics, and of decor. In these responses we have surely a microcosm of the site of the African in contemporary—which is, then, surely to say, postmodern—America.

I have given so much of David Rockefeller not to emphasize the familiar fact that questions of what we call "aesthetic" value are crucially bound up with market value; not even to draw attention to the fact that this is known by those who play the art market. Rather, I want to keep clearly before us the fact that David Rockefeller is permitted to say *anything at all* about the arts of Africa because he is a *buyer* and because he is at the *center,* while Lela Kouakou, who merely makes art and who dwells at the margins, is a poor African whose words count only as parts of the commodification—both for those of us who constitute the museum public and for collectors, like Rockefeller—of Baule art. I want to remind you, in short, of how important it is that African art is a *commodity.*

But the cocurator whose choice will set us on our way is James Baldwin—the only cocurator who picked a piece that was not in the mold of the Africa of the exhibition Primitivism, a sculpture that will be my touchstone, a piece labeled by the museum *Yoruba Man with a Bicycle.* Here is some of what Baldwin said about it:

> This is something. This has got to be contemporary. He's really going to town. It's very jaunty, very authoritative. His errand might prove to be impossible. He is challenging something—or something has challenged him. He's grounded in immediate reality by the bicycle. . . . He's apparently a very proud and silent man. He's dressed sort of polyglot. Nothing looks like it fits him too well.

Baldwin's reading of this piece is, of course and inevitably, "in terms of [his] own . . . criteria," a reaction contextualized only by the knowledge that bicycles are new in Africa and that this piece, anyway, does not look anything like the works he recalls seeing from his earliest childhood at the Schomburg museum in Harlem. And his response torpedoes Vogel's argument for her notion that the only "authentically traditional" African—the only one whose responses, as she says, could have been found a century ago—must be refused a choice among Africa's art cultures because he, unlike the rest of the cocurators, who are Americans and the European-educated Africans, will use his "own . . . criteria." This Baule diviner, this authentically African villager, the message is, does not know what *we,* authentic postmodernists, now know: that the first and last mistake is to judge the Other on one's own terms. And so, in the name of this, the relativist insight, we impose our judgment that Lela Kouakou may not judge sculpture from beyond the Baule culture zone because he will—like all the other African "informants" we have met in the field—read them as if they were meant to meet those Baule standards.

Worse than this, it is nonsense to explain Lela Kouakou's responses as deriving from an ignorance of other traditions—if indeed he is, as he is no doubt supposed to be, like most "traditional" artists today, if he is like, for example, Francis Akwasi of Kumasi. Kouakou may judge other artists by his own standards (what on earth else could he, could anyone, do, save make no judgment at all?), but to suppose that he is unaware that there are other standards within Africa (let alone without) is to ignore a piece of absolute basic cultural knowledge, common to most precolonial as to most colonial and postcolonial cultures on the continent—the piece of cultural knowledge that explains why the people we now call "Baule" exist at all. To be Baule, for example, is, for a Baule, not to be a white person, not to be Senufo, not to be French. The ethnic groups—Lele Kouakou's Baule "tribe," for example—within which all African aesthetic life apparently occurs, are the products of colonial and postcolonial articulations. And someone who knows enough to make himself up as a Baule for the twentieth century surely knows that there are other kinds of art.

But Baldwin's *Yoruba Man with a Bicycle* does more than give the lie to Vogel's strange footnote; it provides us with an image of an object that can serve as a point of entry to my theme: a piece of contemporary African art that will allow us to explore the articulation of the postcolonial and the postmodern. *Yoruba Man with a Bicycle* is described as follows in the catalog:

> Page 124
> Man with a Bicycle
> Yoruba, Nigeria 20th century
> Wood and paint H. 35¾ in.
> The Newark Museum
>
> The influence of the Western world is revealed in the clothes and bicycle of this neo-traditional Yoruba sculpture which probably represents a merchant en route to market.

And it is this word *neotraditional*—a word that is almost right—that provides, I think, the fundamental clue.

But I do not know how to explain this clue without saying first how I keep my bearings in the shark-infested waters around the semantic island of the postmodern. And since narratives, unlike metanarratives, are allowed to proliferate in these seas, I shall begin with a story about my friend the late Margaret Masterman. Sometime in the midsixties Margaret as asked to participate at a symposium, chaired by Karl Popper, at which Tom Kuhn was to read a paper and then she, J. M. W. Watkins, Stephen Toulmin, L. Pearce Williams, Imre Lakatos, and Paul Feyerabend would engage in discussion of Kuhn's work. Unfortunately for Margaret, she developed infective hepatitis in the period leading up to the symposium and she was unable, as a result, to prepare a paper. Fortunately for all of us, though, she was able to sit in her hospital bed—in Block 8, Norwich hospital, to whose staff the paper she finally did write is dedicated—and create a subject index to *The Structure of Scientific Revolutions.* In the course of working through the book with index cards, Margaret identified no "less than twenty-one senses, possibly more, not less" in which Kuhn uses the word *paradigm.* After her catalog of these twenty-one uses, she remarks laconically that "not all these senses of 'paradigm' are inconsistent with one another"; and she continues:

> Nevertheless, given the diversity, it is obviously reasonable to ask: "Is there anything in common between all these senses? Is there, philosophically speaking, anything definite or general about the notion of a paradigm which Kuhn is tying to make clear? Or is he just a histori-

> an-poet describing different happenings which have occurred in the history of science, and referring to them all by using the same word "paradigm"?

The relevance of this tale hardly needs explication. And the task of chasing the word postmodernism through the pages of Lyotard and Jameson and Habermas, in and out of the *Village Voice* and the *T.L.S.* and even the *New York Times Book Review,* makes the task of pinning down Kuhn's paradigm look like work for a minute before breakfast.

Nevertheless, there *is*, I think, a story to tell about all these stories—or, of course, I should say, there are many, but this, for the moment, is mine—and, as I tell it, the Yoruba bicyclist will eventually come back into view.

Let me begin with the most-obvious and surely one of the most-often-remarked features of Jean-Francois Lyotard's account of postmodernity: the fact that it is a metanarrative of the end of metanarratives. To theorize certain central features of contemporary culture as *post* anything, is, of course, inevitably to invoke a narrative, and, from the Enlightenment on, in Europe and European-derived cultures, that "after" has also meant "above and beyond" and to step forward (in time) has been ipso facto to *pro*gress. Brian McHale announces in his recent *Postmodernist Fiction:*

> As for the prefix POST, here I want to emphasize the element of logical and historical *consequence* rather than sheer temporal *posteriority.* Postmodernism follows *from* modernism, in some sense, more than it follows *after* modernism. . . . Postmodernism is the posterity of modernism, that is tautological.

My point, then, is not the boring logical point that Lyotard's view—in which, in the absence of "grand narratives of legitimation," we are left with only local legitimations, imminent in our own practices—might seem to presuppose a "grand narrative of legitimation" of its own, in which justice turns out to reside, unexcitingly, in the institutionalization of pluralism. It is rather that his analysis seems to feel the need to see the contemporary condition as over against an immediately anterior set of practices and as going beyond them. Lyotard's postmodernism—his theorization of contemporary life as postmodern—is *after* modernism because it rejects aspects of modernism. And in this repudiation of one's immediate predecessors (or, more especially, of their theories of themselves) it recapitulates a crucial gesture of the historic avant-garde: indeed, it recapitulates the crucial gesture of the modern "artist"; in that sense of modernity characteristic of sociological usage in which it denotes "an era that was ushered in via the Renaissance, rationalist philosophy, and the Enlightenment, on the one hand, and the transition from the absolutist state to bourgeois democracy, on the other"; in that sense of "artist" to be found in Trilling's account of Arnold's *Scholar Gypsy,* whose "existence is intended to disturb us and make us dissatisfied with our habitual life in culture."

This straining for a contrast—a modernity or a modernism to be *against*—is extremely striking given the lack of any plausible account of what distinguishes the modern from the postmodern that is distinctively formal. In a recent essay, Fredric Jameson grants at one point, after reviewing recent French theorizings (Deleuze, Baudrillard, Debord) that it is difficult to distinguish formally the postmoderm from high modernism:

> Indeed, one of the difficulties in specifying postmodernism lies in its symbiotic or parasitical relationship to [high modernism]. In effect with the canonization of a hitherto scandalous, ugly, dissonant, amoral, antisocial, bohemian high modernism offensive to the middle classes, its promotion to the very figure of high culture generally, and perhaps most importantly, its enshrinement in the academic institution, postmodernism emerges as a way of making creative space for artists now oppressed by those henceforth hegemonic categories of irony, complexity, ambiguity, dense temporality, and particularly, aesthetic and utopian monumentality.

Jameson's argument in this essay is that we must characterize the distinction not in formal terms—in terms, say, of an "aesthetic of *textuality,*" or of "the eclipse, finally, of all depth, especially historicity itself," or of "the 'death' of the subject," or "the *culture of the simulacrum,*" or "the society of the spectacle"—but in terms of "the social functionality of culture itself."

> High modernism, whatever its overt political content, was oppositional and marginal within a middle-class Victorian or philistine or gilded age culture. Although postmodernism is equally offensive in all the respects enumerated (think of punk rock or pornography), it is no longer at all "oppositional" in that sense; indeed, it constitutes the very dominant or hegemonic aesthetic of consumer society itself and significantly serves the latter's commodity production as a virtual laboratory of new forms and fashions. The argument for a conception of postmodernism as a periodizing category is thus based on the presupposition that, even if *all* the formal features enumerated above were already present in the older high modernism, the very significance of those features changes when they become a cultural *dominant* with a precise socioeconomic functionality.

It is the "'waning" of the "dialectical opposition" between high modernism and mass culture—the commodification and, if I may coin a barbarism, the deoppositionalization, of those cultural forms once constitutive of high modernism—that Jameson sees as key to understanding the postmodern condition.

There is no doubt much to be said for Jameson's theorizing of the postmodern. But I do not think we shall understand what is in common to all the various postmodernisms if we stick within Jameson's omnisubsumptive vision. The commodification of a fiction, a stance, of oppositionality that is saleable precisely because its commodification guarantees for the consumer that it is no substantial threat was, indeed, central to the cultural role of "punk rock" in Europe and America. But what, more than a word and a conversation, makes Lyotard and Jameson competing theorists of the *same* postmodern?

I do not—this will come as no surprise—have a definition of the postmodern to put in the place of Jameson's or Lyotard's. But there is now a rough consensus about the structure of the modern-postmodern dichotomy in the many domains—from architecture to poetry to philosophy to rock to the movies—in which it has been invoked. In each of these domains there is an antecedent practice that laid claim to a certain exclusivity of insight and in each of them postmodernism is a name for the rejection of that claim to exclusivity, a rejection that is almost always more playful—though not necessarily less serious—than the practice it aims to replace. That this will not do as a *definition* of postmodernism follows from the fact that in each domain this rejection of exclusivity takes up a certain specific shape, one that reflects the specificities of its setting.

To understand the various postmodernisms this way is to leave open the question how their theories of contemporary social, cultural, and economic life relate to the actual practices that constitute that life; to leave open, then, the relations between postmodernism and postmodernity. Where the practice is theory—literary or philosophical—postmodernism as a *theory* of postmodernity can be adequate only if it reflects to some extent the realities of that practice, because the practice is itself fully theoretical. But when a postmodernism addresses, say, advertising or poetry, it may be adequate as an account of them even if it conflicts with their own narratives, their theories of themselves. For, unlike philosophy and literary theory, advertising and poetry are not largely *constituted* by their articulated theories of themselves.

It is an important question *why* this distancing of the ancestors should have become so central a feature of our cultural lives. And the answer, surely, has to do with the sense in which art is increasingly commodified. To sell oneself and one's products as art in the marketplace, it is important, above all, to clear a space in which one is distinguished from other producers and products—and one does this by the construction and the marking of differences.

It is this that accounts for a certain intensification of the long-standing individualism of post-Renaissance art production: in the age of mechanical reproduction, aesthetic individualism—the characterization of the artwork as belonging to the oeuvre of an individual—and the absorption of the artist's life into the conception of the work can be seen precisely as modes of identifying objects for the market. The sculptor of the bicycle, by contrast, will not be known by those who buy this object; his individual life will make no difference to its future history. (Indeed, he surely knows this, in the sense in which one knows anything whose negation one has never even considered.) Nevertheless, there is *something* about the object that serves to establish it for the market: the availability of Yoruba culture and of stories about Yoruba culture to surround the object and distinguish it from "folk art" from elsewhere. I shall return to this point.

Let me confirm this proposal by instances:

1. In philosophy, postmodernism is the rejection of the mainstream consensus from Descartes through Kant to logical positivism on foundationalism (there is one route to knowledge, which is exclusivism in epistemology) and of metaphysical realism (there is one truth, which is exclusivism in ontology), each underwritten by a unitary notion of reason; it thus celebrates such figures as Nietzsche (no metaphysical realist) and Dewey (no foundationalist). The modernity that is opposed here can thus be Cartesian (in France), Kantian (in Germany), and logical positivist (in America).

2. In architecture, postmodernism is the rejection of an exclusivism of function (as well as the embrace of a certain taste for pastiche). The modernity that is opposed here is the "monumentality," "elitism," and "authoritarianism" of the international style of Le Corbusier or Mies.

3. In "literature," postmodernism reacts against the high seriousness of high modernism, which mobilized "difficulty" as a mode of privileging its own aesthetic sensibility and celebrated a complexity and irony appreciable only by a cultural elite. Modernity here is, say, and in no particular order, Proust, Eliot, Pound, Woolf.

4. In political theory, finally, postmodernism is the rejection of the monism of Big-M Marxist (though not of the newer little-m marxist) and liberal conceptions of justice, and their overthrow by a conception of politics as irreducibly plural, with every perspective essentially contestable from other perspectives. Modernity here is the great nineteenth-century political narratives, of Marx and Mill but includes, for example, such latecomers as John Rawls's reconstruction of *The Liberal Theory of Justice.*

These sketchy examples are meant to suggest how we might understand the family resemblance of the various postmodernisms as governed by a *loose* principle. They also suggest why it might be that the high theorists of postmodernism—Lyotard, Jameson, Habermas, shall we say—can seem to be competing for the same territory: Lyotard's privileging of a certain philosophical antifoundationalism could surely be seen as underwriting (though not, I think, plausibly, as causing) each of these moves; Jameson's characterization of postmodernism as the logic of late capitalism—with the commodification of "cultures" as a central feature—might well account for many features of each of these transitions also; and Habermas's project is surely intended (though in the name of a most un-Lyotardian metanarrative) to provide a modus operandi in a world in which pluralism is, so to speak, a fact waiting for some institutions.

Postmodern culture is the culture in which all of the postmodernisms operate, sometimes in synergy, sometimes in competition. And because contemporary culture is, in certain senses to which I shall return, transnational, postmodern culture is global—though that does not by any means mean that it is the culture of every person in the world.

If postmodernism is the project of transcending some species of modernism—which is to say some relatively self-conscious self-privileging project of a privileged modernity—our *neotraditional* sculptor of the *Yoruba Man with a Bicycle* is presumably to be understood, by contrast, as premodern, that is, traditional. (I am supposing, then, that being neotraditional is a way of being traditional; what work the "neo" does is matter for a later moment). And the sociological and anthropological narratives of tradition through which he or she came to be so theorized is dominated, of course, by Weber.

Weber's characterization of traditional (and charismatic) authority *in opposition* to rational authority is in keeping with his general characterization of modernity as the rationalization of the world, and he insisted on the significance of this characteristically Western process for the rest of humankind. The introduction to *The Protestant Ethic* begins:

> A product of modern European civilization, studying any problem of universal history, is bound to ask himself to what combination of circumstances the fact should be attributed that in Western civilization, and in Western civilization only, cultural phenomena have appeared which (as we like to think) lie in a line of development having universal significance and value.

There is certainly no doubt that Western modernity now has a universal *geographical* significance. The Yoruba bicyclist—like Sting and his Amerindian chieftains of the Amazon rain forest or Paul Simon and the Mbaqanga musicians of *Graceland*—is testimony to that. But, if I may borrow someone else's borrowing, the fact is that the Empire of Signs strikes back. Weber's "as we like to think" reflects his doubts about whether the Western imperium over the world was as clearly of universal value as it was certainly of universal *significance,* and postmodernism surely fully endorses his resistance to this claim. The bicycle enters our museums to be valued by us (David Rockefeller tells us *how* it is to be valued). But just as the *presence* of the object reminds us of this fact, its *content* reminds us that the trade is two-way.

I want to argue that to understand our—our human—modernity we must first understand why the rationalization of the world can no longer be seen as the tendency either of the West or of history; why, simply put, the modernist characterization of modernity must be challenged. To understand our world is to reject Weber's claim for the rationality of what he called rationalization and his projection of its inevitability; it is, then, to have a radically post-Weberian conception of modernity.

We can begin with a pair of familiar and helpful caricatures: Thomas Stearns Eliot is against the soullessness and the secularization of modern society, the reach of Enlightenment rationalism into the whole world. He shares Weber's account of modernity and more straightforwardly deplores it. Le Corbusier is in favor of rationalization—a house is a "machine for living in"—but he, too, shares Weber's vision of modernity. And, of course, the great rationalists—the believers in a transhistorical reason triumphing in the world—from Kant on are the source of Weber's Kantian vision. Modernism in literature and architecture and philosophy (the account of modernity that, on my model, *post*modernism in these domains seeks to subvert) may be for reason or against it: but in each domain rationalization—the pervasion of reason—is seen as the distinctive dynamic of contemporary history.

But the beginning of postmodern wisdom is to ask whether Weberian rationalization is in fact what has happened. For Weber, charismatic authority—the authority of Stalin, Hitler, Mao, Guevara, Nkrumah—is antirational, yet modernity has been dominated by just such charisma. Secularization seems hardly to be proceeding: religions grow in all parts of the world; more than 90 percent of North Americans still avow some sort of theism; what we call "fundamentalism" is as alive in the West as it is in Africa and the Middle and Far East; Jimmy Swaggart and Billy Graham have business in Louisiana and California as well as in Costa Rica and Ghana.

What we can see in all these cases, I think, is not the triumph of Enlightenment capital-R Reason—which would have entailed exactly the end of charisma and the universalization of the secular—not even the penetration of a narrower instrumental reason into all spheres of life, but what Weber mistook for that: namely, the incorporation of all areas of the world and all areas of even formerly "private" life into the money economy. Modernity has turned every element of the real into a sign, and the sign reads "for sale"; this is true even in domains like religion where instrumental reason would recognize that the market has at best an ambiguous place.

If Weberian talk of the triumph of instrumental reason can now be seen to be a mistake, what Weber thought of as the disenchantment of the world—that is, the penetration of a scientific vision of things—describes at most the tiny, and in the United States quite marginal, world of the higher academy and a few islands of its influence. The world of the intellectual *is*, I think, largely disenchanted (even theistic academics largely do not believe in ghosts and ancestor spirits), and fewer people (though still very many) suppose the world to be populated by the multitudes of spirits of earlier religion. Still, what we have seen in recent times in the United States is not secularization—the end of religions—but their commodification; with that commodification, religions have reached further and grown—their markets have expanded—rather than dying away.

Postmodernism can be seen, then, as a new way of understanding the multiplication of distinctions that flows from the need to clear oneself a space; the need that drives the underlying dynamic of cultural modernity. Modernism saw the economization of the world as the triumph of reason, postmodernism rejects that claim, allowing in the realm of theory the same multiplication of distinctions we see in the cultures it seeks to understand.

I anticipate that objection that the Weber I have been opposing is something of a caricature. And I would not be unhappy to admit that there is some truth in this. Weber foresaw, for example, that the rationalization of the world would continue to be resisted, and his view that each case of charisma needed to be "routinized" was not meant to rule out the appearance of new charismatic leaders in our time as in earlier ones: our politics of charisma would, perhaps, not have surprised him. Certainly, too, his conception of reason involved far more than instrumental calculation. Since much of what I have noticed here would have been anticipated by him, it may be as well to see this as a rejection of a narrow (if familiar) misreading of Weber than an argument against what is best in the complex and shifting views of Weber himself.

But I think we could also construe this misreading—which we find, perhaps, in Talcott Parsons—as in part a consequence of a problem with Weber's own work. For part of the difficulty with Weber's work is that, despite the wealth of historical detail in his studies of religion, law, and economics, he often mobilizes theoretical terms that are of a very high level of abstraction. As a result, it is not always clear that there really are significant commonalities among the various social phenomena he assimilates under such general concepts as "rationalization" or charisma. "(This is one of the general problems posed by Weber's famous reliance on "ideal types.") Reinhard Bendix, one of Weber's most important and sympathetic interpreters, remarks at one point in his discussion of one of Weber's theoretical distinctions (the distinction, as it happens, between patrimonialism and feudalism) that "this distinction is clear only so long as it is formulated in abstract terms." In reading Weber it is a feeling that one has over and over again. The problem is exemplified in Weber's discussion of "charisma" in *The Theory of Social and Economic Organization:*

> The term "charisma" will be applied to a certain quality of an individual personality by virtue of which he is set apart from ordinary men and treated as endowed with supernatural, superhuman, or at least specifically exceptional powers or qualities. These are . . . regarded as of divine origin or exemplary, and on the basis of them the individual concerned is treated as a leader.

Notice how charisma is here defined disjunctively as involving *either* magical ("supernatural, superhuman," "of divine origin") capacities, on the one hand, *or* merely "exceptional" or "exemplary" qualities on the other. The first disjunct in each case happily covers the many cases of priestly and prophetic leadership that Weber discusses, for example, in his study *Ancient Judaism.* But it is the latter, presumably, that we should apply in seeking to understand the political role of Hitler, Stalin, or Mussolini, who though no doubt "exceptional" and "exemplary" were not regarded as having "supernatural" powers "of divine origin." The point is that much of what Weber has to say in his general discussion of charisma in *The Theory of Social and Economic Organization* and in the account of "domination" in *Economy and Society* requires that we take its magical aspect seriously. When, however, we do take it seriously, we find his theory fails to apply to the instances of charisma that fall under the second disjunct of his definition. In short, Weber's account of charisma assimilates too closely phenomena—such as the leadership of Stalin, at one end of the spectrum, and of King David or the emperor Charlemagne, at the other—in which magico-religious ideas seem, to put it mildly, to play remarkably different roles. If we follow out the logic of this conclusion by redefining Weberian charisma in such a way as to insist on its magical component, it will follow, by definition, that the disenchantment of the world—the decline of magic—leads to the end of charisma. But we shall then have to ask ourselves how correct it is to claim, with Weber, that magical views increasingly disappear with modernity. And if he is right in this, we shall also have to give up the claim that Weber's sociology of politics—in which charisma plays a central conceptual role—illuminates the characteristic political developments of modernity.

There is a similar set of difficulties with Weber's account of rationalization. In *The Protestant Ethic and the Spirit of Capitalism,* Weber wrote: "If this essay makes any contribution at all, may it be to bring out the complexity of the only superficially simple concept of the rational." But we may be tempted to ask whether our understanding of the genuine complexities of the historical developments of the last few centuries of social, religious, economic, and political history in Western Europe is truly deepened by making use of a concept of rationalization that brings together a supposed increase in means-end calculation (instrumental rationality); a decline in appeal to "mysterious, incalculable forces" and a correlative increasing confidence in calculation (disenchantment or intellectualization); and the growth of "value rationality," which means something like an increasing focus on maximizing a narrow range of ultimate goals. Here, seeking to operate at this high level of generality, assimilating under one concept so many, in my view, distinct and independently intelligible processes, Weber's detailed and subtle appreciation of the dynamics of many social processes is obscured by his theoretical apparatus; it is, I think, hardly surprising that those who have been guided by his theoretical writings have ascribed to him a cruder picture than is displayed in his historical work.

I have been exploring how modernity looks from the perspective of the Euro-American intellectual. But how does it look from the postcolonial spaces inhabited by the *Yoruba Man with a Bicycle?* I shall speak about Africa, with confidence *both* that some of what I have to say will work elsewhere in the so-called Third World *and* that, in some places, it will certainly not. And I shall speak first about the producers of these so-called neotraditional artworks and then about the case of the African novel, because I believe that to focus exclusively on the novel (as theorists of contemporary African cultures have been inclined to do) is to distort the cultural situation and the significance within it of postcoloniality.

I do not know when the *Yoruba Man with a Bicycle* was made or by whom; African art has, until recently, been collected as the property of "ethnic" groups, not of individuals and workshops, so it is not unusual that not one of the pieces in the Perspectives show was identified in the "Checklist" by the name of an individual artist, even though many of them are twentieth-century; (and no one will have been surprised, by contrast, that most of them *are* kindly labeled with the name of the people who own the largely private collections where they now live). As a result I cannot say if the piece is literally postcolonial, produced after Nigerian independence in 1960. But the piece belongs to a genre that has certainly been produced since then: the genre that is here called *neotra-*

ditional. And, simply put, what is distinctive about this genre is that it is produced for the West.

I should qualify. Of course, many of the buyers of first instance live in Africa, many of them are juridically citizens of African states. But African bourgeois consumers of neotraditional art are educated in the Western style, and, if they want African art, they would often rather have a "genuinely" traditional piece—by which I mean a piece that they believe to be made precolonially, or at least in a style and by methods that were already established precolonially. And these buyers are a minority. Most of this art, which is *traditional* because it uses actually or supposedly precolonial techniques, but is *neo*—this, for what it is worth, is the explanation I promised earlier—because it has elements that are recognizably from the colonial or postcolonial in reference, has been made for Western tourists and other collectors.

The incorporation of these works in the West's world of museum culture and its art market has almost nothing, of course, to do with postmodernism. By and large, the ideology through which they are incorporated is modernist: it is the ideology that brought something called "Bali" to Artaud, something called "Africa" to Picasso, and something called "Japan" to Barthes. (This incorporation as an official Other was criticized, of course, from its beginnings: Oscar Wilde once remarked that "the whole of Japan is a pure invention. There is no such country, no such people.") What *is* postmodernist is Vogel's muddled conviction that African art should not be judged "in terms of [someone else's] traditional criteria." For modernism, primitive art was to be judged by putatively *universal* aesthetic criteria, and by these standards it was finally found possible to value it. The sculptors and painters who found it possible were largely seeking an Archimedean point outside their own cultures for a critique of a Weberian modernity. For postmoderns, by contrast, these works, however they are to be understood, cannot be seen as legitimated by culture- and history-transcending standards.

What is useful in the *neotraditional* object as a model—despite its marginality in most African lives—is that its incorporation in the museum world (while many objects made by the same hands—stools, for example—live peacefully in non-bourgeois homes) reminds one that in Africa, by contrast, the distinction between high culture and mass culture, insofar as it makes sense at all, corresponds by and large to the distinction between those with and those without Western-style formal education as cultural consumers.

The fact that the distinction is to be made this way—in most of sub-Saharan Africa excluding the Republic of South Africa—means that the opposition between high culture and mass culture is available only in domains where there is a significant body of Western formal training, and this excludes (in most places) the plastic arts and music. There are distinctions of genre and audience in African musics, and for various cultural purposes there is something that we call "traditional" music that we still practice and value. But village and urban dwellers alike, bourgeois and nonbourgeois, listen, through discs and, more importantly, on the radio, to reggae, to Michael Jackson, and to King Sonny Adé.

And this means that by and large the domain in which it makes most sense is the one domain where that distinction is powerful and pervasive—namely, in African writing in Western languages. So that it is here that we find, I think, a place for consideration of the question of the *post*coloniality of contemporary African culture.

Postcoloniality is the condition of what we might ungenerously call a comprador intelligentsia: of a relatively small, Western-style, Western-trained, group of writers and thinkers who mediate the trade in cultural commodities of world capitalism at the periphery. In the West they are known through the Africa they offer; their compatriots know them both through the West they present to Africa and through an Africa they have invented for the world, for each other, and for Africa.

All aspects of contemporary African cultural life—including music and some sculpture and painting, even some writings with which the West is largely not familiar—have been influenced, often powerfully, by the transition of African societies *through* colonialism, but they are not all in the relevant sense *post*colonial. For the *post* in postcolonial, like the *post* in postmodern is the *post* of the space-clearing gesture I characterized earlier: and many areas of contemporary African cultural life—what has come to be theorized as popular culture, in particular—are not in this way concerned with transcending, with going beyond, coloniality. Indeed, it might be said to be a mark of popular culture that its borrowings from international cultural forms are remarkably insensitive to—not so much dismissive of as blind to—the issue of neocolonialism or "cultural imperialism." This does not mean that theories of postmodernism are irrelevant to these forms of culture: for the internationalization of the market and the commodification of artworks are both central to them. But it *does* mean that these artworks are not understood by their producers or their consumers in terms of a postmodern*ism*: there is no antecedent practice whose claim to exclusivity of vision is rejected through these artworks. What is called "syncretism" here is made possible by the international exchange of commodities, but is not a consequence of a space-clearing gesture.

Postcolonial intellectuals in Africa, by contrast, are almost entirely dependent for their support on two institutions: the African university—an institution whose intellectual life is overwhelmingly constituted as Western—and the Euro-American publisher and reader. (Even when these writers seek to escape the West—as Ngugi wa Thiong'o did in attempting to construct a Kikuyu peasant drama—their theories of their situation are irreducibly informed by their Euro-American formation. Ngugi's conception of the writer's potential in politics is essentially that of the avant-garde, of Left modernism.)

Now this double dependence on the university and the European publisher means that the first generation of modern African novels—the generation of Achebe's *Things Fall Apart* and Laye's *L'Enfant noir*—were written in the context of notions of politics and culture dominant in the French and British university and publishing worlds in the fifties and sixties. This does not mean that they were like novels written in

Western Europe at that time: for part of what was held to be obvious both by these writers and by the high culture of Europe of the day was that new literatures in new nations should be anticolonial and nationalist. These early novels seem to belong to the world of eighteenth- and nineteenth-century literary nationalism; they are theorized as the imaginative recreation of a common cultural past that is crafted into a shared tradition by the writer; they are in the tradition of Scott, whose *Minstrelsy of the Scottish Border* was intended, as he said in the preface, to "contribute somewhat to the history of my native country; the peculiar features of whose manners and character are daily melting and dissolving into those of her sister and ally." The novels of this first stage are thus realist legitimations of nationalism: they authorize a "return to traditions" while at the same time recognizing the demands of a Weberian rationalized modernity.

From the later sixties on, these celebratory novels of the first stage become rarer: Achebe, for example, moves from the creation of a usable past in *Things Fall Apart* to a cynical indictment of politics in the modern sphere in *A Man of the People.* But I should like to focus on a francophone novel of the later sixties, a novel that thematizes in an extremely powerful way many of the questions I have been asking about art and modernity: I mean, of course, Yambo Ouologuem's *Le Devoir de Violence.* This novel. like many of this second stage, represents a challenge to the novels of this first stage: it identifies the realist novel as part of the tactic of nationalist legitimation and so it is (if I may begin a catalog of its ways-of-being-*post*-this-and-that) *postrealist.*

Now postmodernism is, of course, postrealist also. But Ouologuem's postrealism is surely motivated quite differently from that of such postmodern writers as, say, Pynchon. Realism naturalizes: the originary "African novel" of Chinua Achebe (*Things Fall Apart*) and of Camara Laye (*L'Enfant noir*) is "realist." So Ouologuem is against it, rejects—indeed, assaults—the conventions of realism. He seeks to delegitimate the forms of the realist African novel, in part, surely, because what it sought to naturalize was a nationalism that, by 1968, had plainly failed. The national bourgeoisie that took on the baton of rationalization, industrialization, bureaucratization in the name of nationalism, turned out to be a kleptocracy. Their enthusiasm for nativism was a rationalization of their urge to keep the national bourgeoisies of other nations—and particularly the powerful industrialized nations—out of their way. As Jonathan Ngaté has observed, "*Le Devoir de Violence . . . deal[s] with a world in which the efficacy* of the call to the Ancestors as well as the Ancestors themselves is seriously called into question." That the novel is in this way postrealist allows its author to borrow, when he needs them, the techniques of modernism, which, as we learned from Fred Jameson, are often also the techniques of postmodernism. (It is helpful to remember at this point how Yambo Ouologuem is described on the back of the Éditions Du Seuil first edition: "Né en 1940 au Mali. Admissible à l'École normale supérieure. Licencié ès Lettres. Licencié en Philosophie. Diplômé d'études supérieures d'Anglais. Prépare une thèse de doctorat de Sociologie." Borrowing from European modernism is hardly going to be difficult for someone so qualified, to be a Normalien is indeed, in Christopher Miller's charming formulation, "roughly equivalent to being baptized by Bossuet.")

Christopher Miller's discussion—in *Blank Darkness*—of *Le Devoir de Violence* focuses usefully on theoretical questions of intertextuality raised by the novel's persistent massaging of one text after another into the surface of its own body. The book contains, for example, a translation of a passage from Graham Greene's 1934 novel *It's a Battlefield* (translated and improved, according to some readers!) and borrowings from Maupassant's *Boule de suif* (hardly an unfamiliar work for francophone readers; if this latter is a theft, it is the adventurous theft of the kleptomaniac, who dares us to catch him at it).

And the book's first sentence artfully establishes the oral mode—by then an inevitable convention of African narration—with words that Ngaté rightly describes as having the "concision and the striking beauty and power of a proverb," and mocks us in this moment because the sentence echoes the beginning of André Schwartz-Bart's decidedly un-African 1959 holocaust novel *Le Dernier des justes,* an echo that more substantial later borrowings confirm.

> Our eyes drink the flash of the sun, and, conquered, surprise themselves by weeping. Maschallah! oua bismillah! . . . An account of the bloody adventure of the niggertrash—dishonor to the men of nothing—could easily begin in the first half of this century; but the true history of the Blacks begins very much earlier, with the Saïfs, in the year 1202 of our era, in the African kingdom of Nakem. . . .
>
> Our eyes receive the light of dead stars. A biography of my friend Ernie could easily begin in the second quarter of the 20th century; but the true history of Ernie Lévy begins much earlier, in the old anglican city of York. More precisely: on the 11 March 1185.

The reader who is properly prepared will expect an African holocaust, and these echoes are surely meant to render ironic the status of the rulers of Nakem as descendants of Abraham El Héït, "le Juif noir."

The book begins, then, with a sick joke at the unwary reader's expense against nativism: and the assault on realism is—here is my second signpost—postnativist; this book is a murderous antidote to a nostalgia for *Roots.* As Wole Soyinka has said in a justly well-respected reading, "the Bible, the Koran, the historic solemnity of the griot are reduced to the histrionics of wanton boys masquerading as humans." It is tempting to read the attack on history here as a repudiation not of roots but of Islam, as Soyinka does when he goes on to say:

> A culture which has claimed indigenous antiquity in such parts of Africa as have submitted to its undeniable attractions is confidently proven to be imperialist; worse, it is demonstrated to be essentially hostile to the indigenous culture. . . . Ouologuem pronounces the Moslem incursion into black Africa to be corrupt, vicious, decadent, elitist and insensitive. At the least such a work functions as a wide swab in the deck-clearing operation for the commencement of racial retrieval.

But it seems to me much clearer to read the repudiation as a repudiation of national history; to see the text as postcolonially postnationalist as well as anti- (and thus, of course, post-) nativist. (Indeed, Soyinka's reading here seems to be driven by his own equally representative tendency to read Africa as race and place into everything.) Raymond Spartacus Kassoumi—who, if anyone, is the hero of this novel—is, after all, a son of the soil, but his political prospects by the end of the narrative are less than uplifting. More than this, the novel explicitly thematizes, in the anthropologist Shrobenius—an obvious echo of the name of the German Africanist Frobenius, whose work is cited by Senghor—the mechanism by which the new elite has come to invent its traditions through the "science" of ethnography:

> Saïf made up stories and the interpreter translated, Madoubo repeated in French, refining on the subtleties to the delight of Shrobenius, that human crayfish afflicted with a groping mania for resuscitating an African universe—cultural autonomy, he called it, which had lost all living reality; . . . he was determined to find metaphysical meaning in everything . . . African life, he held, was pure art . . . .

At the start we have been told that "there are few written accounts and the versions of the elders diverge from those of the griots, which differ from those of the chroniclers." Now we are warned off the supposedly scientific discourse of the ethnographers.

Because this is a novel that seeks to delegitimate not only the form of realism but the content of nationalism, it will to that extent seem to us misleadingly to be postmodern. *Mis*leadingly, because what we have here is not postmodern*ism* but postmoderni*zation*; not an aesthetics but a politics, in the most literal sense of the term. After colonialism, the modernizers said, comes rationality; that is the possibility the novel rules out. Ouologuem's novel is typical of this second stage in that it is not written by someone who is comfortable with and accepted by the new elite, the national bourgeoisie. Far from being a celebration of the nation, then, the novels of the second stage—the postcolonial stage—are novels of delegitimation; rejecting the Western imperium, it is true, but also rejecting the nationalist project of the postcolonial national bourgeoisie. And, so it seems to me, the basis for that project of delegitimation is very much not the postmodernist one: rather, it is grounded in an appeal to an ethical universal; indeed it is based, as intellectual responses to oppression in Africa largely are based, in an appeal to a certain simple respect for human suffering, a fundamental revolt against the endless misery of the last thirty years. Ouologuem is hardly likely to make common cause with a relativism that might allow that the horrifying new-old Africa of exploitation is to be understood—legitimated—in its own local terms.

Africa's postcolonial novelists—novelists anxious to escape neocolonialism—are no longer committed to the nation, and in this they will seem, as I have suggested, misleadingly postmodern. But what they have chosen instead of the nation is not an older traditionalism but Africa—

the continent and its people. This is clear enough, I think, in *Le Devoir de Violence,* at the end of which Ouologuem writes:

> Often, it is true, the soul desires to dream the echo of happiness, an echo that has no past. But projected into the world, one cannot help recalling that Saïf, mourned three million times, is forever reborn to history beneath the hot ashes of more than thirty African republics.

If we are to identify with anyone, *in fine,* it is with "la négraille"—the niggertrash, who have no nationality. For these purposes one republic is as good—which is to say as bad—as any other. If this postulation of oneself as African—and neither as of this or that allegedly precolonial ethnicity nor of the new nation-states—is implicit in *Le Devoir de Violence,* in the important novels of V. Y. Mudimbe, *Entre les eaux, Le Bel immonde*—recently made available in English as *Before the Birth of the Moon*—and *L'Écart,* this postcolonial recourse to Africa is to be found nearer the surface and over and over again.

There is a moment in *L'Écart,* for example, when the protagonist, whose journal the book is, recalls a conversation with the French girlfriend of his student days—the young woman on whom he reflects constantly as he becomes involved with an African woman.

> "You can't know, Isabelle, how demanding Africa is."
>
> "It's important for you, isn't it?"
>
> "To tell you the truth, I don't know . . . I really don't . . . I wonder if I'm not usually just playing around with it."
>
> "Nara . . . I don't understand. For me, the important thing is to be myself. Being european isn't a flag to wave."
>
> "You've never been wounded like . . ."
>
> "Your dramatizing, Nara. You carry your african-ness like a martyr. . . . That makes one wonder. . . . I'd be treating you with contempt if I played along with you."
>
> "The difference is that Europe is above all else an idea, a juridical institution . . . while Africa . . ."
>
> "Yes? . . ."
>
> "Africa is perhaps mostly a body, a multiple existence. . . . I'm not expressing myself very well."

This exchange seems to me to capture the essential ambiguity of the postcolonial African intellectual's relation to Africa. But let me pursue Africa, finally, in Mudimbe's first novel, *Entre les eaux,* a novel that thematizes the question most explicitly.

In *Entre les eaux*—a first-person narrative—our protagonist is an African Jesuit, Pierre Landu, who has a "doctorat ne théologie et [une] licence en droit canon" acquired as a student in Rome. Landu is caught between his devotion to the church and, as one would say in more Protestant language, to Christ; the latter leads him to repudiate the official Roman Catholic hierarchy of his homeland and join with a group of Marxist guerrillas, intent on removing the corrupt postindependence state. When he first tells his immediate superior in the hierarchy, Father

Howard, who is white, of his intentions, the latter responds immediately and remorselessly that this will be treason.

> "You are going to commit treason," the father superior said to me when I informed him of my plans.
>
> "Against whom?"
>
> "Against Christ."
>
> "Father, isn't it rather the West that I'm betraying. Is it still treason? Don't I have the right to dissociate myself from this christianity that has betrayed the Gospel?"
>
> "You are a priest, Pierre."
>
> "Excuse me, Father, I'm a black priest."

It is important, I think, not to see the blackness here as a matter of race. It is rather the sign of Africanity. To be a black priest is to be a priest who is also an African and thus committed, nolens-volens, to an engagement with African suffering. This demand that Africa makes has nothing to do with a sympathy for African cultures and traditions; reflecting—a little later—on Father Howard's alienating response, Landu makes this plain.

> Father Howard is also a priest like me. That's the tie that binds us. Is it the only one? No. There's our shared tastes.
>
> Classical music. Vivaldi. Mozart. Bach. . . .
>
> And then there was our reading. The books, we used to pass each other. Our shared memories of Rome. Our impassioned discussions on the role of the priest, and on literature and on the mystery novels that we each devoured. I am closer to Father Howard than I am to my compatriots, even the priests.
>
> Only one thing separates us: the color of our skins.

In the name of this "couleur de la peau," which is precisely the *sign* of a solidarity with Africa, Landu reaches from Roman Catholicism to Marxism, seeking to gather together the popular revolutionary energy of the latter and the ethical—and religious—vision of the former; a project he considers in a later passage, where he recalls a long-ago conversation with Monseigneur Sanguinetti in Rome. "The Church and Africa," the Monseigneur tells him, "are counting on you." Landu asks in the present:

> Could the church really still count on me? I would have wished it and I wish it now. The main thing meanwhile is that Christ counts on me. But Africa? Which Africa was Sanguinetti speaking of? That of my black confrères who have stayed on the straight and narrow, or that of my parents whom I have already betrayed? Or perhaps he was even speaking of the Africa that we defend in this camp?

Whenever Landu is facing a crucial decision, it is framed for him as a question about the meaning of Africa.

After he is accused of another betrayal—this time by the rebels, who have intercepted a letter to his bishop (a letter in which he appeals to him to make common cause with the rebels, to recover them for Christ)—Landu is condemned to death. As he awaits execution, he remembers something an uncle had said to him a decade earlier about "the ancestors."

> "You'll be missed by them . . . ," my uncle had said to me, ten years ago. I had refused to be initiated. What did he mean? It is I who miss them. Will that be their curse? The formula invaded me, at first unobtrusively, but then it dazzled me, stopping me from thinking: "Wait till the ancestors come down. Your head will burn, your throat will burst, your stomach will open and your feet will shatter. Wait till the ancestors come down. . . ." They had come down. And I had only the desiccation of a rationalized Faith to defend myself against Africa.

The vision of modernity in this passage is not, I think, Weberian. In being postcolonial, Pierre Landu is against the rationalizing thrust of Western modernity (that modernity here, in this African setting, is represented by Catholicism confirms how little modernity has ultimately to do with secularization). And even here, when he believes he is facing his own death, the question "What does it mean to be an African?" is at the center of his mind.

A raid on the camp by government forces saves Pierre Landu from execution; the intervention of a bishop and a brother powerfully connected within the modern state saves him from the fate of a captured rebel. He retreats from the world to take up the life of a monastic with a new name—no longer Peter-on-whom-I-will-build-my-church but Mathieu-Marie de L'Incarnation—in a different, more contemplative order. As we leave him his last words, the last of the novel, are "l'humilité de ma bassesse, quelle gloire pour l'homme!" Neither Marx nor Saint Thomas, the novel suggests—neither of the two great political energies of the West in Africa—offers a way forward. But this retreat to the otherworldly cannot be a political solution. Postcoloniality has, also, I think, become a condition of pessimism.

Postrealist writing; postnativist politics; a *transnational* rather than a *national* solidarity. And pessimism: a kind of *post*optimism to balance the earlier enthusiasm for *The Suns of Independence.* Postcoloniality is *after* all this: and its *post,* like postmodernism's, is also a *post* that challenges earlier legitimating narratives. And it challenges them in the name of the suffering victims of "more than thirty republics." But it challenges them in the name of the ethical universal; in the name of *humanism,* "le gloire pour l'homme." And on that ground it is not an ally for Western postmodernism but an agonist, from which I believe postmodernism may have something to learn.

For what I am calling humanism can be provisional, historically contingent, antiessentialist (in other words, postmodern), and still be demanding. We can surely maintain a powerful engagement with the concern to avoid cruelty and pain while nevertheless recognizing the contingency of that concern. Maybe, then, we can recover within postmodernism the postcolonial writers' humanism—the concern for human suffering, for the victims of the postcolonial state (a concern we find everywhere: in Mudimbe, as we have seen; in Soyinka's *A Play of Giants;* in Achebe, Farrah, Gordimer, Labou Tansi—the list is difficult to complete)—while still rejecting the master narratives of modernism. This human impulse—an impulse that transcends obligations to churches and to nations—I propose we learn from Mudimbe's Landu.

But there is also something to reject in the postcolonial adherence to Africa of Nara, the earlier protagonist of Mudimbe's *L'Écart:* the sort of Manicheanism that makes Africa *"a body"* (nature) against Europe's juridical reality (culture) and then fails to acknowledge—even as he says it—the full significance of the fact that Africa is also *"a multiple existence." Entre les eaux* provides a powerful postcolonial critique of this binarism: we can read it as arguing that if you postulate an either-or choice between Africa and the West, there is no place for you in the real world of politics, and your home must be the otherworldly, the monastic retreat.

If there is a lesson in the broad shape of this circulation of cultures, it is surely that we are all already contaminated by each other, that there is no longer a fully autochthonous *echt*-African culture awaiting salvage by our artists (just as there is, of course, no American culture without African roots). And there is a clear sense in some postcolonial writing that the postulation of a unitary Africa over against a monolithic West—the binarism of Self and Other—is the last of the shibboleths of the modernizers that we must learn to live without.

Already in *Le Devoir de Violence,* in Ouologuem's withering critique of "Shrobéniusologie," there were the beginnings of this postcolonial critique of what we might call "alteritism," the construction and celebration of oneself as Other. Ouologuem writes, ". . . henceforth Negro art was baptized 'aesthetic' and hawked in the imaginary universe of 'vitalizing exchanges.'" Then, after describing the phantasmic elaboration of some interpretative mumbo jumbo "invented by Saïf," he announces that ". . . Negro art found its patent of nobility in the folklore of mercantile intellectualism, oye, oye, oye . . ." Shrobenius, the anthropologist, as apologist for "his" people; a European audience that laps up this exoticized other; African traders and producers of African art, who understand the necessity to maintain the "mysteries" that construct their product as "exotic"; traditional and contemporary elites who require a sentimentalized past to authorize their present power: all are exposed in their complex and multiple mutual complicities.

> Witness the splendor of its art—the true face of Africa is the grandiose empires of the Middle Ages, a society marked by wisdom, beauty, prosperity, order, nonviolence, and humanism, and it is here that we must seek the true cradle of Egyptian civilization.
>
> Thus drooling, Shrobenius derived a twofold benefit on his return home: on the one hand, he mystified the people of his own country who in their enthusiasm raised him to a lofty Sorbonnical chair, while on the other hand he exploited the sentimentality of the coons, only too pleased to hear from the mouth of a white man that Africa was "the womb of the world and the cradle of civilization."
>
> In consequence the niggertrash donated masks and art treasures by the ton to the acolytes of "Shrobeniusology."

A little later, Ouologuem articulates more precisely the interconnections of Africanist mystifications with tourism, and the production, packaging, and marketing of African artworks.

> An Africanist school harnessed to the vapors of magico-religious, cosmological, and mythical symbolism had been born: with the result that for three years men flocked to Nakem—and what men!—middlemen, adventurers, apprentice bankers, politicians, salesmen, conspirators—supposedly 'scientists,' but in reality enslaved sentries mounting guard before the 'Shrobeniusological' monument of Negro pseudosymbolism.
>
> Already it had become more than difficult to procure old masks, for Shrobenius and the missionaries had had the good fortune to snap them all up. And so Saïf—and the practice is still current—had slapdash copies buried by the hundredweight, or sunk into ponds, lakes, marshes, and mud holes, to be exhumed later on and sold at exorbitant prices to unsuspecting curio hunters. These three-year-old masks were said to be charged with the weight of four centuries of civilization.

Ouologuem here forcefully exposes the connections we saw earlier in some of David Rockefeller's insights into the international system of art exchange, the international art world: we see the way in which an ideology of disinterested aesthetic value—the "baptism" of "Negro art" as "aesthetic"—meshes with the international commodification of African expressive culture, a commodification that requires, by the logic of the space-clearing gesture, the manufacture of Otherness. (It is a significant bonus that it also harmonizes with the interior decor of modem apartments.) Shrobenius, "ce marchand-confectionneur d'idéologie," the ethnographer allied with Saïf—image of the "traditional" African ruling caste—has invented an Africa that is a body over against Europe, the juridical institution, and Ouologuem is urging us vigorously to refuse to be thus Other.

Sara Suleri has written recently, in *Meatless Days,* of being treated as an "Otherness-machine"—and of being heartily sick of it. If there is no way out for the post-colonial intellectual in Mudimbe's novels, it is, I suspect, because *as* intellectuals—a category instituted in black Africa by colonialism—we are always at risk of becoming Otherness-machines. It risks becoming our principal role. Our only distinction in the world of texts to which we are latecomers is that we can mediate it to our fellows. This is especially true when postcolonial meets postmodern, for what the postmodern reader seems to demand of its Africa is all too close to what modernism—as documented in William Rubin's Primitivism exhibit of 1985—demanded of it. The role that Africa, like the rest of the Third World, plays for Euro-American post-modernism—like its better-documented significance for modernist art—must be distinguished from the role postmodernism might play in the Third World. What that might be it is, I think, too early to tell. And what happens will happen not because we pronounce upon the matter in theory but out of the changing everyday practices of African cultural life.

For all the while, in Africa's cultures, there are those who will not see themselves as Other. Despite the overwhelming reality of economic decline; despite unimaginable poverty; despite wars, malnutrition, disease, and political instability, African cultural productivity grows apace: popular literatures, oral narrative and poetry, dance, drama, music, and visual art all thrive. The contemporary cultural production of many African societies—and the many traditions whose evidences so vigorously remain—is an antidote to the dark vision of the postcolonial novelist.

And I am grateful to James Baldwin for his introduction to the *Yoruba Man with a Bicycle*—a figure who is, as Baldwin so rightly saw, polyglot, speaking Yoruba and English, probably some Hausa and a little French for his trips to Cotonou or Cameroon; someone whose "clothes do not fit him too well." He and the other men and women among whom he mostly lives suggest to me that the place to look for hope is not just to the postcolonial novel—which has struggled to achieve the insights of a Ouologuem or Mudimbe—but to the all-consuming vision of this less-anxious creativity. It matters little who it was made *for;* what we should learn from is the imagination that produced it. *The Man with a Bicycle* is produced by someone who does not care that the bicycle is the white man's invention—it is not there to be Other to the Yoruba Self; it is there because someone cared for its solidity; it is there because it will take us further than our feet will take us; it is there because machines are now as African as novelists—and as fabricated as the kingdom of Nakem.

Africana Womanism: An Historical, Global Perspective for Women of African Descent[1]

Clenora Hudson-Weems

> WOMEN WHO ARE CALLING THEMSELVES BLACK FEMINISTS NEED ANOTHER WORD THAT DESCRIBES WHAT THEIR CONCERNS ARE. BLACK FEMINISM IS NOT A WORD THAT DESCRIBES THE FLIGHT OF BLACK WOMEN.
>
> JULIA HARE, *BLACK ISSUES*

The crucial role of the Africana woman within the constructs of the modern feminist movement is a critical and controversial issue in both the academy and the community today. Given that female subjugation and exploitation are real issues that unquestionably must be combated, many Africana academicians have uncritically accepted the concept of feminism, which emphasizes female empowerment, to reflect the level of struggle or concerns of Africana women. Because of the erroneous assumption that gender issues are an exclusive for the feminist, it must be noted that confronting a patriarchal system that oppresses women in general is a concern for all people, and thus addressing gender problems in our society does not necessarily translate into feminism. It appears that for many of those who embrace feminism, they do so because of its theoretical and methodological legitimacy in the academy and their desire to be a legitimate part of that community. Moreover, they do so because of the absence of a suitable existing framework for their individual needs as Africana women. Be that as it may, while many have accepted the label, more and more Africana women today are beginning to reevaluate feminism and its applicability, particularly in terms of its agenda and its historical realities. Even some White women have become disenchanted with feminist theory, and have concluded that the modern feminist movement does not accurately reflect their reality or level of struggle.

The key to understanding true feminism is to acknowledge its historical and current female-centered agenda, which does not accurately reflect the true agenda of Africana women, who are instead family-centered by necessity as a result of a racist society which has wreaked havoc on their male counterparts as well as their children globally. Observe the position of a South African activist, Ruth Mompati, in her response to witnessing the decomposed bodies of so many children victims of apartheid:

> The South African woman, faced with the above situation, finds the order of her priorities in her struggle for human dignity and her rights as a woman dictated by the general political struggle of her people as a whole. The national liberation of the black South African is a prerequisite to her own liberation and emancipation as a woman and a worker. (Mompati, 112–113)

Clearly, as Daphne Williams Ntiri, editor of *One Is Not a Woman*, asserts, "Human discrimination transcends sex discrimination . . . [and] the costs of human suffering are high when compared to a component, sex obstacle (Ntiri, 6).

It is also significant to understand the venomous origins of feminism and its exclusivity, which left Africana women and women in general not of the educated upper middle class on the outside, let alone on the fringes, of the pre-feminist (suffragist) movement. One of the leading conservative suffragists, Carrie Chapman Catt, advocated Anglo-Saxon purity and insisted that White men understand "the usefulness of woman suffrage as a counterbalance to the foreign vote, and as a means of legally preserving White supremacy in the South" (Carol and Nobel, 296). The following quotation from *Africana Womanism* presents the overall picture of feminism in its embryonic stage:

> Feminism, earlier called the Woman's Suffrage Movement, started when a group of liberal White women, whose concerns then were for the abolition of slavery and equal rights for all people regardless of race, class and sex, dominated the scene among women on the national level. . . . However, in 1870 the Fifteenth Amendment to the Constitution of the United States ratified the voting rights of Africana men, leaving women, White women in particular, and their desire for the same rights, unaddressed. . . . The result was a racist reaction to the Amendment and Africans in particular. Thus, from the 1880s on, an organized movement among White women shifted the pendulum to a radically conservative posture on the part of White women in general (Hudson-Weems, 20–21).

Africana women, on the other hand, have historically demonstrated their concern for the entire Africana community, and hence race empowerment has been and continues to be a number one priority, with class and gender following. Consider, for example, historical Africana women activists like Sojourner Truth (abolitionist and universal suffragist, respectively), Harriet Tubman (Underground Railroad conductor), and Ida B. Wells (anti-lynching crusader), all of whom have been erroneously misclaimed as pre-feminists. Truth's "And Aren't I a Woman" oration demonstrates the primacy of overcoming racial obstacles before addressing the absurdity of female subjugation; Tubman risked her life time and again in freeing Africana men, women, and children in slavery, thereby establishing her commitment to racial parity; Wells's investigations into the lynching of Africana men reveal that most of these lynchings resulted from the perception that Africana men were a threat to the established economic system. Clearly racial equality rather than female empowerment was the primary concern of all these women, a substantive commentary on the prioritizing of race, class, and gender as a key feature of Africana Womanism. According to Filomina Chioma Steady in *The Black Woman Cross-Culturally,*

> Regardless of one's position, the implications of the feminist movement for the black woman are complex. . . . Several factors set the black woman apart as having a different order of priorities. She is oppressed not simply because of her sex but ostensibly because of her race and, for the majority, essentially because of their class. Women belong to different socio-economic groups and do not represent a universal category. Because the majority of black women are poor, there is likely to be some alienation from the middle-class aspect of the women's movement which perceives feminism as an attack on men rather than a system which thrives on inequality. (Steady, 23–24)

Thus, as noted Africana sociologist Joyce Ladner postulates, "black women do not perceive their enemy to be black men, but rather the enemy is considered to be oppressive forces in the larger society which subjugate black men, women and children" (277–278).

Another key feature of this concept is the importance of self-naming, which takes us back to Africa. In African cosmology, the proper naming—*nommo*—is crucial to existence. Hence,

> Africana women might begin by naming and defining their unique movement "Africana Womanism." . . . In refining this terminology into a theoretical framework and methodology, "Africana Womanism" identifies the participation and the role of Africana women in the struggle, but does not suggest that female subjugation is the most critical issue they face in their struggle for parity. Like Black feminism, Africana Womanism acknowledges societal gender problems as critical issues to be resolved; however, it views feminism, the suggested alternative to these problems, as a sort of inverted White patriarchy, with the White feminist now in command and on top. Mainstream feminism is women's co-opting themselves into mainstream patriarchal values (Hudson-Weems, 38)

Merely appending *Black* to *feminism* (Black feminism) is insufficient, particularly since Africana women have served as models for White women in their struggle to move from home to workplace and to break silence. Africana women have always been in the workplace next to their male counterparts. Moreover, they have always been verbal, voicing their opinions about issues in general. Hence, naming ourselves after White women and concentrating most of our energies on their agenda would be an act of duplicating a duplicate.

The agenda of Africana women is uniquely our own. It is one that is grounded in our history and culture, and hence Africana women cannot risk sacrificing our struggle for the benefit of someone else's struggle. A very objective and insightful White feminist, Bettina Aptheker, understands the reality of the White feminist and the Africana woman as two separate issues:

> When we place women at the center of our thinking we are going about the business of creating an historical and cultural matrix from which women may claim autonomy and independence over their own lives. For women of color, such autonomy cannot be achieved in conditions of racial oppression and cultural genocide. . . . In short, "feminist," in the modern sense, means the empowerment of women. For women of color, such an equality, such an empowerment, cannot take place unless the communities in which they live can successfully establish their own racial and cultural integrity. (Aptheker, 13)

Africana Womanism as a theoretical concept and methodology defines a new paradigm, which offers an alternative to all forms of feminism. It is a terminology and concept that consider both ethnicity (Africana) and gender (Womanism), which I coined and defined in the mid-1980s. In 1989, the *Western Journal of Black Studies* published my article "Cultural and Agenda Conflicts in Academia: Critical Issues for Africana Women's Studies," which is the second chapter of *African Womanism.* It was later established that the concept is

> neither an outgrowth nor an addendum to feminism, . . . Black feminism, African feminism, or Walker's womanism that some Africana women have come to embrace. Africana Womanism is an ideology created and designed for all women of African descent. It is grounded in African culture, and therefore, it necessarily focuses on the unique experiences, struggles, needs, and desires of Africana women. It critically addresses the dynamics of the conflict between the mainstream feminist, the Black feminist, the African feminist, and the Africana womanist. (Hudson-Weems, 24)

Eighteen descriptors represent the agenda of the Africana woman, all of which are present in her life to varying degrees. First, the Africana womanist *names* and *defines* herself and her movement: hence, Africana Womanism. She is *family-centered*—the Africana womanist is more concerned with her entire family than with just herself and her sisters—even though genuine *sisterhood* is also very important in her reality. The Africana womanist also welcomes *male presence and participation* in her struggle, as her destiny is often intertwined with his in their broader struggle for humanity and liberation for Africana people. She has demonstrated and continues to demonstrate enormous *strength,* both in a physical and a psychological sense. Moreover, the Africana womanist desires *positive male companionship.* However, her *role as homemaker,* as it has always been, is much relaxed. She demands *respect* and *recognition* in her incessant search for *wholeness* and *authenticity.* Her authentic point of reference emphasizes her tremendous sense of *spirituality.* From this perspective, she acknowledges the existence of spiritual reality, which brings into account the power of comprehension, healing, and the unknown. She *respects and appreciates elders,* insisting that her young do likewise. Finally, demanding *no separate space* for nourishing her individual needs and goals, even though *ambition* is an aspect of her character, the Africana womanist is committed to the art of *mothering* and *nurturing* her own in particular and humankind in general.

If Africana Womanism is allowed to reach its full potential, that of reclaiming Africana women via identifying our own collective struggle and acting upon it, then Africana people the world over will be better for it. Moreover, whenever a people takes control over its struggle, tailoring it to meet the collective needs and demands, success is almost invariably inevitable. When success in one's goals is realized, it makes for a more peaceful reality for all concerned. Further, one is more inclined to a wholesome and amicable relationship with others, knowing that the concerns of the people are respected and met.

In conclusion, Africana Womanism establishes that, above all else, "the primary goal of Africana women . . . is to create their own criteria for assessing their realities, both in thought and in action." (Hudson-Weems, 50)

Bibliography

Aptheker, Bettina. "Strong Is What We Make Each Other: Unlearning Racism within Women's Studies." *Women's Studies Quarterly* 1, No. 4 (Winter 1981).

Hare, Julia. Quoted in "Feminism in Academe: The Race Factor" by Eileen Crawford, in *Black Issues in Higher Education* 10, No. 1 (March 11, 1993).

Hudson-Weems, Clenora. *Africana Womanism: Reclaiming Ourselves,* 2d rev. ed. Troy, Mich.: Bedford, 1994.

Ladner, Joyce. *Tomorrow's Tomorrow: The Black Woman.* Garden City, N.Y.: Anchor, 1972.

Mompati, Ruth. "Women and Life Under Apartheid." In *One Is Not A Woman, One Becomes: The African Woman in a Transitional Society,* ed. Daphne Williams Ntiri. Troy, Mich.: Bedford, 1982.

Ntiri, Daphne Williams, ed. *One Is Not A Woman, One Become: The African Woman in a Transitional Society.* Troy, Mich.: Bedford, 1982.

Steady, Filomina Chioma, ed. *The Black Woman Cross-Culturally.* Cambridge, Mass.: Schenkman, 1981.

Note

1. A previously unpublished essay based on concepts in Clenora Hudson-Weem's book *Africana Womanism: Reclaiming Ourselves* (1993).

The Classroom as Curriculum

American Association of University Women

In 1992, the American Association of University Women (AAUW) published a study entitled How Schools Shortchange Girls: A Study of Major Findings on Girls and Education. *The authors say their report is "a synthesis of all the available research on the subject of girls in school. . . . [and] presents compelling evidence that girls are not receiving the same quality, or even quantity, of education as their brothers." The AAUW report gained instant, nationwide attention and prompted several U.S. Senators to introduce legislation protecting girls from unequal treatment in publicly funded schools. Chapter 2 of the report, "The Classroom as Curriculum," examines what students from preschool to college learn about gender from the ways in which teachers interact differently with girls and boys.*

Students can learn as much from what they experience in school as they can from the formal content of classroom assignments. Classroom interactions, both with the teacher and other students, are critical components of education. These interactions shape a school. They determine in large measure whether or not a school becomes a community: a place where girls and boys can learn to value themselves and others, where both the rights and the responsibilities of citizens are fostered.

Teacher-Student Interactions

Whether one looks at preschool classrooms or university lecture halls, at female teachers or male teachers, research spanning the past twenty years consistently reveals that males receive more teacher attention than do females.[1] In preschool classrooms boys receive more instructional time, more hugs, and more teacher attention.[2] The pattern persists through elementary school and high school. One reason is that boys demand more attention. Researchers David and Myra Sadker have studied these patterns for many years. They report that boys in one study of elementary and middle school students called out answers eight times more often than girls did. When boys called out, the typical teacher reaction was to listen to the comment. When girls called out, they were usually corrected with comments such as, "Please raise your hand if you want to speak."[3]

It is not only the attention demanded by male students that explains their greater involvement in teacher-student exchanges. Studies have found that even when boys do not volunteer, the teacher is more likely to solicit their responses.[4]

The issue is broader than the inequitable distribution of teacher *contacts* with male and female students; it also includes the inequitable *content* of teacher comments. Teacher remarks can be vague and superficial or precise and penetrating. Helpful teacher comments provide students with insights into the strengths and weaknesses of their answers. Careful and comprehensive teacher reactions not only affect student learning, they can also influence student self-esteem.[5]

The Sadkers conducted a three-year study of more than 100 fourth, sixth- and eighth-grade classrooms. They identified four types of teacher comments: praise, acceptance, remediation, and criticism.

They found that while males received more of all four types of teacher comments, the difference favoring boys was greatest in the more useful teacher reactions of praise, criticism, and remediation. When teachers took the time and made the effort to specifically evaluate a student's performance, the student receiving the comment was more likely to be male.[6] These findings are echoed in other investigations, indicating that boys receive more precise teacher comments than females in terms of both scholarship and conduct.[7]

The differences in teacher evaluations of male and female students have been cited by some researchers as a cause of "learned helplessness," or lack of academic perseverance, in females. Initially investigated in animal experiments, "learned helplessness" refers to a lack of perseverance, a debilitating loss of self-confidence.[8] This concept has been used to explain why girls sometimes abandon while boys persistently pursue academic challenges for which both groups are equally qualified.[9]

One school of thought links learned helplessness with attribution theory. While girls are more likely to attribute their success to luck, boys are more likely to attribute their success to ability. As a result of these different causal attributions, boys are more likely to feel mastery and control over academic challenges, while girls are more likely to feel powerless in academic situations.[10]

Studies also reveal that competent females have higher expectations of failure and lower self-confidence when encountering new academic situations than do males with similar abilities.[11] The result is that female students are more likely to abandon academic tasks.[12]

However, research also indicates that the concepts of learned helplessness and other motivation constructs are complex. Psychologist Jacquelynne Eccles and her colleagues have found that there is a high degree of variation within each individual in terms of motivational constructs as one goes across subject areas. New evidence indicates that it is too soon to state a definitive connection between a specific teacher behavior and a particular student outcome.[13] Further research on the effects of teacher behavior and student performance and motivation is needed.

The majority of studies on teacher-student interaction do not differentiate among subject areas. However, there is some indication that the teaching of certain subjects may encourage gender-biased teacher behavior while others may foster more equitable interactions. Sex differences in attributing success to luck versus effort are more likely in subject areas where teacher responses are less frequent and where single precise student responses are less common.[14]

Two recent studies find teacher-student interactions in science classes particularly biased in favor of boys.[15] Some mathematics classes have less biased patterns of interaction overall when compared to science classes, but there is evidence that despite the more equitable overall pattern, a few male students in each mathematics class receive particular attention to the exclusion of all other students, male and female.[16]

Research on teacher-student interaction patterns has rarely looked at the interaction of gender with race, ethnicity, and/or social class. The limited data available indicate that while males receive more teacher attention than females, white boys receive more attention than boys from various racial and ethnic minority groups.[17]

Evidence also suggests that the attention minority students receive from teachers may be different in nature from that given to white children. In elementary school, black boys tend to have fewer interactions overall with teachers than other students and yet they are the recipients of four to ten times the amount of qualified praise ("That's good, but . . .") as other students.[18] Black boys tend to be perceived less favorably by their teachers and seen as less able than other students.[19] The data are more complex for girls. Black girls have less interaction with teachers than white girls, but they attempt to initiate interaction much more often than white girls or than boys of either race. Research indicates that teachers may unconsciously rebuff these black girls, who eventually turn to peers for interaction, often becoming the class enforcer or go-between for other students.[20] Black females also receive less reinforcement from teachers than do other students, although their academic performance is often better than boys.[21]

In fact, when black girls do as well as white boys in school, teachers attribute their success to hard work but assume that the white boys are not working up to their full potential.[22] This, coupled with the evidence that blacks are more often reinforced for their social behavior while whites are likely to be reinforced for their academic accomplishments, may contribute to low academic self-esteem in black girls.[23] Researchers have found that black females value their academic achievements less than black males in spite of their better performance.[24] Another study found that black boys have a higher science self-concept than black girls although there were no differences in achievement.[25]

The Design of Classroom Activities

Research studies reveal a tendency beginning at the preschool level for schools to choose classroom activities that will appeal to boys' interests and to select presentation formats in which boys excel or are encouraged more than are girls.[26] For example, when researchers looked at lecture versus laboratory classes, they found that in lecture classes teachers asked males academically related questions about 80 percent more often than they questioned females; the patterns were mixed in laboratory classes.[27] However, in science courses, lecture classes remain more common than laboratory classes.

Research indicates that if pupils begin working on an activity with little introduction from the teacher, everyone has access to the same experience. Discussion that follows after all students have completed an activity encourages more participation by girls.[28] In an extensive multistate study, researchers found that in geometry classes where the structure was changed so that students read the book and did problems *first* and *then* had classroom discussion of the topic, girls outperformed boys in two of five tests and scored equally in the other three. Girls in the experimental class reversed the general trend of boys' dominance on applications, coordinates, and proof taking, while they remained on par with boys on visualizations in three dimensions and transformations. In traditional classes where topics were introduced by lecture first and then students read the book and did the problems, small gender differences favoring boys remained.[29]

Successful Teaching Strategies

There are a number of teaching strategies that can promote more gender-equitable learning environments. Research indicates that science teachers who are successful in encouraging girls share several strategies.[30] These included using more than one textbook, eliminating sexist language, and showing fairness in their treatment and expectations of both girls and boys.

Other research indicates that classrooms where there are no gender differences in math are "girl friendly," with less social comparison and competition and an atmosphere students find warmer and fairer.[31]

In their 1986 study, *Women's Ways of Knowing,* Belenky, Clinchy, Goldberger, and Tarule point out that for many girls and women, successful learning takes place in an atmosphere that enables students to empathetically enter into the subject they are studying, an approach the authors term "connected knowing." The authors suggest that an acceptance of each individual's personal experiences and perspectives facilitates students' learning. They argue for classrooms that emphasize collaboration and provide space for exploring diversity of opinion.[32]

Few classrooms foster "connected learning," nor are the majority of classrooms designed to encourage cooperative behaviors and collaborative efforts. The need to evaluate, rank, and judge students can undermine collaborative approaches. One recent study that sampled third-, fifth-, and seventh-grade students found that successful students reported fewer cooperative attitudes than did unsuccessful students. In this study the effects of gender varied as a function of grade level. Third-grade girls were more cooperative than their male peers, but by fifth grade the gender difference had disappeared.[33] Other studies do not report this grade level-gender interaction, but rather indicate that girls tend to be more cooperative than boys but that cooperative attitudes decline for all students as they mature.[34]

Some educators view the arrival of new classroom organizational structures as a harbinger of more effective and more equitable learning environments. "Cooperative learning" has been viewed as one of these potentially more successful educational strategies. Cooperative learning is designed to eliminate the negative effects of classroom competition while promoting a cooperative spirit and increasing heterogeneous and cross-race relationships. Smaller cooperative work groups are designed to promote group cohesion and interdependence, and mobilize these positive feelings to achieve academic objectives.[35] Progress and academic performance are evaluated on a group as well as an individual basis; the group must work together efficiently or all its members will pay a price.[36] A number of positive results have been attributed to cooperative learning groups, including increasing cross-race friendships, boosting academic achievement, mainstreaming students with disabilities, and developing mutual student concerns.[37]

However, positive cross-sex relationships may be more difficult to achieve than cross-race friendships or positive relationships among students with and without disabilities. First, as reported earlier in this report, there is a high degree of sex-segregation and same-sex friendships in elementary and middle school years.[38] Researchers have found that the majority of elementary students preferred single-sex work groups.[39] Second, different communication patterns of males and females can be an obstacle to effective cross-gender relationships. Females are more indirect in speech, relying often on questioning, while more direct males are more likely to make declarative statements or even to interrupt.[40] Research indicates that boys in small groups are more likely to receive requested help from girls; girls' requests, on the other hand, are more likely to be ignored by the boys.[41] In fact, the male sex may be seen as a status position within the group. As a result, male students may choose to show their social dominance by not readily talking with females.[42]

Not only are the challenges to cross-gender cooperation significant, but cooperative learning as currently implemented may not be powerful enough to overcome these obstacles. Some research indicates that the infrequent use of small, unstructured work groups is not effective in reducing gender stereotypes, and, in fact, increases stereotyping. Groups often provide boys with leadership opportunities that increase their self-esteem. Females are often seen as followers and are less likely to want to work in mixed-sex groups in the future.[43]Another study indicates a decrease in female achievement when females are placed in mixed-sex groups.[44] Other research on cooperative education programs has reported more positive results.[45] However, it is clear that merely providing an occasional group learning experience is not the answer to sex and gender differences in classrooms.

Problems in Student Interactions

The ways students treat each other during school hours is an aspect of the informal learning process, with significant negative implications for girls. There is mounting evidence that boys do not treat girls well. Reports of student sexual harassment—the unwelcome verbal or physical conduct of a sexual nature imposed by one individual on another—among junior high school and high school peers are increasing. In the majority of cases a boy is harassing a girl.[46]

Incidents of sexual harassment reveal as much about power and authority as they do about sexuality; the person being harassed usually is less powerful than the person doing the harassing. Sexual harassment is prohibited under Title IX, yet sex-biased peer interactions appear to be permitted in schools, if not always approved. Rather than viewing sexual harassment as serious misconduct, school authorities too often treat it as a joke.

When boys line up to "rate" girls as they enter a room, when boys treat girls so badly that they are reluctant to enroll in courses where they may be the only female, when boys feel it is good fun to embarrass girls to the point of tears, it is no joke. Yet these types of behaviors are often viewed by school personnel as harmless instances of "boys being boys."

The clear message to both girls and boys is that girls are not worthy of respect and that appropriate behavior for boys includes exerting power over girls—or over other, weaker boys. Being accused of being in any way like a woman is one of the worst insults a boy can receive. As one researcher recently observed:

> *"It is just before dismissal time and a group of very active fourth-graders are having trouble standing calmly in line as they wait to go to their bus. Suddenly one of the boys grabs another's hat, runs to the end of the line, and involves a number of his buddies in a game of keep-away. The boy whose hat was taken leaps from his place in line, trying to intercept it from the others, who, as they toss it back and forth out of his reach, taunt him by yelling, 'You woman! You're a woman!' When the teacher on bus duty notices, she tells the boys that they all have warnings for not waiting in line properly. The boys resume an orderly stance but continue to mutter names—'Woman!' 'Am not.' 'Yes, you are.'—under their breath."*
>
> Margaret Stubbs, October 1990

Harassment related to sexual orientation or sexual preference has received even less attention as an equity issue than heterosexual sexual harassment.[47] Yet, examples of name calling that imply homophobia, such as "sissy," "queer," "gay," "lesbo," are common among students at all levels of schooling. The fourth-grade boys who teased a peer by calling him a "woman" were not only giving voice to the sex-role stereotype that women are weaker than and therefore inferior to men; they were also challenging their peer's "masculinity" by ascribing feminine characteristics to him in a derogatory manner. Such attacks often prevent girls, and sometimes boys, from participating in activities and courses that are traditionally viewed as appropriate for the opposite sex.

When schools ignore sexist, racist, homophobic, and violent interactions between students, they are giving tacit approval to such behaviors. Environments where students do not feel accepted are not environments where effective learning can take place.

Implications

Teachers are not always aware of the ways in which they interact with students. Videotaping actual classrooms so that teachers can see themselves in action can help them to develop their own strategies for fostering gender-equitable education. The use of equitable teaching strategies should be one of the criteria by which teaching performance is evaluated.

Research studies indicate that girls often learn and perform better in same-sex work groups than they do in mixed-sex groupings. Additional research is needed, however, to better understand the specific dynamics of these interactions, particularly the circumstances under which single-sex groupings are most beneficial. Single-sex classes are illegal under Title IX, but usually single-sex work groups within coed classes are not. Teachers should be encouraged to "try out" many different classroom groupings, not only in mathematics and science classes but across a wide range of subject matter. It is critical that they carefully observe the impact of various groupings and write up and report their findings.

Notes

1. See for example, J. Brophy and T. Good, *Teacher-Student Relationships: Causes and Consequences* (New York: Holt, Rinehart, and Winston, 1974); M. Jones, "Gender Bias in Classroom Interactions," *Contemporary Education* 60 (Summer 1989):216–22; M. Lockheed, *Final Report: A Study of Sex Equity in Classroom Interaction* (Washington, DC: National Institute of Education, 1984); M. Lockheed and A. Harris, *Classroom Interaction and Opportunities for Cross-Sex Peer Learning in Science,* paper presented at the Annual Meeting of the American Educational Research Association, New York, April 1989; M. Sadker and D. Sadker, "Sexism in the Classroom: From Grade School to Graduate School," *Phi Delta Kappan* 68 (1986):512; R. Spaulding, *Achievement, Creativity and Self-Concept Correlates of Teacher-Pupil Transactions in Elementary School* (Cooperative Research Project No. 1352), (Washington, DC: U.S. Department of Health Education and Welfare, 1963).
2. L. Serbin et al., "A Comparison of Teacher Responses to the Pre-Academic and Problem Behavior of Boys and Girls," *Child Development* 44 (1973):796–804; M. Ebbeck, "Equity for Boys and Girls: Some Important Issues," *Early Child Development and Care* 18 (1984): 119–31.
3. D. Sadker, M. Sadker, and D. Thomas, "Sex Equity and Special Education," *The Pointer* 26 (1981):33–38.
4. D. Sadker and M. Sadker, "Is the OK Classroom OK?" *Phi Delta Kappan* 55 (1985):358–67.
5. J. Brophy, "Teacher Praise: A Functional Analysis," *Review of Educational Research* 51 (1981):5–32; A. Gardner, C. Mason, and M. Matyas, "Equity, Excellence and 'Just Plain Good Teaching!'" *The American Biology Teacher* 51 (1989):72–77.
6. M. Sadker and D. Sadker, *Year 3: Final Report, Promoting Effectiveness in Classroom Instruction* (Washington, DC: National Institute of Education, 1984).

7. D. Baker, "Sex Differences in Classroom Interactions in Secondary Science," *Journal of Classroom Interaction* 22 (1986):212–18; J. Becker, "Differential Treatment of Females and Males in Mathematics Classes," *Journal for Research in Mathematics Education* 12 (1981):40–53; L. Berk and N. Lewis, "Sex Role and Social Behavior in Four School Environments," *Elementary School Journal* 3 (1977):205–21; L. Morse and H. Handley, "Listening to Adolescents: Gender Differences in Science Classroom Interaction," in *Gender Influences in Classroom Interaction,* L. Wilkerson and C. Marrett, eds., (Orlando, FL: Academic Press, 1985), pp. 37–56.

8. M. Seligman and S. Maier, "Failure to Escape Traumatic Shock," *Journal of Experimental Psychology* 74 (1967):1–9.

9. C. Dweck and N. Repucci, "Learned Helplessness and Reinforcement Responsibility in Children," *Journal of Personality and Social Psychology* 25 (1973):109–16; C. Dweck and T. Goetz, "Attributions and Learned Helplessness," in *New Directions in Attribution Research,* J. Harvey, W. Ickes, and R. Kidd, eds., (Hillsdale, NJ: Erlbaum, 1978).

10. See for example, K. Deaux, "Sex: A Perspective on the Attribution Process," in *New Directions in Attribution Research;* C. Dweck and E. Bush, "Sex Differences in Learned Helplessness: I. Differential Debilitation with Peer and Adult Evaluators," *Developmental Psychology* 12 (1976):147–56; C. Dweck, T. Goetz, and N. Strauss, "Sex Differences in Learned Helplessness: IV. An Experimental and Naturalistic Study of Failure Generalization and Its Mediators," *Journal of Personality and Social Psychology* 38 (1980):441–52; L. Reyes, *Mathematics Classroom Processes,* paper presented at the Fifth International Congress on Mathematical Education, Adelaide, Australia, August 1984; P. Wolleat et al., "Sex Differences in High School Students' Causal Attributions of Performance in Mathematics," *Journal for Research in Mathematics Education* 11 (1980):356–66; D. Phillips, "The Illusion of Incompetence among Academically Competent Children," *Child Development* 55 (1984):200–16; E. Fennema et al., "Teachers' Attributions and Belief about Girls, Boys, and Mathematics," *Educational Studies in Mathematics* 21 (1990): 55–69.

11. E. Maccoby and C. Jacklin, *The Psychology of Sex Differences* (Stanford, CA: Stanford University Press, 1974); E. Lenney," Women's Self-Confidence in Achievement Settings," *Psychological Bulletin* 84 (1977): 1–13; J. Parsons and D. Ruble," The Development of Achievement-Related Expectancies," *Child Development* 48 (1977): 1075–79; Dweck, Goetz, and Strauss," Sex Differences in Learned Helplessness: IV" ; J. Goetz," Children's Sex Role Knowledge and Behavior: An Ethnographic Study of First Graders in the Rural South," *Theory and Research in Social Education* 8 (1981): 31–54.

12. W. Shepard and D. Hess, "Attitudes in Four Age Groups Toward Sex Role Division in Adult Occupations and Activities," *Journal of Vocational Behavior* 6 (1975): 27–39; C. Dweck and E. Elliot," Achievement Motivation," in *Handbook of Child Psychology,* vol. 4, P. Mussen and E. Hefthierington, eds., (New York: Wiley, 1983); R. Felson," The Effect of Self-Appraisals of Ability on Academic Performance," *Journal of Personality and Social Psychology* 47 (1984): 944–52; M. Stewart and C. Corbin, "Feedback Dependence among Low Confidence Preadolescent

Boys and Girls," Research. *Quarterly for Exercise and Sport* 59 (1988): 160–64.

13. J. Eccles-Parsons et al., "Sex Differences in Attributions and Learned Helplessness," *Sex Roles* 8 (1982): 421–32; J. Eccles-Parsons, C. Kaczala, and J. Meece," Socialization of Achievement Attitudes and Beliefs: Classroom influences," *Child Development* 53 (1982): 322–39; J. Eccles," Expectancies, Values and Academic Behaviors," in *Achievement and Achievement Motives,* J. Spence, ed., (San Francisco, CA: W. H. Freeman and Co., 1983); J. Eccles, *Understanding Motivation: Achievement Beliefs, Gender-Roles and Changing Educational Environments,* address before American Psychological Association, New York, 1987.
14. B. Licht, S. Stader, and C. Swenson," Children's Achievement Related Beliefs: Effects of Academic Area, Sex, and Achievement Level," *Journal of Educational Research* 82 (1989): 253–60.
15. J. Kahle," Why Girls Don't Know," in *What Research Says to the Science Teacher—the Process of Knowing,* M. Rowe, ed., (Washington, DC: National Science Testing Association, 1990), pp. 55–67; V. Lee," Sexism in Single-Sex and Coeducational Secondary School Classrooms," paper presented at the annual meeting of the American Sociological Association, Cincinnati, OH, August 8, 1991.
16. J. Eccles, "Bringing Young Women to Math and Science," in *Gender and Thought: Psychological Perspectives,* M. Crawford and M. Gentry, eds., (New York: Springer-Verlag, 1989), pp. 36–58; Licht et al., "Children's Achievement Related Beliefs."
17. Sadker and Sadker, *Year* 3; L. Grant, "Race-Gender Status, Classroom Interaction and Children's Socialization in Elementary School," in *Gender Influences in Classroom Interaction,* L Wilkinson and C. Marrett, eds., (Orlando, FL: Academic Press, 1985), pp. 57–75.
18. Grant, "Race-Gender Status," p. 66.
19. C. Cornbleth and W. Korth, "Teacher Perceptions and Teacher-Student Interaction in Integrated Classrooms," *Journal of Experimental Education* 48 (Summer 1980): 259–63; B. Hare, *Black Girls: A Comparative Analysis of Self-Perception and Achievement by Race, Sex and Socioeconomic Background,* Report No. 271, (Baltimore, MD: John Hopkins University, Center for Social Organization of Schools, (1979).
20. S. Damico and E. Scott, "Behavior Differences Between Black and White Females in Desegregated Schools," *Equity and Excellence* 23 (1987): 63–66; Grant," Race-Gender Status." See also J. Irvine," Teacher-Student Interactions: Effects of Student Race, Sex, and Grade Level," *Journal of Educational Psychology* 78 (1986): 14–21.
21. Damico and Scott," Behavior Differences;" Hare, "Black Girls: A Comparative Analysis."
22. Damico and Scott," Behavior Differences;" L. Grant," Black Females 'Place' in Integrated Classrooms," *Sociology of Education* 57 (1984): 98–111.
23. Damico and Scott," Behavior Differences."

24. D. Scott-Jones and M. Clark," The School Experience of Black Girls: The Interaction of Gender, Race and Socioeconomic Status," *Phi Delta Kappan* 67 (March 1986): 20–526.

25. V. Washington and J. Newman," Setting Our Own Agenda: Exploring the Meaning of Gender Disparities Among Blacks in Higher Education," *Journal of Negro Education* 60 (1991): 19–35.

26. E. Fennema and P. Peterson, "Effective Teaching for Girls and Boys: The Same or Different?" in *Talks to Teachers,* D. Berliner and B. Rosenshine, eds., (New York: Random House, 1987), pp. 111–25; J. Stallings, "School Classroom and Home Influences on Women's Decisions to Enroll in Advanced Mathematics Courses," in *Women and Mathematics: Balancing the Equation,* S. Chipman, L. Brush, and D. Wilson, eds., (Hillsdale, NJ: Erlbaum, 1985), pp. 199–224; S. Greenberg," Educational Equity in Early Education Environments," in *Handbook for Achieving Sex Equity Through Education,* S. Klein, ed., (Baltimore, MD: Johns Hopkins University Press, 1985), pp. 457–69.

27. D. Baker," Sex Differences in Classroom Interactions in Secondary Science," *Journal of Classroom Interaction* 22 (1986): 212–18.

28. D. Jorde and A. Lea," The Primary Science Project in Norway," in *Proceedings of Growth GSAT Conference,* J. Kahle, J. Daniels, and J. Harding, eds., (West Lafayette, IN: Purdue University, 1987), pp. 68–72.

29. P. Flores, "How Dick and Jane Perform Differently in Geometry: Test Results on Reasoning, Visualization and Affective Factors," paper presented at American Educational Research Association Meeting, Boston, MA, April 1990.

30. J. Kahle, *Factors Affecting the Retention of Girls in Science Courses, and Careers: Case Studies of Selected Secondary Schools* (Reston, VA: The National Association of Biology Teachers, October 1983).

31. Eccles, "Bringing Young Women to Math and Science."

32. M. Belenky et al., *Women's Ways of Knowing—The Development of Self, Body, and Mind* (New York: Basic Books; s, Inc., 1986).

33. G. Engelhard and J. Monsaas, "Academic Performance, Gender and the Cooperative Attitudes of Third, Fifth and Seventh Graders," *Journal of Research and Development in Education* 22 (1989): 13–17.

34. A. Ahlgren and D. Johnson, "Sex Differences in Cooperative and Competitive Attitudes from 2nd Through the 12th Grades," *Developmental Psychology* 15 (1979): 45–49; B. Herndon and M. Carpenter, "Sex Differences in Cooperative and Competitive Attitudes in a Northeastern School," *Psychological Reports* 50 (1982): 768–70; L Owens and R. Straton, "The Development of Co-operative Competitive and Individualized Learning Preference Scale for Students," *Journal of Educational Psychology* 50 (1980): 147–61.

35. S. Sharon et al., eds., *Cooperation in Education* (Provo, UT: Brigham Young University Press, 1980); S. Bossert, *Task Structure and Social Relationships* (Cambridge, MA: Harvard University Press, 1979); W. Shrum, N. Cheek, and S. Hunter," Friendship in the School: Gender and Racial Homophily," *Sociology of Education* 61 (1988): 227–39.

36. E. Aronson, *The Jigsaw Classroom* (Beverly Hills, CA: Sage, 1978); D. DeVries and K. Edwards, "Student Teams and Learning Games: Their Effects on Cross-Race and Cross-Sex Interaction," *Journal of Educational Psychology* 66 (1974): 741–49; P Okebukola, "Cooperative Learning and Students' Attitude to Laboratory Work," *Social Science and Mathematics* 86 (1986): 582–90; R. Slavin, "How Student Learning Teams Can Integrate the Desegregated Classroom," *Integrated Education* 15 (1977): 56–58.

37. N. Blaney et al., "Interdependence in the Classroom: A Field Study," *Journal of Educational Psychology* 69 (1977): 121–28; D. DeVries and K. Edwards, "Student Teams and Learning Games: Their Effects on Cross-Race and Cross-Sex Interaction," *Journal of Educational Psychology* 66 (1974): 741–49; Sharon et al., *Cooperation in Education;* R. Slavin, "Cooperative Learning," *Review of Educational Research* 50 (1980): 315–42; R. Slavin, "Cooperative Learning and Desegregation," in W. Hawley ed., *Effective School Desegregation* (Berkeley, CA: Sage, 1981); R. Weigle, P. Wiser, and S. Cook, "The Impact of Cooperative Learning Experiences on Cross-Ethnic Relations and Attitude," *Journal of Social Issues* 3 (1975): 219–44.

38. M. Hailman, *The Evolution of Children's Friendship Cliques (*ERIC Document Reproduction Service no. ED 161556, 1977); R. Best, *We've All Got Scars: What Boys and Girls Learn in Elementary School* (Bloomington, IN: University Press, 1983); J. Eccles-Parsons," Sex Differences in Mathematics Participation," in M. Steinkamp and M. Machr, eds., *Women in Science* (Greenwich, CT: JAI Press, 1984); M. Hallman and N. Tumma, "Classroom Effects on Change in Children," *Sociology of Education* 51 (1978): 170–282.

39. M. Lockheed, K. Finklestein, and A. Harris, *Curriculum and Research for Equity: Model Data Package* (Princeton, NJ: Educational Testing Service, 1979).

40. B. Eakins and R. Eakins, "Sex Roles, Interruptions, and Silences in Conversation," in B. Thorne and N. Henley, eds., *Sex Differences in Human Communication* (Boston, MA: Houghton Mifflin, 1978); N. Henley and B. Thorne, "Women Speak and Men Speak: Sex Differences and Sexism in Communications, Verbal and Nonverbal," in A. Sargent, ed., *Beyond Sex Roles* (St. Paul, MN: West Publishing Company, 1977); R. Lakoff, *Languages and Women's Place (*New York; Harper Colophon Books, 1976); D. Tannen, *You Just Don't Understand: Women and Men in Conversation* (New York: William Morrow, 1990).

41. L. Wilkinson, J. Lindow, and C. Chiang, "Sex Differences and Sex Segregation in Students' Small-Group Communication," in L. Wilkinson and C. Marret, eds., *Gender Influences in Classroom Interaction* (Orlando, FL: Academic Press, 1985), pp. 185–207.

42. J. Berger, T. Conner, and M. Fisek, eds., *Expectation States Theory: A Theoretical Research Program* (Cambridge, MA: Winthrop, 1974); M. Lockheed and A. Harris, "Cross-Sex Collaborative Learning in Elementary Classrooms," *American Educational Research Journal* 21 (1984):275–94.

43. Lockheed and Harris, "Cross-Sex Collaborative Learning."

44. C. Weisfeld et at., "The Spelling Bee; A Naturalistic Study of Female Inhibitions in Mixed-Sex Competitions," *Adolescence* 18 (1983):695–708.
45. For example, M. Lockheed and A. Harris, "Classroom Interaction and Opportunity for Cross-Sex Peer Learning in Science," *Journal of Early Adolescence* (1982): 135–43.
46. S. Strauss, "Sexual Harassment in the School: Legal Implications for Principals," *National Association of 'Secondary School Principals Bulletin* (1988):93–97; N. Stein, ed. *Who's Hurt and Who's Liable: 'Sexual Harassment in Massachusetts Schools,* 4th ed. (Quincy, MA: Massachusetts Department of Education, 1986).
47. D. Grayson, "Emerging Equity Issues Related to Homosexuality," *Peabody Journal of Education* 64 (1989):132–45.

How It Feels to Be Colored Me

Zora Neale Hurston

I am colored but I offer nothing in the way of extenuating circumstances except the fact that I am the only Negro in the United States whose grandfather on the mother's side was *not* an Indian chief.

I remember the very day that I became colored. Up to my thirteenth year I lived in the little Negro town of Eatonville, Florida. It is exclusively a colored town. The only white people I knew passed through the town going to or coming from Orlando. The native whites rode dusty horses, the Northern tourists chugged down the sandy village road in automobiles. The town knew the Southerners and never stopped cane chewing when they passed. But the Northerners were something else again. They were peered at cautiously from behind the curtains by the timid. The more venturesome would come out on the porch to watch them go past and got just as much pleasure out of the tourists as the tourists got out of the village.

The front porch might seem a daring place for the rest of the town, but it was a gallery seat for me. My favorite place was atop the gate-post. Proscenium box[1] for a born first-nighter. Not only did I enjoy the show, but I didn't mind the actors knowing that I liked it. I usually spoke to them in passing. I'd wave at them and when they returned my salute, I would say something like this: "Howdy-do-well-I-thank-you-where-you-goin'?" Usually automobile or the horse paused at this, and after a queer exchange of compliments, I would probably "go a piece of the way" with them, as we say in farthest Florida. If one of my family happened to come to the front in time to see me, of course negotiations would be rudely broken off. But even so, it is clear that I was the first "welcome-to-our-state" Floridian, and I hope the Miami Chamber of Commerce will please take notice.

During this period, white people differed from colored to me only in that they rode through town and never lived there. They liked to hear me "speak pieces" and sing and wanted to see me dance the parse-me-la, and gave me generously of their small silver for doing these things, which seemed strange to me for I wanted to do them so much that I needed bribing to stop. Only they didn't know it. The colored people gave no dimes. They deplored any joyful tendencies in me, but I was their Zora nevertheless. I belonged to them, to the nearby hotels, to the county—everybody's Zora.

But changes came in the family when I was thirteen, and I was sent to school in Jacksonville. I left Eatonville, the town of the oleanders, as Zora. When I disembarked from the river-boat at Jacksonville, she was no more. It seemed that I had suffered a sea change. I was not Zora of Orange County anymore, I was now a little colored girl. I found it out in certain ways. In my heart as well as in the mirror, I became a fast brown—warranted not to rub nor run.

But I am not tragically colored. There is no great sorrow dammed up in my soul, nor lurking behind my eyes. I do not mind at all. I do not belong to the sobbing school of Negrohood who hold that nature somehow has given them a low-down dirty deal and whose feelings are all hurt about it. Even in the helter-skelter skirmish that is my life, I have seen that the world is to the strong regardless of a little pigmentation more or less. No, I do not weep at the world—I am too busy sharpening my oyster knife.

Someone is always at my elbow reminding me that I am the granddaughter of slaves. It fails to register depression with me. Slavery is sixty years in the past. The operation was successful and the patient is doing well, thank you. The terrible struggle that made me an American out of a potential slave said "On the line!" The Reconstruction said "Get set!"; and the generation before said "Go!" I am off to a flying start and I must not halt in the stretch to look behind and weep. Slavery is the price I paid for civilization, and the choice was not with me. It is a bully adventure and worth all that I have paid through my ancestors for it. No one on earth ever had a greater chance for glory. The world to be won and nothing to be lost. It is thrilling to think—to know that for any act of mine, I shall get twice as much praise or twice as much blame. It is quite exciting to hold the center of the national stage, with the spectators not knowing whether to laugh or to weep.

The position of my white neighbor is much more difficult. No brown specter pulls up a chair beside me when I sit down to eat. No dark ghost thrusts its leg against mine in bed. The game of keeping what one has is never so exciting as the game of getting.

I do not always feel colored. Even now I often achieve the unconscious Zora of Eatonville before the Hegira.[2] I feel most colored when I am thrown against a sharp white background.

For instance at Barnard. "Besides the waters of the Hudson" I feel my race. Among the thousand white persons, I am a dark rock surged upon, and overswept, but through it all, I remain myself. When covered by the waters, I am; and the ebb but reveals me again.

Sometimes it is the other way around. A white person is set down in our midst, but the contrast is just as sharp for me. For instance, when I sit in the drafty basement that is The New World Cabaret[3] with a white person, my color comes. We enter chatting about any little nothing that we have in common and are seated by the jazz waiters. In the abrupt way that jazz orchestras have, this one plunges into a number. It loses no time in circumlocutions, but gets right down to business. It constricts the thorax and splits the heart with its tempo and narcotic harmonies. This orchestra grows rambunctious, rears on its hind legs and attacks the tonal veil with primitive fury, rending it, clawing it until it breaks through to the jungle beyond. I follow those heathen—follow them exultingly. I dance wildly inside myself; I yell within, I whoop; I shake my assegai[4] above my head, I hurl it true to the mark *yeeeeooww!* I am in the jungle and living in the jungle way. My face is painted red and yellow and my body is painted blue. My pulse is throbbing like a war drum. I want to slaughter something—give pain, give death to what, I do not know. But the piece ends. The men of the orchestra wipe their lips and rest their fingers. I creep back slowly to the veneer we call civilization with the last tone and find the white friend sitting motionless in his seat—smoking calmly.

"Good music they have here," he remarks, drumming the table with his fingertips.

Music. The great blobs of purple and red emotion have not touched him. He has only heard what I felt. He is far away and I see him but dimly across the ocean and the continent that have fallen between us. He is so pale with his whiteness then and I am *so* colored.

At certain times I have no race, I am *me*. When I set my hat at a certain angle and saunter down Seventh Avenue, Harlem City, feeling as snooty as the lions in front of the Forty-Second Street Library, for instance. So far as my feelings are concerned, Peggy Hopkins Joyce on the Boule Mich[5] with her gorgeous raiment, stately carriage, knees knocking together in a most aristocratic manner, has nothing on me. The cosmic Zora emerges. I belong to no race nor time. I am the eternal feminine with its string of beads.

I have no separate feeling about being an American citizen and colored. I am merely a fragment of the Great Soul that surges within the boundaries. My country, right or wrong.

Sometimes, I feel discriminated against, but it does not make me angry. It merely astonishes me. How *can* any deny themselves the pleasure of my company? It's beyond me.

But in the main, I feel like a brown bag of miscellany propped against a wall. Against a wall in company with other bags, white, red and yellow. Pour out the contents, and there is discovered a jumble of small things priceless and worthless. A first-water diamond, an empty spool, bits of broken glass, lengths of string, a key to a door long since crumbled away, a rusty knife blade, old shoes saved for a road that never was and never will be, a nail bent under the weight of things too heavy for any nail, a dried flower or two still a little fragrant. In your hand is the brown bag. On the ground before you is the jumble it held—so much like the jumble in the bags, could they be emptied, that all might be dumped in a single heap and the bags refilled without altering the content of any greatly. A bit of colored glass more or less would not matter. Perhaps that is how the Great Stuffer of Bags filled them in the first place—who knows?

Footnotes

1. Box at the very front of the auditorium, closest to the stage.
2. Forced march of Mohammed from Mecca to Medina in A.D. 622; hence any forced flight or journey for safety.
3. Popular Harlem nightclub in the 1920s.
4. A slender spear used by some South African tribes.
5. Boulevard St. Michel, a street on the Left Bank of Paris running near the Sorbonne University and through the Latin Quarter. It has never been a particularly fashionable street, but it is filled with cafés that were—and still are—much frequented by Americans in Paris. "Joyce": a socialite and heiress, much photographed.

Black Youths' Rate of Suicide Rising Sharply

Pam Belluck

Chicago, March 19—Suicide among black teen-agers, once quite rare, has increased dramatically over the last two decades, a troubling rise that might reflect the strain some black families feel in making the transition to middle-class life, according to a study released today by the Federal Centers for Disease Control and Prevention.

The study found that the rate of black suicide for youths ages 15 to 19 had more than doubled, to 8.1 per 100,000 in 1995, from 3.6 per 100,000 in 1980. It found a similar pattern, with far smaller numbers for ages 10 to 14.

Researchers also found that the gap between black and white teenage suicides was narrowing—in 1980 the white rate for ages 10 to 19 was more than twice the black rate, but in 1995 the white rate was 42 percent higher.

And a 1995 study of high school students across the country found that black students were as likely as whites to try suicide.

Suicide is now the third leading cause of death for black 15- to 24-year-olds, matching the rate in the general population. The leading cause of death for black youths is homicide, followed by accidents.

Dr. Tonji Durant, the author of the centers' study, said the researchers hypothesized that one factor in the increase might be the growth in the black middle class.

"One theory," Dr. Durant said, "is that upwardly mobile families may experience some stress with their social environments and for some reason they may, in the process of adapting to a new class, adopt some of the behaviors of the larger society, including the ways of dealing with hopelessness and depression."

Several experts on black youth suicide said today that the move to middle-class life might be accompanied by a splintering of community and family support networks, a weakening of bonds to religion, and the pressure of trying to compete in historically white-dominated professions and social circles.

The centers' study did not look at the economic status of the suicide victims, and Dr. Durant said more research was needed to better understand the patterns. But a 1995 study at Columbia University of teen-age suicides in the New York metropolitan region found that unlike white and Hispanic teen-agers, black teen-agers who committed suicide tended to come from higher socio-economic backgrounds than blacks in the general population.

"What we figured is that the greater economic and educational advances that blacks were able to make, the less protection they were able to give their children from the stresses that lead to suicide," said Dr. David Shaffer, a professor of child psychiatry and pediatrics at Columbia who conducted that study, which looked in detail at each of 140 teen-age suicides in New York City and its suburbs from 1984 to 1996. "One theory is that when you live in poorer circumstances, there is a kind of stress inoculation. If you're exposed to a lot of stress then you're able to deal with many problems. It's a bit like being immunized."

Les Franklin, who formed a foundation to help black teen-agers after his 16-year-old son shot himself in the head in 1990 in their affluent Denver neighborhood, said his own experience and others he knew of supported that analysis.

"You don't find a lot of gang members committing suicide." Mr. Franklin said. His son Shaka grew up in a large house with a swimming pool out back and a Porsche in the four-car garage, a life style Mr. Franklin, then working for I.B.M., said was a huge step up from his own upbringing in an apartment crowded with aunts and grandparents.

"His mother and I were divorced—he never forgave us for that," Mr. Franklin said, "I know that he had low self-esteem. He was a kid living in pain." But, Mr. Franklin added; "he didn't have the hard life. He didn't have the preparation that I got to learn how to survive."

At a time when young black males are murdered at an alarming rate—111 per 100,000, according to 1995 figures—the suicide numbers are especially troubling, people who work with black youths say.

The centers' study of the 3,030 suicides of blacks ages 10 to 19 between 1980 and 1995 found a stark increase in the suicide rate among black males between the ages of 15 and 19, which jumped 146 percent from 1980 to 1995. Suicide among white males in that age group increased 22 percent during that time. Suicides among black teen-age girls increased also, but, like whites, those rates were much lower than for boys.

Nearly three quarters of black male teenagers who killed themselves used guns, the most common weapon for white males as well.

David C. Clark, a professor of psychiatry at Rush Medical College in Chicago, who is working on a study of all Chicago suicides for ages 15 to 19 from 1991 to 1996, said preliminary findings had suggested that "black adolescents are more likely to kill themselves in the presence of somebody else, in the presence of a girlfriend or in the presence of someone who they hang out with."

That means, Professor Clark said, that "for every black adolescent suicide we have a greater likelihood that there are other teens that are exposed to the shock."

Some experts said that in urban neighborhoods with a lot of guns, some young men, feeling hopeless or depressed, might indirectly commit suicide by putting themselves in situations where they are likely to be shot. The centers' study did not include these kinds of deaths.

Professor Clark said the increase in black teen-age suicide might also reflect the paucity of mental health services in many black communities and a cultural distrust of mental health professionals.

"It certainly has been a taboo in the African-American community to talk about suicide," said Kenya Napper Bello, president of Free Mind Generation, a nonprofit organization in Atlanta dedicated to preventing suicide and depression among blacks. "It is often viewed as a sin."

Ms. Bello's husband, Razak, killed himself at the age of 27 by jumping out the 32nd-floor window of a downtown Atlanta hotel. Mr. Bello, a rising young executive at Coca-Cola, had been told by doctors that he had manic-depressive illness, but was so uncomfortable with the idea that he needed help that, two months before his death, he stopped taking medication and seeing a counselor.

"It was unfortunate that he suffered from a condition that he felt he had to hide, including from me," said Ms. Bello, who also started a newsletter called "Black Men Don't Commit Suicide," after a myth that she said some people still subscribe to.

"We have prided ourselves as a community in being resilient and being able to deal with adversity," Ms. Bello said. "The white community seems to be far more advanced in talking about it, getting psychological support and coming up with answers. We're just not there."

A Black Experience

Leonard C. Archer

Sometimes when I am conversing with my colleagues in the academic profession, and the subject is about doctoral dissertations, I am asked, "Archer, what did you write about?" I reply, "The NAACP and the American Theatre." Then come the questions and remarks: Why did you write about that? I didn't know the NAACP had an interest in the American theatre. Did someone select that subject for you? The NAACP is a protest organization that uses legal actions. Where did you get your Ph.D.?

The answer to the last question, The Ohio State University. Thus follows a black experience. From 1947 to 1957, I was director of a theatre at Central State College, Wilberforce, Ohio, which is about fifty-five miles from Columbus, Ohio. Annually, the Black president of that predominantly Black college sounds a warning to the faculty members who do not have doctorates that they may suffer reductions in salaries or even lose their jobs. "Some of you are too lazy to drive fifty-five miles to Ohio State," he chided. I had no doctorate so I heeded the president's stern warning and began attending Ohio State in 1950—summers, part-time during the school year, then full-time to meet residence requirements. I reached the candidacy for the doctor's degree in 1956, having worn out two automobiles in the process.

When I had completed the courses for a major in theatre and minors in dramatic literature, radio, and television, my adviser asked, "Leonard, what will you write about for your dissertation?" Now in those days, and maybe still in these, certain universities and departments guided Black students into writing on Black subjects. Instinctively, I backed into the subject of this study. First, I suggested to my adviser a study of nineteenth-century novels which had been adapted for the stage as melodramas. He rejected this because he thought the subject would be too boring for me. Remembering that he had seen me play the part of a runaway slave and dancer in *The 17th Star* (1953), a symphonic drama written by Paul Green for the Ohio Sesquicentennial, I offered a study of symphonic dramas, such as *Unto These Hills* which is presented every summer in Cherokee, North Carolina. He didn't think much of symphonic dramas.

At this point, I pulled out my ace in the hole—"The National Association of Colored People and the American Theatre." "Now that's a good subject," he admitted. It is ironic that in those days many Black students snickered when Black students wrote on Black subjects.

For several reasons this subject interested me most. First, it would be a continuation of my master's thesis at the University of Toronto (1940), "Negro Life as a Folk Basis for American Drama." My first adviser left Ohio State for another position. My second adviser kept me under very close supervision while I wrote the final draft; he explained that a local chapter of the NAACP had picketed one of his theatrical productions at another university. Second, the materials and resources of this study have been a part of my own Black experience. Who am I? Those of my generation think we know who we are. We are descendants of African-Americans who were later dehyphenated by other Americans, segregated and called "Negroes." And now, young Black revolutionaries have discarded "Negro," an invention of white America, in favor of Black.

I was born at Morehouse College, Atlanta, Georgia, where my father was successively football coach, Greek and mathematics professor, dean, and president from 1905 to 1938. My mother taught at a Baptist mission for the education for Negro men. Dr. John Hope was president of Morehouse then. Only Booker T. Washington had greater influence than Hope's in the cause of higher education for Negroes.

In Atlanta, there has been, and still is, a quality of life called "the Negro exposure," primarily due to the city's five Negro colleges and their schools of divinity: Atlanta University, Congregationalist; Clark College, Northern Methodist; Morris Brown College, African-Methodist Episcopal; Morehouse and Spelman College, Northern Baptist. During 1928 and 1929, Hope was the guiding spirit in a courageous experimental movement to bring these institutions, with their varying religious denominations, together as one University. The old, prestigious Atlanta University became the graduate school. All-male Morehouse and all-female Spelman continued as undergraduate schools. These three institutions became known as the Atlanta University Affiliation. Clark and Morris Brown cooperated to form a federation of colleges.[1]

As president of the new Atlanta University, Hope assembled a distinguished faculty; he invited his friend Dr. W. E. B. Du Bois to join the faculty and Du Bois accepted in 1934. From 1897 to 1910, when Du Bois was a professor of history and economics at Atlanta University, the two men formed a lasting friendship. Both were active in the Niagara Movement and the NAACP. Du Bois returned to Atlanta determined to educate the "Talented Tenth" of Negroes. My wife Alice was one of Du Bois's students who studied for a master's in sociology. My father Samuel Howard Archer and Hope were close friends, both having been trained for the Baptist mission to Negro education—Hope at Brown University, 1890–1894; Archer at Colgate University, 1898–1902. When Hope accepted the presidency of Atlanta, my father succeeded him as president of Morehouse.

My interest in the theatre began when a tall, slender, vocally and physically attractive lady, Miss Anne Cooke, joined the Spelman faculty as a Professor of Speech and Drama and Director of the University Players. I shifted some of my interest from English and athletics to acting with the University Players. John Hope had a great love for the theatre, and he was concerned about the limited opportunity for Negroes in the South to witness great plays by great artists because of the policy of segregation. He discussed the problem with Miss Cooke and the result was the Atlanta University Summer Theatre. Another Negro thespian was added to the staff to assist Miss Cooke. He was John McLin Ross, M.F.A., Yale University. Beginning in 1934, the University Players presented five plays during the season, with three performances of each play. The first Black theatre established in Atlanta, the project was an immediate success.[2]

Besides Du Bois, two other Negroes were part of this Atlanta exposure. James Weldon Johnson graduated from Atlanta University in 1894 and became field secretary for the NAACP in 1916; Walter White graduated in 1916 and became assistant executive secretary of the NAACP in 1918. These two men along with Du Bois were very active in the Negro Renaissance of the 1920s and 1930s; they were literary spokesmen. Du Bois, as editor of *The Crisis*, the NAACP monthly magazine, compiled records and opinions concerning the theatre, literature, and art.

If anyone wonders why it takes a Black student so long to earn a Ph.D., the first reason is financial. My adviser suggested that I go to New York City, at my own expense, and interview producers, writers, and performers. So I wrote a letter to Dr. Du Bois, who was living in Brooklyn in 1957, and asked for an appointment. He replied by letter to grant the appointment, and here is a part of what he wrote:

> I remember your father as one of my best friends during my years of teaching at Atlanta University. . . . From the first, the NAACP deliberately limited its efforts to securing civil, political and social rights for American Negroes. On the other hand, of course, it was impossible for the NAACP to neglect entirely so great a cultural force as the theatre. Indirectly, therefore, the NAACP did a good deal. James Weldon Johnson, who was chief executive for seventeen years, had close connections with the theatre.
>
> *The Crisis,* under my editorship, was interested in the theatre, published some plays and started the Krigwa Little Theatre. Most of our connections with the theatre will be found in the pages of *The Crisis.*

Today, the prestige and vitality of the NAACP may be questioned when one looks back to Malcolm X and Martin Luther King, Jr. However, in 1944, Gunnar Myrdal, the Swedish sociologist, made this evaluation of the Association's intellectual potential: "Few similar organizations have reached the organizational stability and the membership the size of the NAACP. It should also be stressed that while the lack of mass following is a weakness, the high intellectual quality of the membership is an asset."[3]

My Black experiences have allied me to this study—my Black world surrounded by a white world, my twoness of souls turning from white to Black, my Black instinct always adjusting my thoughts and actions in times and places. From an education in the humanities and the Judaeo-

Christian faith and from such Black experience should come a Black image which would be more acceptable to the white world.

Negro intellectuals like Du Bois, James Weldon Johnson, and Walter White have always been seriously concerned about the Black image, individual and racial. The Negro middle class have desired and promoted Black images which they thought would be favorable to the white world and especially to influential white Americans. They reasoned that as each individual Black person is accepted by white people, there would be less white repression upon Black people, and eventually the entire Black middle class would be accepted. Today, the Black revolutionaries have termed this type of reasoning as one which accommodates white people—accommodation rather than confrontation for equal social justice. Dr. Nathan Hare, a Black sociologist, has written rather critically of the Negro middle class in his book, *The Black Anglo-Saxons.* He devoted one chapter to this social class as Image Makers: "They are obsessed by their belief that all Negroes have to do to break down discrimination is to impress the white community with proper public manners and the sincerity of the Negroes' quest for integration. Accordingly, they are burdened with two self images: a Negro self and a white self."[4]

Since the Supreme Court of the United States judged that separate education is not equal education in 1954, the boycott of buses and stores in Montgomery, Alabama, in 1955, the sit-ins, kneel-ins, wade-ins, Freedom Rides and the March on Washington, the Black images have become more visible in America than ever before. Before these demonstrations, Black images were almost invisible.

After the turn of the century, Dr. Du Bois and other Black intellectuals became interested about favorable and unfavorable Black images as they appeared in theatres as well as literature. The NAACP has been surprisingly active with campaigns and protests against unfavorable representations of Negro life appearing on stage, in motion pictures, on the radio, and on television. These campaigns and protests involved Black performers from Paul Robeson to Uncle Tom, from Lena Horne to Aunt Dinah, from Duke Ellington to Rag-Time Joe, from Shakespeare's Othello to Amos 'n' Andy.

Notes

1. Clarence A. Bacote, *The Story of Atlanta University,* (Princeton, New Jersey: Princeton University Press, 1969), p. 256.
2. *Ibid.,* pp. 296–297.
3. Gunnar Myrdal, *An American Dilemma,* II (New York: Harper and Brothers, 1944), p. 821.
4. Nathan Hare, *The Black Anglo-Saxons,* (New York: Marzani and Munsell, Publishers Inc., 1965), p. 37.

Address: Democratic National Convention, San Francisco, July 17, 1984

Jesse Jackson

Our flag is red, white and blue, but our Nation is a rainbow—Red, Yellow, Brown, Black and White—we're all precious in God's sight.

America is not like a blanket—one piece of unbroken cloth, the same color, the same texture, the same size. America is more like a quilt—many patches, many pieces, many colors, many sizes, all woven and held together by a common thread. The White, the Hispanic, the Black, the Arab, the Jew, the woman, the Native American, the small farmer, the businessperson, the environmentalist, the peace activist, the young, the old, the lesbian, the gay and the disabled make up the American quilt

Even in our fractured state, all of us count and fit somewhere. We have proven that we can survive without each other. But we have not proven that we can win and make progress without each other. We must come together.

From Fannie Lou Hamer in Atlantic City in 1964 to the Rainbow Coalition in San Francisco today; from the Atlantic to the Pacific, we have experienced pain but progress as we ended America's apartheid laws, we got public accommodations, we secured voting rights, we obtained open housing, as young people got the right to vote. We lost Malcolm, Martin, Medgar, Bobby and John and Viola. The team that got us here must be expanded, not abandoned.

Twenty years ago, tears welled up in our eyes as the bodies of Schwerner, Goodman, and Chaney were dredged from the depths of a river in Mississippi. Twenty years later, our communities, Black and Jewish, are in anguish, anger and in pain. Feelings have been hurt on both sides.

There is a crisis in communications. Confusion is in the air, but we cannot afford to lose our way. We may agree to agree or agree to disagree on issues; we must bring back civility to the tensions.

We are co-partners in a long and rich religious history—the Judeo-Christian traditions. Many Blacks and Jews have a shared passion for social justice at home and peace abroad. We must seek a revival of the spirit inspired by a new vision and new possibilities. We must return to higher ground.

We are bound by Moses and Jesus, but also connected with Islam and Mohammed. These three great religions—Judaism, Christianity, and Islam—were all born in the revered and Holy City of Jerusalem.

We are bound by Dr. Martin Luther King, Jr., and Rabbi Abraham Heschel, crying out from their graves for us to reach common ground. We are bound by shared blood and shared sacrifices. We are much too intelligent; much too bound by our Judeo-Christian heritage; much too victimized by racism, sexism, militarism, and anti-Semitism; much too threatened as historical scapegoats to go on divided one from another. We must turn from finger-pointing to clasped hands. We must share our burdens and our joys with each other once again. We must turn to each other and not on each other, and choose higher ground.

Twenty years later, we cannot be satisfied by just restoring the old coalition. Old wine skins must make room for new wine. We must heal and expand. The Rainbow Coalition is making room for Arab Americans. They, too, know the pain and hurt of racial and religious rejection. They must not continue to be made pariahs. The Rainbow Coalition is making room for Hispanic Americans who this very night are living under the threat of the Simpson-Mazzoli bill. And [for] farm workers from Ohio who are fighting the Campbell Soup Company with a boycott to achieve legitimate workers' rights.

The Rainbow is making room for the Native American, the most exploited people of all, a people with the greatest moral claim amongst us. We support them as they seek the restoration of land and water rights, as they seek to preserve their ancestral homelands and the beauty of a land that was once all theirs. They can never receive a fair share for all they have given us. They must finally have a fair chance to develop their great resources and to preserve their people and their culture.

The Rainbow Coalition includes Asian Americans now being killed in our streets, scapegoats for the failures of corporate, industrial and economic policies.

The Rainbow is making room for young Americans. . . .

The Rainbow includes disabled veterans. The color scheme fits in the Rainbow. The disabled have their handicap revealed and their genius concealed; while the able-bodied have their genius revealed and their disability concealed. But ultimately, we must judge people by their values and their contribution. Don't leave anybody out. I would rather have Roosevelt in a wheelchair than Reagan on a horse.

The Rainbow is making room for small farmers. . . .

The Rainbow includes lesbians and gays. No American citizen ought to be denied equal protection under the law.

We must be unusually committed and caring as we expand our family to include new members. All of us must be tolerant and understanding as the fears and anxieties of the rejected . . . express themselves in many different ways. Too often, what we call hate, as if it were some deeply rooted philosophy or strategy, is simply ignorance, anxiety, paranoia, fear and insecurity.

In 1984, my heart is made to feel glad, because I know there is a way out—justice. The requirement for rebuilding America is justice. The linchpin of progressive politics in our Nation will not come from the North. . . . [It] in fact will come from the South.

That is why I argue over and over again. We look from Virginia around to Texas. There is only one Black Congressperson out of 115. Nineteen years later, we are locked out of the Congress, the Senate, and the Governor's Mansion.

What does this large black vote mean? Why do I fight to win second primaries and fight gerrymandering and annexation and at-large elections? Why do we fight over that? Because I tell you, you cannot hold someone in the ditch unless you linger there with them. . . .

If you want a change in this Nation, you enforce that Voting Rights Act. We will get 12 to 20 Black, Hispanic, female and progressive congresspersons from the South. We can save the cotton, but we have got to fight the boll weevils. We have got to make a judgment. . . .

It is not enough to hope ERA will pass. How can we pass ERA? If Blacks vote in great numbers, progressive Whites win. It is the only way progressive Whites win. If Blacks vote in great numbers, Hispanics win. When Blacks, Hispanics, and progressive Whites vote, women win. When women win, children win. When women and children win, workers win. We must all come together. We must come up together.

I have a message for our youth. I challenge them to put hope in their brains and not dope in their veins. I told them that like Jesus, I, too, was born in the slum, and just because you are born in the slum does not mean the slum is born in you, and you can rise above it if your mind is made up. I told them in every slum there are two sides. When I see a broken window, that is the slummy side. Train some youth to become a glazier; that is the sunny side. When I see a missing brick, that is the slummy side. Let that child in the union and become a brick mason and build; that is the sunny side. When I see a missing door, that is the slummy side. Train some youth to become a carpenter; that is the sunny side. And when I see the vulgar words and hieroglyphics of destitution on the walls, that's the slummy side. Train some youth to become a painter, an artist; that is the sunny side.

We leave this place looking for the sunny side because there is a brighter side somewhere. I am more convinced than ever that we can win. We will vault up the rough side of the mountain. We can win. I just want young America to do me one favor. . . .

Exercise the right to dream. You must face reality—that which is; but then dream of the reality that ought to be—that must be. Live beyond the pain of reality with the dream of a bright tomorrow. Use hope and imagination as weapons of survival and progress. Use love to motivate you and obligate you to serve the human family.

Young America, dream. Choose the human race over the nuclear race. Bury the weapons and don't burn the people. Dream—dream of a new value system.

Teachers who teach for life and not just for a living, teach because they can't help it. Dream of lawyers more concerned about justice than a judgeship. Dream of doctors more concerned about public health than personal wealth. Dream of preachers and priests who will prophesy and not just profiteer. Preach and dream! Our time has come. Our time has come.

Suffering breeds character, character breeds faith, and faith will not disappoint. Our time has come. Our faith, hopes and dreams will prevail. Our time has come. Weeping has endured for night, but now joy cometh in the morning.

Our time has come. No grave can hold our body down. Our time has come. No lie can live forever. Our time has come. We must leave the racial battleground and find the economic common ground and moral higher ground. America, our time has come.

We come from disgrace to Amazing Grace. Our time has come. Give me your tired, give me your poor, your huddled masses who yearn to breathe free, and come November there will be a change because our time has come.

Thank you and God bless you.

Imagination and Truth: An Interview with James Alan McPherson

Nell Beram

Ralph Ellison died in the spring of 1994, when I was attending the Iowa Writers' Workshop and studying with James Alan McPherson, author of the short story collections *Hue and Cry* (1969) and *Elbow Room* (1977), winner of the Pulitzer Prize. If memory serves, he had initially decided not to attend Ellison's funeral in New York, but on the way to class that Tuesday afternoon I learned from the secretary that McPherson had appointed me to direct the workshop in his stead. My peers mutinied: so treasured were his insights that it was inconceivable that we hold class without him. I told myself that my impeachment was nothing personal.

The two-hour makeup workshop we instigated was distinguished by McPherson's presence in body only; as I recall, he did not utter a single word throughout. "You exercised your rights in a democracy" he explained at the end of class, "and I have exercised mine." For years I have felt bad about having been party to robbing him of a chance to mourn his late friend and mentor without penalty, but it was while reading his stunning new offering, *Crabcakes: A Memoir* (Simon & Schuster), that I came to understand that his silent protest had been predicated not on a sense of having been personally mistreated but on a larger conviction that, out of respect for our very humanness, we must be permitted to honor grief and other wearying emotions with full capacity, as we would joy.

I spoke with McPherson over the telephone in late January while he was touring with his new book.

Beram: I want to ask you a few questions about teaching. I had a great experience in your workshop, and during my two years at Iowa, people would pass around your insights like mantras. One was, "Don't be afraid to move people with your writing." That sounds very basic but somehow rather daunting.

McPherson: If I said it, I meant, Why write if it can't resonate beyond your own feelings and touch the feelings of other people? I thought there was an awful lot of shyness in the workshop, and also a sense that you were there to learn conventions about writing, that as long as you've satisfied the conventions, you've created something that was formally okay but with no content that was meaningful to other people.

Beram: I believe you told us one day that every time someone sits down to write a story, he or she is trying to convince him or herself not to commit suicide.

McPherson: No, I never said that. My friend [the novelist] Ernest Gaines told me that he always puts a gun on his desk when he starts to write—that's Hemingway kind of stuff. I'm not that kind of macho person.

Beram: I thought you were making the point that in a successful piece of fiction there isn't a single extraneous or meaningless detail, whereas life can seem without design: a killer will win the lottery and a saint will get run over by a bus. By writing a story, we're saying there's a grand design to it all.

McPherson: There's a book coming out in May from Beacon Press called *Fathering Daughters.* DeWitt Henry [founding editor of *Ploughshares*] and I solicited a lot of authors to write essays about daughters. My piece is called "Disneyland." I'm not pushing it, but there's something that's related to what you said that I wrote to my daughter. I sent her an e-mail asking, "If you were king for a day, how would you go about having your way?" And she wrote back this beautiful thing. I was amazed by the reach and vitality of Rachel's imagination as she attempted to reorder the world according to her own sense of happiness.

I sent her another e-mail, saying, "You've learned that God created the world in six days, and that on the seventh day He rested. But this is not entirely true. I've heard it said that on the seventh day God was not yet pleased with His creations, so on that day He created imagination, and he gave this thing dominion over all else He had created. The person who is blessed with this is able to stand at his own place and at the same time project himself into another person's place, and see through the eyes of that person and understand the world from that person's point of view. This gift is called compassion, and it is a very, very rare thing indeed."

I say that now because, if I said something earlier about a gun, I was talking about the same kind of honesty and compassion that God practices. With imagination, you undertake obligations to tell the truth, be emotionally honest, leave yourself open to the person you're writing about.

Beram: One more question about teaching. As I understand it, you're slightly wary of the current emphasis on multiculturalism in the arts.

McPherson: Once we had black and white as insufficient categories to define human beings in their complexity, and now we generalize from that so we've got Asian, Spanish, black, white, female, male, smokers, nonsmokers. I worry about that because it leads to a fragmentation of a communal sense, a fragmentation of a common moral undergirding, so to speak. It leads to what I'll call moral dandyism—I stole the phrase from Lewis Lapham [editor of *Harper's*].

In California, where I live now, they just passed a law that you can't smoke in bars. Now, from Los Angeles to San Francisco you see all these homeless people—no one cares about homeless people, they care about smoking at bars! And that's moral dandyism, you see? You find a cause that you can invest with meaning and value, and that becomes a substitute for meaningful action that will cost you something. I think the same is true for putting people in categories: it's easier to deal with them that way. And you still get a sense of righteousness and authenticity because your category is deciding your whole world for you. That's a dangerous thing, I think.

Beram: So are you uncomfortable with self-identifying as black, or a person of color? I believe that you describe yourself as a "Negro."

McPherson: I use that word in an ironic way. I also say that I've been in Iowa twenty years trying to pass as a Nordic youth. I make fun of it because I think you can't get very far. . . . There are a certain number of good people in the world: you're not going to find them all within your own precinct or group, so don't integrate just to integrate; integrate because you might see somebody good across the line. Do you understand? I've always done that. You can't say all the good, beautiful, truthful people are in my group, my category, and the rest of them are invalid.

I remember Ellison told me, Never separate yourself, and never assume that you know it all, because somebody out there always knows more than you. He told me a story about how when he was at Tuskegee he majored in classical music, taking courses in classical trumpet. He had a teacher there named Hazel Harrison who had studied with the composer Busoni in Italy before the Fascists came to power, and then she came back to this country as a black woman and taught at Tuskegee. So, Ellison once got lower marks on an exam, and he went to her complaining about the incompetence of his other teachers, and she said to him, "You can't do that." She said, "In this country, especially, never underestimate the competence of anybody. The man who sweeps out the waiting room at Kehow Station, the train station downtown, could know more about classical trumpet than you." He said, "That's bullshit."

Later, as a young man in Harlem, Ellison canvassed for the socialists; he was knocking on doors in a dark apartment building and he heard some black men arguing about who was the best ballerina at the Met—"And they were arguing like black people do, saying, "Man, that bitch can *jeté* like a motherfucker!" All the technical terms, they knew. So Ellison knocked on the door, and they opened it; they were sitting around a table with an oil lamp on it. They signed the petition. And Ellison said, "How did you guys learn so much about classical ballet?" And they said, "Well, during the week we work as stevedores, but on weekends we work as Egyptians at the Met." I'm saying multiculturalism threatens to turn everybody inward, saying, "*My* group contains it all," and decreases awareness of what goes on outside one's own special category.

Beram: I hear you. I'm half Syrian and I'm leery of the idea of wearing that badge more proudly than I do because I was brought up by my WASP mother. I think about this a lot: how do I play this out? It doesn't seem true to my experience in any way.

McPherson: If they gave you some affirmative action you might want to go with it.

Beram: We'll see what my tax return looks like this year.

McPherson: Well, the black American middle class has stolen all the goodies from affirmative action, and the poor people are not better off. [The middle class are] the biggest nationalists about blackness, you know, because they get rewarded for it. Years ago, they were passing as white!

Beram: What do you think about Ebonics?

McPherson: I was out in Oakland last January, and that whole thing came up. I made myself a business card that said, "James 'Biff' McPherson, Freelance Public Intellectual. You sit, I bullshit. Special discount on Ebonics." The thing about Ebonics is, it's another hustle. Somebody can say years from now, I came up from Ebonics, instead of up from slavery. The bureaucrats have a vested interest in lowering the basement floor, so to speak. They get more publicity and business out of it.

Beram: Living in Iowa City, you've been fairly ensconced in so-called whiteness, whatever that is. Do you have a definition for it, or do you think, as some intellectuals are arguing, that whiteness doesn't exist, except as an absence of something else?

McPherson: About Iowa City: I never once felt put-upon because of my race in Iowa City. I'll tell you a story. When I first went there, I was very reticent. I remember one time I went to the instant bank teller booth on the corner of Clinton, right across from the mall. And there was a black guy there ahead of me. He left in a hurry. I went to put my card in, but he'd left his. So I pulled it out, and it had a Spanish woman's name on it. So I thought, This brother has ripped off this woman, he's taking her money, and if I leave it someone else will find it. But if I give it to the police, I've got to describe him. And I worried about that.

I finally went down to the police station and I gave the card to a policeman and said, "Somebody left this in a teller machine." He said, "Did you see anybody in there before you?" I said, "It's a black male," and I described him. I felt so bad because I felt I'd betrayed him, a brother. So I went home and that night the policeman called and said, "He wants to thank you: it was his wife's card." And that's the other side of it, you see. You live in an environment in which you *can* do the right thing. And it has the result that you find more human space. That's all you need: human space.

Beram: That's an amazing story.

McPherson: I'm just trying to say that whiteness is an abstraction. Ellison always lectured me on how race is an abstraction. If you're going to talk about the sociological advantages that you get out of being white in terms of structures, I can go with that and that's what the teachers of "white studies" are doing, probably. But to make it a source of narcissism like the black stuff, you know? That's a dangerous thing. I just wish that whiteness wouldn't take on the romanticism and narcissism that blackness and ethnic studies has.

Beram: One of the many pleasures of your new book, *Crabcakes,* is the fact that it calls into question the definition of a memoir. In fact, "Grant Hall," the essay that you wrote for *DoubleTake* this winter, contains more factual information about your life than the entire memoir does. We don't learn in *Crabcakes* that you've gone to Harvard Law School or that you have a daughter. How did you decide what information to include and what information to exclude from your memoir?

McPherson: You recall I used to make jokes in class?

Beram: Yes. I miss them.

McPherson: I wasn't trying to write a memoir. I was trying to create a new form. But they're marketing it as a memoir. I was trying to create "a Negro," because I "stole" everything in the book. Remember that Henry James observed that "form *takes*!"

Beram: I was wondering if you were making a comment on the genre as a whole.

McPherson: When I called it a Negro, I meant to say that I stole everything in the thing, and Negroes steal. I was trying to make something new. That's all.

Beram: Were you trying to be provocative?

McPherson: No, not about the memoir. I was trying to distance myself from the contemporary memoir form when I quoted from Rousseau toward the end. This morning I saw a girl on a television talk show confessing to a crowd of people that she has two boyfriends and she's pregnant and doesn't know which one's the father of the baby. And I was thinking, that's so private: why do you want to share it with a studio audience and with the whole world? That the idea behind the memoir is that it's something private between you and what I call the *intermundia,* from a Latin word meaning "the place between the worlds." So I tried my best to keep away from the contemporary memoir form, but the publisher is marketing it that way.

Beram: There are so many references in *Crabcakes* to forces of destruction, from the Kobe earthquake to your neighbor's dying son to the Hindenburg explosion and various other catastrophes, and yet I'm left with a hopeful feeling at the end. One would think that the sum of the book's parts would lead the reader elsewhere. Where does that hope come from?

McPherson: Hope grows out of despair. Last December, when I was in Japan, I went with some friends to a ceremony sponsored by one of the most prestigious publishers in Japan. He'd spent twenty-five years translating all of Faulkner's work into Japanese—I mean even his unpublished work, his letters, everything. They were having a ceremony about that, with about five hundred Faulknerians in the audience, professors from the University of Tokyo, people giving presentations on Faulkner. Now this publisher was going out of business. He'd put everything he had into this one effort, and he was closing his doors after that. There's something in Japanese aesthetics called *wabi-sabi*: that means beauty and sadness. Have you ever seen a rock garden? Where you have beautiful flowers coming out of something that looks decayed and old? That's the idea. Elegance and grime. But something beautiful comes out of it. Much like the blues, I think.

Nell Beram is assistant editor of the *Hungry Mind Review*. Her work has appeared recently in *American Short Fiction, Long Shot,* and *Z* magazine.

To Light the Fire of Science, Start with Some Fantasy and Wonder

Chet Raymo

Chet Raymo is a professor of physics at Stonehill College in Easton, Massachusetts. This essay, the argumentative claim for which appears both in the title and (twice) in the selection itself, was first published in Raymo's regular column, "Science Musings," in the Boston Globe *on December 21, 1992. The charm of this argument lies partially in Raymo's taking apparently unconnected pursuits—the reading of children's literature and the practice of good science—and supporting their connection in convincing ways. In the process, he makes his own love of science and of children's literature clear to all.*

Every year about this time I am asked by friends and colleagues to recommend good science books for kids, to fill the remaining hollows in Santa's pack.

There are lots of terrific science books out there, and a good place to find them is the children's book section of the Museum of Science shop. But my advice to parents is: Don't be overly worried about providing science books for your kids. Expose them to good children's literature, and the science will take care of itself.

Over the years I have often referred to children's books in this column, including the nonsense books of Dr. Seuss, Antoine de Saint-Exupery's *The Little Prince*, Lewis Carroll's *Alice in Wonderland* and *Through the Looking-glass*, L. Frank Baum's *Wizard of Oz*, Kenneth Grahame's *The Wind in the Willows* and Felix Salten's *Bambi*.

All of these in essays about science.

What's the connection?

Children's books capture the curious, willing-to-be-surprised, "let's pretend" quality of good science. Children's books are full of the playfulness that is part of all first-rate science. And, like science, children's books insist upon the reality of the unseen.

Science books for children are packed full of interesting information. What most of these books do not convey is the story of how the information was obtained, why we understand it to be true, and how it might embellish the landscape of the mind. For many children—and adults, too—science is information, a mass of facts. But facts are not science any more than a table is carpentry.

Science is an attitude toward the world—curious, skeptical, undogmatic and sensitive to beauty and mystery. The best science books for children are the ones that convey these attitudes. They are not necessarily the books labeled "science."

From a "science" book we might learn that a flying bat might snap up 15 insects per minute, or that the frequency of its squeal can range as high as 50,000 cycles per second. Useful information, yes.

But consider the information in this poem from Randall Jarrell's "The Bat-Poet":

A bat is born
Naked and blind and pale.
His mother makes a pocket of her tail
And catches him. He clings to her long fur
By his thumbs and toes and teeth.
And then the mother dances through the night
Doubling and looping, soaring, and somersaulting—
Her baby hangs on underneath.

Oh what wondrous information! Even the rhythm of the poem ("naked and blind and pale"; "thumbs and toes and teeth") mimics the flight of mother and child, doubling and looping in the night.

More:

The mother eats the moth and gnats she catches
In full flight; in full flight
The mother drinks the water of the pond
She skims across. Her baby hangs on tight.

That wonderful line—"In full flight; in full flight"—conveys the single most important fact about bats: their extraordinary aviator skills. Jarrell's repeated phrase conveys useful facts about chiropteran dining; it also lets the child feel in her bones what it is to be a bat.

In Jarrell's book, the Bat-Poet recites his poem about bats to a chipmunk. Afterwards, he asks, "Did you like the poem?"

The chipmunk replies, "Oh, of course. Except I forgot it was a poem. I just kept thinking how queer it must be to be a bat."

The Bat-Poet says, "No, it's not queer. It's wonderful."

That's what good children's literature does—makes us feel the queerness and wonderfulness of nature. It's the best possible introduction to science.

Albert Einstein wrote: "When I examine myself and my methods of thought, I come to the conclusion that the gift of fantasy has meant more to me than any talent for abstract, positive thinking." The best time—perhaps the only time—to acquire the gift of fantasy is childhood. We live in an age of information. Too much information can swamp the boat of wonder, especially for a child. What children need is not more information, but fantasy.

Einstein also write: "The most beautiful experience we can have is the mysterious. It is the fundamental emotion which stands at the cradle of all true art and true science." At first, this might seem a strange thought. We are frequently asked to believe that science is the antithesis of mystery. Nothing could be further from the truth. Mystery invites the attention of the curious mind. Unless we perceive the world as mysterious—queer and wonderful—we will never be curious about what makes it tick.

Deconstructing Wordsworth's "A Slumber Did My Spirit Seal"

Geoffrey Hartman

It does not matter whether you interpret the second stanza (especially its last line) as tending toward affirmation, or resignation, or a grief verging on bitterness. . . .

That [the poem] is a kind of epitaph is relevant, of course. We recognize, even if genre is not insisted on, that Wordsworth's style is laconic, even lapidary. There may be a mimetic or formal motive related to the ideal of epitaphic poetry. But the motive may also be, in a precise way, meta-epitaphic. The poem, first of all, marks the closure of a life that has never opened up. Lucy is likened in other poems to a hidden flower or the evening star. Setting overshadows rising, and her mode of existence is inherently inward, westering. I will suppose then, that Wordsworth was at some level giving expression to the traditional epitaphic wish: Let the earth rest lightly on the deceased. If so, his conversion of this epitaphic formula is so complete that to trace the process of conversion might seem gratuitous. The formula, a trite if deeply grounded figure of speech, has been catalyzed out of existence. Here it is formula itself, or better, the adjusted words of the mourner that lie lightly on the girl and everyone who is a mourner.

I come back, then, to the "aesthetic" sense of a burden lifted, rather than denied. A heavy element is made lighter. One may still feel that the term "elation" is inappropriate in this context; yet elation is, as a mood, the very subject of the first stanza. For the mood described is love or desire when it *eternizes* the loved person, when it makes her a star-like being that "could not feel/The touch of earthly years." This *naive* elation, this spontaneous movement of the spirit upward, is reversed in the downturn or catastrophe of the second stanza. Yet this stanza does not close out the illusion; it preserves it within the elegiac form. The illusion is elated, in our use of the word: *aufgehoben* seems the proper term. For the girl is still, and all the more, what she seemed to be: beyond touch, like a star, if the earth in its daily motion is a planetary and erring rather than a fixed star, and if all on this star of earth must partake of its sublunar, mortal, temporal nature. . . .

To sum up: In Wordsworth's lyric the specific gravity of words is weighed in the balance of each stanza; and this balance is as much a judgment on speech in the context of our mortality as it is a meaningful response to the individual death. At the limit of the medium of words, and close to silence, what has been purged is not concreteness, or the empirical sphere of the emotions—shock, disillusion, trauma, recognition, grief, atonement—what has been purged is a series of flashy schematisms and false or partial mediations: artificial plot, inflated consolatory rhetoric, the coercive absolutes of logic or faith.

Characteristics of Contemporary Native American Literature

Craig Lesley

Craig Lesley, professor of English at Clackamas College in Portland, Oregon, and board member of the Northwest Native American Writers Association, has published both academic and creative work. He is well known for his scholarly work on Native American literature and for his novels. His 1984 novel Winterkill *received the Western Writers of America Golden Spur Award for best novel of 1984, the Medicine Pipe Bearer's Award for best first novel and the Pacific Northwest Booksellers Association Award for the best book of 1984. He published a second highly-acclaimed novel,* River Song, *in 1989; He is also the editor of* Talking Leaves: Contemporary Native American Short Stories *(1991). The essay which follows first appeared in a 1979 National Council of Teachers of English publication,* New Students in Two-Year College: Twelve Essays.

When N. Scott Momaday was awarded the Pulitzer Prize in 1969 for *House Made of Dawn*, contemporary Native American literature gained national recognition. The excellent Indian anthologies and novels published since Momaday's work demonstrate that Native American literature continues to be a major literary force. Nonetheless, many readers remain puzzled about the new Indian writers because their works cannot be explicated fully according to familiar literary criteria.

This bewilderment decreases if readers bear in mind that the Indian writer's perceptions, values, and culture are different from those familiar to the Anglo. Once these differences are accounted for, we can better understand contemporary Indian writing.

By studying Native American literature we learn that there are other ways of perceiving, other values operating; in the process, we increase our understanding of the Native American's perceptions and values. We also gain another perspective from which to view our own idea of the way things are.

Importance of the Land

An investigation of contemporary Native American writing reveals the close relationship between the writer's work and the land. Native

Americans understand themselves in relation to the landscape. There is reciprocity between the Native American and the land, a participation of the Indian in the landscape. This strong connection is readily apparent in most works by contemporary Native American authors. Momaday explains, "Man understands himself in relation to the tree over here and the mountain over here and the river and naturally operates out of that environment, operates immediately out of it. . . . Man understands that he is obligated in certain ways to the landscape, that he is responsible for it, that he shares in the spirit of place" ("A Conversation with N. Scott Momaday," 1976, p. 19).

Leslie Silko, a Laguna Pueblo writer, demonstrates an understanding of Momaday's concept as she explains, "I grew up at Laguna Pueblo. . . . This place that I am from is everything I am as a writer and human being." She understands also Momaday's discussion of man's relation to a tree or a mountain. In the introduction to her book *Laguna Woman* (1974), Silko describes her great-grandmother's place at Laguna and the large cottonwood tree there. In her short story "The Man to Send Rain Clouds," Silko uses the cottonwood as a focal point: "They found him under a big cottonwood tree. His Levi jacket and pants were faded a light-blue so that he had been easy to find. The big cottonwood tree stood apart from a small grove of winterbare cottonwoods which grew in the wide, sandy arroyo" (Rosen, 1974, p. 3).

Search for the Center

This close relationship to the land enables Native Americans to have a sense of a "center," a place where they belong. Frequently, the center is pictured as a hoop. In "The Great Vision" chapter of *Black Elk Speaks* (Neihardt, 1961), the Oglala Sioux holy man reveals the significance of the center when he recalls the Voice which told him in his youth, "Behold the circle of the nation's hoop, for it is holy, being endless, and thus all powers shall be one power in the people without end" (p. 35). At the conclusion of his work, after assessing the slaughter at Wounded Knee, Black Elk states, "And I, to whom so great a vision was given in my youth,—you see me now a pitiful old man who has done nothing, for the nation's hoop is broken and scattered. There is no center any longer, and the sacred tree is dead" (p. 276).

In spite of Black Elk's disillusionment, the idea of the center remains an integral part of Native American perception and a major characteristic of their literature. If, as many have noted, the theme of twentieth-century literature is the search for the self and the country of the self, the Native American's search ends at the center. Much of contemporary Indian literature attempts to express the reality of the center and to encourage discovery of the center.

Silko's novel *Ceremony* (1977) traces the attempts of Tayo, an Indian World War II veteran, to regain health and mental equilibrium after his discharge and return to the Southwest. A contemporary medicine man finally counteracts the illness with a ceremony designed to return Tayo's awareness of the center. The main image of this ceremony is the hoop. The medicine man chants (p. 143):

I will bring you through my hoop,
I will bring you back.

. . . .

Come home, happily
Return belonging to your home
Return to long life and happiness again. . . .

After the ceremony has been performed, Tayo regains his health, his awareness of the center, and his relationship to his land: "The magnetism of the center spread over him smoothly like rainwater down his neck and shoulders; the vacant cool sensation glided over the pain like feather-down wings. It was pulling him back, close to the earth, where the core was cool and silent as mountain stone, . . . " (p. 201).

Momaday's *House Made of Dawn* (1968) also includes a passage that expresses the importance of the center and its relation to the land. His speaker, Abel, remembers: "And that night your grandfather . . . told you stories in the firelight. And you were little and right there in the center of everything, the sacred mountains, the snow-covered mountains and the hills, the gullies and the flats, the sundown and the night, everything—where you were little, where you were and had to be" (p. 157).

Abel's closeness to the center depends on his ties with the land. That closeness is enhanced by the presence of his grandfather. Relatives, clan members, adopted families, tribal elders, and medicine men all give the individual a sense of place in the tribal system, but the influence of the grandparents or great-grandparents is frequently the most important. Silko's close relationship with her great-grandmother indicates this importance: "My mother had to work, so I spent most of my time with my great-grandma, following her around her yard while she watered the holly-hocks and blue morning glories. When I got older I carried the coal bucket inside for her. Her name was Maria Anaya and she was born in Paguate village, north of Old Laguna. . . . She took care of me and my sisters and she told us about how things were when she was a little girl" (Silko, 1974, pp. 34–35).

Relationship to the Past

The closeness of Native Americans to their grandparents, great-grandparents, or great-great-grandparents suggests another part of the Indian's perception that makes the center approachable. The Indian writer works out of a concept of time that is cyclical rather than linear. This involves a concentric continuity whereby the speaker becomes closer to the future (and the past) rather than further removed. Momaday ("A Conversation with N. Scott Momaday," 1976) expresses the Indian's relation to heritage and time as follows:

> I think the storyteller in Indian tradition understands that he is dealing in something that is timeless. He has a sense of its projection into the past. And it's an unlimited kind of projection. I am speaking, I am telling a story, I am doing something that my father's father's father's father's father's father's father did. That kind of understanding of the past and of a continuity in the human voice is a real element in the oral

> tradition. And it goes forward in the same way. I am here and what I am doing is back here and it will be here. (p. 21)

In *House Made of Dawn,* Momaday uses three distinct narrative voices to emphasize a sense of continuity and closeness with the past. A mythical voice describes the rituals and the Native American's relation to the land. A historical voice records the life of Abel's grandfather. A contemporary voice indicates the protagonist's (Abel's) perceptions. By interweaving the three voices, Momaday demonstrates the integral dynamic relationships between Abel and his heritage and tribal history.

Because Native American writers understand that they are a part of the past and the past is part of them, they may merge their voices with voices from the past. One example occurs when Charles Ballard, a contemporary Quapaw-Cherokee, takes on the voice of White Antelope in his poem "Sand Creek." White Antelope, a Cheyenne chief who was slaughtered in the Sand Creek massacre, reportedly folded his arms and sang his death chant "Nothing lives long/Only the earth and the mountain" until he was shot down by the advancing cavalry. Ballard (Rosen, 1975, p. 123) integrates that chant with his poem as his voice and the voice of a historical figure become one:

And on this day too old to run am I
Too old for the land of the young

Black Kettle raises the flag
The air is crisp and cold
Here at my home I sing
Nothing lives long
But the earth and the mountain

White Antelope is my name

The Indian culture is based on the land, the tribe, and the past. Each contributes to the Native American's sense of self and close relation to the center. Contemporary Indian writers' awareness of the center and their belief in their culture save them from the estrangement and bewilderment that permeate the works of their Anglo counterparts—what Theodore Roszak terms "the dispiriting conviction of cosmic absurdity." While Anglo writers find themselves in conflict with their disintegrating culture, where technology and shifting values constantly erode the sense of self, Native American writers embrace their culture and strengthen the self in the process.

Bitterness Toward White Culture

Although contemporary Native American literature lacks a voice that expresses alienation toward Indian culture, it contains many voices that express an alienation toward the white culture, "the world of stalking white men" according to the unnamed speaker of Welch's *Winter in the Blood.* The bitterness, hostility, and estrangement of many Native American voices indicate that Indians feel a morally and spiritually inferior culture has encroached upon their own.

Anita Probst's poem "Manifest Destiny" (Niatum, 1975, p. 163) expresses her anger for the intruding white culture:

My mother used to say, Brown Child
of the red sand, wash your feet
with river flowers, climb high
upon the rocks and smile out
the stars. Now as a woman,
I remember a man who said
all Indians are rich
they just don't know how to save,
except by cans of beer.
And like the buffalo, you took my brown
skin and hung it on the wall.
I am gentle, but angry:
Is this how you white men
mount your trophies. Tomorrow, I see
my son; in his eyes there is more than quiet pain—
now blood-red flames bloom anger
and he has yet to live.

The images of "red sand" and "river flowers" and the actions of climbing the rocks and smiling at the stars establish the speaker's close relationship to the natural world. The man who says that Indians don't know how to save and the allusion to buffalo slaughter suggest the white man's lack of sensitivity toward and exploitation of the natural world. A more personal exploitation, probably a sexual one, can be inferred from the brown skin hanging on the wall and the reference to white men mounting trophies. Although brief, the poem effectively expresses a principal conflict between two cultures, The Indian respects and lives in harmony with nature. The white man conquers and exploits nature; in addition, he exploits the Indian—in both general and specific situations.

One method of exploitation involved introducing Indians to alcohol, then acquiring their goods. The current effects of Indian alcohol abuse are all too evident. The poem "Eclipse" (Niatum, 1975, p. 178) exemplifies Probst's general bitterness toward the white man's drink. In addition, it illustrates a personal loss of love through the effects of alcohol:

Black Wolf, naked night-hunter,
you crouch in the corner and growl.
Your howling eclipses my pleas
and the broken bottles rip open your mouth
with the quick surge of an eagle's anger.

You dark man, trying to hide the blood
and spitting it at me in rage;
baring your raw lips and black tongue.
on all fours you crawl from my lodge
and try to find the moon.
Once you said it was in my eyes.

Through the effects of alcohol, the once sensitive lover has become a snarling, crawling beast. James Welch's poem "D-Y Bar" (Niatum, 1975. p. 250) contains similar imagery of man degraded into animal:

In stunted light, Bear Child tells a story
to the mirror. He acts his name out,
creeks muscling gorges fill his glass
with gumbo. The bear crawls on all fours
and barks like a dog. Slithering snake-wise
he balances a nickel on his nose

The Native American writer understands that, too frequently, overuse of alcohol results in death. The unnamed speaker of Welch's *Winter in the Blood* (1974) lives with the nightmare memory of his father freezing to death after getting drunk in a white man's bar and driving off the road. Although the speaker frequents bars as his father did, eventually he rejects their corrosive influence: "I had had enough of Havre, enough of town, of walking home, hung over, beaten up, or both. I had had enough of the people, the bartenders, the bars. . . ."

The titles of A. K. Redwing's poems are sufficient to express his alienation from and contempt for the white culture. Some examples include "Chrome Babies Eating Chocolate Snowmen," "Two Hookers," "Written in Unbridled Repugnance Near Sioux Falls, Alabama—April 30, 1974," and "A Lost Mohican Visits Hell's Kitchen."

A Vietnam veteran, Redwing frequently points an accusatory finger at the white political-military establishments (Rosen, 1975, p. 134):

Clarence Shortbull died.
    the bullet was aimed decades ago
by a finger from Washington

and another example (pp. 135–136):

A group of touring politicians is shown an elaborate
    Ball of Chicanery—
"Brilliant," said the Bozo from Wazoo . . .
"A commendable piece of artistry," said another.
    they continue on their tour
of freshly polished commodes . . .
playing their role of blind men at a silent movie

An August eagle floats majestically across the sky,
    He is met by a SAM II . . .
the feathers land selectively in living rooms
    from Maine to Seattle . . .

In deploring racism, Redwing concludes (pp. 137–138):

Bronze statues of ancient rapists
applaud tactical squads crunching skulls
As in the dim light of humanity,
Adam weeps. . . .

Belief in the Power of Words

Although alienated and angered by the white culture, Probst, Welch, Redwing, and other contemporary Indian writers are not content to express criticism of the way things are. Their words are intended to bring about a change, to improve the way things are. In commenting about her work *Ceremony* (1977), Silko indicates the force behind Indian writing: "This novel is essentially about the powers inherent in the process of storytelling. . . . I feel the power that the stories still have to bring us together, especially when there is loss and grief." Silko's novel (1977, p. 2) includes a poem, also entitled "Ceremony," which further emphasizes the power of the story:

I will tell you something about stories,
[he said]
They aren't just entertainment.
Don't be fooled.
They are all we have, you see,
all we have to fight off
illness and death.

You don't have anything
if you don't have the stories.

For the Indian, the word is powerful and sacred. It is powerful enough to change reality. In the introduction to *American Indian Prose and Poetry*, Margot Astrov explains that the word is "the directing agency that stands powerfully behind every 'doing,'" "the reality above all tangible reality" (p. 15). B. L. Whorf's studies among the Hopis indicate that these people believe thought can "determine and direct reality." The corn plant serves to illustrate: "By concentrating his thoughts on the corn plant, [the Hopi] feels he can influence its growth and maturation" (Astrov, 1972, p. 20). In a similar fashion, Crazy Horse dreamed and sang himself into what he felt was another state of consciousness, into another reality. Today, the Indian writer intends to direct his words to effect a change.

Because the word is so powerful for Native Americans, they feel a strong responsibility to use it properly. An ancient song of the Navajo priests reveals the priests' belief in a self within the self, a kind of conscience that ensures the proper use of language (Astrov, 1972, p. 15):

That standing within me
Which speaks with me
Some of these things are always looking at me.
I am never out of sight.
Therefore I must tell the truth.
I hold my word tight to my breast.

This song emphasizes the moral relationship which the Indian feels exists between humans and language. Momaday refers to this relationship in "A Conversation with N. Scott Momaday" (1976). He maintains that "magic, and the idea of magic, is very highly developed. . . . It is everywhere. . . . [The Native American writer] is aware of its power. He understands that by exerting the force of language on the physical world, he can bring about actual change. And that's a marvelous attitude. It insures that people use language responsibly" (p. 20).

Henry Realbird, a Crow from Montana, adds another dimension. According to him, serious language among the Crows is referred to as "real talk." Much of real talk reveals the wisdom of the Great Spirit, and may come to humans from other humans, dreams, ghosts, or animals. Real talk is intended to lead or instruct men, and it is never false. If humans hear real talk and ignore it, they may suffer the consequences since it was intended for their instruction. The leaders in the village—the respected elders, chiefs, and medicine men—are ones who understand and listen to real talk.

Because Indians know the power of words and because they feel an obligation to use language responsibly, one can infer that Native American writers would not use language to lie, betray, trick, or cover up. Momaday suggests that the Indian writer is basically honest when he explains, "You know we have a stereotype of the Indian who speaks the truth, the white man who speaks with the forked tongue. . . . There is a basis to the assumption that in an oral culture one deals in the truth. One has a higher regard for language: one tends to take it more seriously. One tends to have a better understanding of what can happen to him if he uses it carelessly, if he abuses it" ("A Conversation with N. Scott Momaday," 1976, p. 21).

The Indian writer's closeness to nature and sensitivity for responsible use of language result in the use of metaphors derived from natural objects, from observations of natural phenomena. Indian names frequently suggest this closeness and responsibility. For example, Henry Realbird's relatives were purposefully named, as he explains in a composition he wrote for a college class:

> My mother's name, Cow-Necklace, was given to her by a clan uncle, Bird Horse. Working as a cowboy, Bird Horse observed that cows with bells around their necks were more trustworthy and dependable than the rest of the lot. By giving my mother the name, he was wishing she too would be a nice and trustworthy person.
>
> My father has two names. His first name, Horse-Catching-Up, is his childhood name. This is a name given to him by his great-grandfather, Medicine Crow. Medicine Crow had a dream in which an old mare was talking to her colt and said, "Horse-Catching-Up." The name means that there are colts every year to the extent that they are all close together age-wise. This denotes the ease and good fortune of the old mare to foal every year with no real problems. Medicine Crow hoped that my father, too, would lead a life of ease and good fortune.
>
> Then returning from World War II, my father was given the name Bird Shirt, a name of his clan. It is customary for a male to change to an adult name after the first participation in a war party. The parents, along with the clan uncles, are the proper persons to handle the transfer of a particular name change. Names like this one stay in the clan but they are transferred from one member to another as the need arises.

> It is not unusual for an individual to acquire several names in his life. In some instances, a person's name is changed if he is having difficulty in life. The new name will help him find himself.

Henry's explanation indicates the closeness of the Indians to their names and to their environment. In addition, it suggests that a change in name can bring about a change in fortune, attitude, or personality. This echoes the idea that words can change the way things are and offers another reason why the Native American chooses to use words in a responsible manner.

Indian writers frequently use metaphors taken from nature. In her poem "Red Rock Ceremonies" (Niatum, 1975, p. 164), Anita Probst demonstrates an ability to create striking natural metaphors:

With low thunder, with red bushes smooth
as water stones, with the blue-arrowed rain,
its dark feathers curving down
and the white-tailed running deer—
the desert sits, a maiden with obsidian eyes,
brushing the star-tasseled dawn from her lap.

Here, the natural metaphors include "red bushes smooth as water stones," "blue-arrowed rain" with its "dark feathers," "obsidian eyes," and "star-tasseled dawn." Probst's notable personification of the desert as a maiden who brushes "the star-tasseled dawn from her lap" reinforces the closeness of the Indian to the natural world. At times, Indian writers further emphasize that closeness in passages where the speaker merges personal identity with an object or animal from the natural world. Examples are abundant, but some are particularly memorable. In her poem "Indian Song: Survival" (Rosen, 1975, p. 25), Silko writes:

taste me,
  I am the wind
touch me,
  I am the lean gray deer
running on the edge of the rainbow.

Silence as a Part of Indian Literature

Silence is the final characteristic of Native American literature to be discussed in this essay. Perhaps this concept is the most intriguing because it cannot be demonstrated with concrete examples. Nevertheless, it operates in contemporary Indian writing.

In an oral story, we recognize silence as an inherent part of the story. Silence may be used dramatically to build suspense, or it may provide a period of time in which the imagination of the listener can work. But in a written story or poem, silence and the importance of that silence are more difficult to apprehend. In spite of that difficulty, Charles Ballard, in speaking of Indian literature, encourages the reader to listen to the silence and what it is saying.

In part, the silence indicates the complex and highly personal relationship Indian writers have to their heritage and their society. This relationship cannot always be expressed in words. It is to be felt, sensed, intuited.

The silence of contemporary Native American literature also is one manifestation of the awesome silence that reflects the mystery of the Indian culture, a mystery that by its very nature denies expression. At times the Indian's approach to this mystery is through ritual or ceremony; but the concept itself defies articulation. To make a comparison with a concept from our own culture, we might reflect on the variety and intensity of feelings that are lumped together as love. Often love is best expressed in nonverbal ways.

In addition, the silence arises from the Indian's sense of continuity and the unity of all things. Imagine for a moment a great moving wheel that is touching the heavens and the earth. Imagine further that the wheel contains the mystery of creation as well as the ceremony, ritual, and "real talk" necessary to instruct humans and unify them with all creation. Only a section of the wheel touches the earth at a given time. Yet there is a certainty, given the cyclical movement of the wheel, that all sections will touch (or indeed have touched) at some time. We may infer from this that much is already known or much is to be known. In addition, much will be expressed through ritual and ceremony.

Finally, there is the power of the words themselves, and the Indian's desire to use them in a responsible manner. Those with a profound respect for words do not want to use them foolishly or unnecessarily. Indian writers tend to use few words; much of the intensity of their literature derives from this economy. Underlying this economy is the wisdom of knowing which words to use. At times a few will do when many will not. Underlying it also is the knowledge that much is already known. As the Papago singer says, "The song is short because we know so much." In the silence, it seems to me, there is the certainty that much has gone before and much is yet to come—the certainty of unity. Moreover, the silence affords the poetical imagination and the mythological understanding a chance to operate. The silence, in other words, invites the reader as listener to become a participant in the Indian writer's work and world.

Works Cited

Astrov, M., ed. *American Indian Prose and Poetry: An Anthology.* New York: John Day, 1972.

Brown, Dee. *Bury My Heart at Wounded Knee.* New York: Holt, Rinehart & Winston, 1970.

"A Conversation with N. Scott Momaday." *Sun Tracks* Spring 1976: 18–21.

Deloria, Vine, Jr. *Custer Died for Your Sins.* New York: Macmillan, 1969.

Momaday, N. Scott. *House Made of Dawn.* New York: Harper & Row, 1968.

Critical Theory and Debate: The Black Aesthetic or Black Poststructuralism?

Robert Johnson

What's Goin' On?

As previously mentioned, during the late 1960s and early 1970s a group of black scholars and writers developed the sociopoetics called the Black Aesthetic, a black criterion for judging the validity and/or beauty of a work of art. Larry Neal, Amiri Baraka, and the other New Black critics, including Maulana Karenga, Clarence Major, Addison Gayle, Jr., Hoyt Fuller, Stephen Henderson, Carolyn Fowler, and Sarah Webster Fabio, rejected Eurocentric standards applied to literary works by American blacks and instead demanded that African American literature be judged according to an aesthetic grounded in African American culture. Dismissing the notion of Immanuel Kant and his inheritors that form, not content, is the imperative of art, these critics insisted that in order to have value, black art must address the sociopolitical and spiritual needs of African American people. Like W. E. B. DuBois, Richard Wright, Ann Petry, and others before them, they insisted that, above all else, black literature must contribute to the cause of black liberation. But with the new black aestheticians, the old theme of liberation took on new meaning. In short, they raised the process of self-definition—of black racial consciousness and black group solidarity—of the Harlem Renaissance and Reformation periods to the level of revolutionary thought. They called for black art to serve as a revolutionary weapon, not merely in polemics against Euro-American psychological and cultural domination but also in reinterpretation of the African American life experience.

These theorists eschewed "protest" literature, which in previous periods was directed toward a white audience and appealed to white morality. Rather, they insisted upon an art of liberating vision—a black way of seeing the world—because, as Clarence Major maintains in his 1967 essay "A Black Criterion," "seeing the world through white eyes from a black soul causes death." In *Understanding the New Black Poetry* (1972), Stephen Henderson explains, "The present movement [the Black Arts Movement] is different from the Harlem Renaissance in the extent of its attempt to speak directly to Black people *about themselves* in order to move them toward self-knowledge and collective freedom." Thus, placing their stamp of disapproval firmly on the T. S. Eliot-Ezra Pound notion that art should be reserved for the intellectual few, Baraka and Neal, in their landmark black nationalist anthology *Black Fire* (1968), contended that the black literary artist must make black art more meaningful by taking it to the black masses. In the Afterword to the work, Neal declared, "the artist and the political activist are one." Accordingly, Maulana Karenga insisted in his essay "Black Art: Mute Matter Given Force and Function" (1968), black art must be "functional," "collective," and "committed."

This "art for people's sake," all of the New Black critics agreed, should be judged according to its presentation of the traditions and styles stemming from African and African American cultures. Thus, as Neal highlighted in his seminal essay "The Black Arts Movement" (1968), these aestheticians called for "a separate symbolism, mythology, critique, and iconology," which emphasized the cultural heritage of African Americans, particularly African survival and the oral tradition. Understanding that rhythmic assertion has always characterized black cultural assertion, they underscored the need for the black artist to use spirituals, gospels, blues, and jazz rhythms, along with various other aspects of the folk culture, as instruments for the revolution. Not since the 1930s had black art and politics been so tightly joined.

These aesthetic assumptions were immediately challenged during the years of the conservative Nixon administration, not only by the traditional white literary establishment, but also by a new group of black academic critics who, as products of affirmative action programs, entered the faculties at white colleges and universities throughout the country. Desiring acceptance and integration into the mainstream of American life and the financial rewards that accompany it, some of these black academicians, especially from Ivy League schools, rejected what they considered to be the purely racial and sociopolitical approach of Black Aesthetic critics in favor of what, in a Eurocentric sense, is a literary one. For example, Yale University critic Robert Stepto, who co-edited *Afro-American Literature: The Reconstruction of Instruction* (1979) with his colleague Dexter Fisher, proposed that African American texts should be examined by applying the Eurocentric literary method called structuralism. It is a theory built on the literary model advanced by the French scholar Ferdinand de Saussure, who postulated that language, rather than social, political, and historical events and forces, shapes the world. "[N]o ideas are established in advance," Saussure maintains in *Course in General Linguistics* (1916; reprinted London, 1974), "and nothing is distinct before the introduction of linguistic structure." Structuralism is reminiscent of the New Criticism, the school of literary thought advocated in the late 1940s to the 1960s by white conservative Southern

Agrarians Robert Penn Warren, John Crowe Ransom, Allen Tate, and others who believed that literature should be judged on its own formal structure without reference to social, political, or other external values.

Also calling for a structuralist of linguistic approach to the literature is the influential University of Pennsylvania professor Houston A. Baker, the author of *The Journey Back: Issues in Black Literature and Criticism* (1980) and *Blues Ideology, and Afro-American Literature: A Vernacular Theory* (1984). Applying his scientific method, or what he terms an "anthropology of art," Baker attempts in these highly complex, obscure works to establish a middle ground between structuralism and Black Aesthetic assumptions. According to Vincent Leitch, in *American Literary Criticism from the Thirties to the Eighties* (1988), Baker relies on anthropological structuralism because "he envisions culture as a linguistic discourse based on systematic rules, principles, and conventions, all of which regularize the social production of art." Although, "Baker locates black works of literature within *black* American culture," Leitch adds, "[w]hat he sought was a means of depoliticizing, deidealizing, and depersonalizing the powerful premises propounded by the Black Aestheticians."

Further targeting the Black Aesthetic through a structuralist theoretical approach to African American literature has been Harvard University professor Henry Louis "Skip" Gates, Jr. With his critical studies *Black Literature and Literary Theory* (1984) and *The Signifying Monkey: Theory of African American Literary Criticism* (1988), Gates has emerged as the most powerful African American literary critic since Alain Locke. More influential than Locke ever was, Gates has become, in Haki Madhubuti's words in *Claiming Earth: Race, Rage, Rape, Redemption* (1994), "a modern day, 21st century, updated Booker T. Washington." Madhubuti observes,

> Never before in the history of Black-white relationships in the United States has an African American been given such unrestricted, unlimited access to influential white journals, newspapers, magazines, and quarterlies. His articles and reviews are not just filler material, equal to that of other scholars or writers, but are often the cover pieces—giving Mr. Gates a tremendous amount of influence and power in the arena of ideas.

Gates has become virtually an "unchallenged authority" in the area of black studies, who, Madhubuti adds,

> is frequently called upon to pass judgment on the "new negros" . . . assistant professors fighting for tenure, young writers looking for publishing contracts and grants, and first book authors looking for a jacket blurb to help sales . . . and his stamp of approval may be the difference between a fast or slow future.

But nowhere has Gates's influence been more keenly felt than in the field of African American literature. As with Locke during the Harlem Renaissance, Gates has towered over the New Black Renaissance. In recent years, he, along with fellow critic Houston Baker, took over the editorship of two leading black literary journals, *Callaloo* and *African*

American Review (formerly *Black American Literature Forum*), thus controlling both canon and voice of African American critical and literary thought. But unlike Locke, Gates has insisted on an ahistorical view of African American literature and has said that "a literary text is a linguistic event; its application must be an activity of close, textual analysis." In an earlier piece, "Preface to Blackness: Text and Pretext," from Robert Stepto and Dexter Fisher's *Afro-American Literature: The Reconstruction of Instruction* (1979), Gates contends "that the correspondence of content between a writer and his world is less significant to literary criticism than is a correspondence of organization of structure, for a relation of content may be a mere reflection of prescriptive, scriptural canon." He concludes that "black literature is a verbal art like other verbal arts. 'Blackness' is not a material object or an event but a metaphor." According to Gates, "If he [the black writer] does embody a 'Black Aesthetic,' then it can be measured not by 'content,' but by a complex structure of meanings."

As a result of Gates's enormous power over the academic world of African American studies, very few literary critics have publicly criticized him. However, Deborah E. McDowell, professor of English at the University of Virginia, has suggested in her essay "The Changing Same: Generational Connections and Black Women Novelists" (1987), that Gates, in his lack of intertextual aggression in examining works by black women fiction writers, has made a particularly male theory of African American intertexuality, that is, a rather selective male theory of black literary history. But his strongest opposition to date has come from black scholar Joyce A. Joyce, professor of black studies at Chicago State University and author of the 1995 American Book Award-winning text *Warriors, Conjurers, and Priests: Defining African-Centered Literary Criticism* (1994). The heated critical debate between Joyce, on the one side, and Gates and Baker, on the other, that appeared in the pages of the literary journal *New Literary History* (Winter 1987) struck at the very heart of contemporary conceptions of black art and of race-gender-class issues within the black community. In the essays "The Black Canon: Reconstructing Black American Literary Criticism" and "Who the Cap Fit," Joyce assesses the critical practices of both Gates and Baker. A Black Aesthetic critic, Joyce passionately attacks them for their black poststructuralism, which she finds to be not only obfuscated, but, more importantly, pernicious because it is a movement toward operating in a "historical vacuum." As she points out, "The Black creative writer has continuously struggled to assert his or her real self and to establish a connection between the self and the people outside that self." Moreover, she says, she "cannot fathom why a Black critic would trust that the master [the poststructuralist] would provide the African American with tools through which one can seek independence." For Joyce, black poststructuralism is an acceptance of "elitist American values," a literary theory that widens the gap between intellectuals and "those masses of Blacks whose lives are still stifled by oppressive environmental, intellectual phenomena." Above all, she criticizes Gates and Baker for dismissing "race as an important element of literary analysis of Black literature" and is outraged by Gates's position that blackness is a mere metaphor. She insists that these critics are irresponsible in their black political commitment, as demonstrated in their critical practice. Finally, she claims that they are alienated from their own culture. Unlike these Ivy League scholars, she sees the black critic not simply as an explicator of texts but also as "a point of consciousness" for the history of black people.

With equal passion, Gates and Baker responded with patriarchal certainty about their African American place in the mainstream canon of American literature. In his essay "'What's Love Got To Do With It?': Critical Theory, Integrity, and the Black Idiom," Gates says that "what Joyce Joyce erroneously thinks of as our 'race' is our *culture*"—that is, that race is a metaphor but black culture is a reality that he both believes in and loves. Gates maintains that the black critic's responsibility is to "translate [contemporary theory] into the black idiom" and identify "indigenous black principles" of critical discourse. The African American experience, in other words, is essentially a linguistic exercise. Baker, in his essay "In Dubious Battle," defends poststructuralism as a radical way of undercutting Western hegemony, of maintaining "a thoroughgoing critique of Western philosophy and its privileges such as colonialism, slavery, racism, and so on." He insists that this recent criticism seems arcane to Joyce because it is "a new sound."

Certainly, all three scholars have critical shortcomings. Joyce underestimates the importance of language, Gates undermines the primacy of "content" in art, and both Gates and Baker attempt to use a textual model to encompass an entire culture. Equally important, all of them overlook the significance of rhythm as an aesthetic principle. As novelist-theorist Sylvia Wynter writes in her study "Ethnopoetics-Sociopoetics?" (1976), "Rhythm as an aesthetic principle not only synthesizes the dialectic of form and content, but makes impossible the separation of form from sense perception, since form is born out of and concomitant with the experiencing of the senses, rather than being a priori." Put simply, blackness is much more than a book. In particular, Gates does not seem to understand the power of race (or, for that matter, racism)—metaphor or not—in American society. To quote the black philosopher Cornel West, "Race matters."

Moreover, to the black poststructuralist's charge that the New Black Aesthetic is political, one can easily respond that poststructuralism, like the New Criticism, is in its own way political; it is a denial of history, an implicit acceptance of the status quo, and an imposition of Eurocentric ideas of structure. Certainly, as Joyce posits, the Black Aesthetic not only serves a significant sociopolitical and literary function but also constitutes a questioning of basic aesthetic assumptions at a time when they clearly need to be questioned.

In Search of Our Mothers' Gardens

Alice Walker

Alice Walker, prolific African-American essayist, novelist, and poet, is the award-winning author of The Color Purple (1982) *and* The Temple of My Familiar *(1989). The present piece is the title essay in her collection* In Search of Our Mothers' Gardens *(1974). References to "gardens" do not appear until the end of the essay; by the time we have reached that point, Walker has prepared us to understand. She has created a common ground for readers through stories of black "grandmothers and mothers . . . driven to a numb and bleeding madness by the springs of creativity in them for which there was no release." We see the creative "seeds" these women have handed to their daughters and, more particularly, the seeds that Walker's own mother handed to her.*

> I described her own nature and temperament. Told how they needed a larger life for their expression. . . . I pointed out that in lieu of proper channels, her emotions have over-flowed into paths that dissipated them. I talked, beautifully I thought, about an art that would be born, an art that would open the way for women the likes of her. I asked her to hope, and build up an inner life against the coming of that day. . . . I sang, with a strange quiver in my voice, a promise song.
>
> Jean Toomer, "Avey," CANE

The poet speaking to a prostitute who falls asleep while he's talking—

When the poet Jean Toomer walked through the South in the early twenties, he discovered a curious thing: black women whose spirituality was so intense, so deep, so *unconscious,* that they were themselves unaware of the richness they held. They stumbled blindly through their lives: creatures so abused and mutilated in body, so dimmed and confused by pain, that they considered themselves unworthy even of hope. In the selfless abstractions their bodies became to the men who used them, they became more than "sexual objects," more even than mere women: they became "Saints." Instead of being perceived as whole persons, their bodies became shrines: what was thought to be their minds became temples suitable for worship. These crazy Saints stared out at the world, wildly, like lunatics—or quietly, like suicides; and the "God" that was in their gaze was as mute as a great stone.

Who were these Saints? These crazy, loony, pitiful women?

Some of them, without a doubt, were our mothers and grandmothers.

In the still heat of the post-Reconstruction South, this is how they seemed to Jean Toomer: exquisite butterflies trapped in an evil honey, toiling away their lives in an era, a century, that did not acknowledge them, except as "the *mule* of the world." They dreamed dreams that no one knew—not even themselves, in any coherent fashion—and saw visions no one could understand. They wandered or sat about the countryside crooning lullabies to ghosts, and drawing the mother of Christ in charcoal on courthouse walls.

They forced their minds to desert their bodies and their striving spirits sought to rise, like frail whirlwinds from the hard red clay. And when those frail whirlwinds fell, in scattered particles, upon the ground, no one mourned. Instead, men lit candles to celebrate the emptiness that remained, as people do who enter a beautiful but vacant space to resurrect a God.

Our mothers and grandmothers, some of them: moving to music not yet written. And they waited.

They waited for a day when the unknown thing that was in them would be made known; but guessed, somehow in their darkness, that on the day of their revelation they would be long dead. Therefore to Toomer they walked, and even ran, in slow motion. For they were going nowhere immediate, and the future was not yet within their grasp. And men took our mothers and grandmothers, "but got no pleasure from it." So complex was their passion and their calm.

To Toomer, they lay vacant and fallow as autumn fields, with harvest time never in sight: and he saw them enter loveless marriages, without joy; and become prostitutes, without resistance; and become mothers of children, without fulfillment.

For these grandmothers and mothers of ours were not Saints, but Artists; driven to a numb and bleeding madness by the springs of creativity in them for which there was no release. They were Creators, who lived lives of spiritual waste, because they were so rich in spirituality—which is the basis of Art—that the strain of enduring their unused and unwanted talent drove them insane. Throwing away this spirituality was their pathetic attempt to lighten the soul to a weight their work-worn, sexually abused bodies could bear.

What did it mean for a black woman to be an artist in our grandmothers' time? In our great-grandmothers' day? It is a question with an answer cruel enough to stop the blood.

Did you have a genius of a great-great-grandmother who died under some ignorant and depraved white overseer's lash? Or was she required to bake biscuits for a lazy backwater tramp, when she cried out in her soul to paint watercolors of sunsets, or the rain falling on the green and peaceful pasturelands? Or was her body broken and forced to bear children (who were more often than not sold away from her)—eight, ten, fifteen, twenty children—when her one joy was the thought of modeling heroic figures of rebellion, in stone or clay?

How was the creativity of the black woman kept alive, year after year and century after century, when for most of the years black people have been in America it was a punishable crime for a black person to read or write? And the freedom to paint, to sculpt, to expand the mind with action did not exist. Consider, if you can bear to imagine it, what might have been the result if singing, too, had been forbidden by law. Listen to the voices of Bessie Smith, Billie Holiday, Nina Simone, Roberta Flack, and Aretha Franklin, among others, and imagine those voices muzzled for life. Then you may begin to comprehend the lives of our "crazy," "Sainted" mothers and grandmothers. The agony of the lives of women who might have been Poets, Novelists, Essayists, and Short-Story Writers (over a period of centuries), who died with their real gifts stifled within them.

And, if this were the end of the story, we would have cause to cry out in my paraphrase of Okot p'Bitek's great poem:

O, my clanswomen
Let us all cry together!
Come,
Let us mourn the death of our mother,
The death of a Queen
The ash that was produced
By a great fire!
O, this homestead is utterly dead
Close the gates
With *lacari* thorns,
For our mother
The creator of the Stool is lost!
And all the young women
Have perished in the wilderness!

But this is not the end of the story, for all the young women—our mothers and grandmothers, *ourselves*—have not perished in the wilderness. And if we ask ourselves why, and search for and find the answer, we will know beyond all efforts to erase it from our minds, just exactly who, and of what, we black American women are.

One example, perhaps the most pathetic, most misunderstood one, can provide a backdrop for our mothers' work: Phillis Wheatley, a slave in the 1700s.

Virginia Woolf, in her book *A Room of One's Own*, wrote that in order for a woman to write fiction she must have two things, certainly: a room of her own (with key and lock) and enough money to support herself.

What then are we to make of Phillis Wheatley, a slave, who owned not even herself? This sickly, frail black girl who required a servant of her own at times—her health was so precarious—and who, had she been white, would have been easily considered the intellectual superior of all the women and most of the men in the society of her day.

* Virginia Woolf wrote further, speaking of course not of our Phillis, that "any woman born with a great gift in the sixteenth century [insert "eighteenth century," insert "black woman," insert "born or made a slave"] would certainly have gone crazed, shot herself, or ended her days in some lonely cottage outside the village, half witch, half wizard [insert "Saint"], feared and mocked at. For it needs little skill and psychology to be sure that a highly gifted girl who had tried to use her gift for poetry would have been so thwarted and hindered by contrary instincts [add "chains, guns, the lash, the ownership of one's body by someone else, submission to an alien religion"], that she must have lost her health and sanity to a certainty."

The key words, as they relate to Phillis, are "contrary instincts." For when we read the poetry of Phillis Wheatley—as when we read the novels of Nella Larsen or the oddly false-sounding autobiography of that freest of all black women writers, Zora Hurston—evidence of "contrary instincts" is everywhere. Her loyalties were completely divided, as was, without question, her mind.

But how could this be otherwise? Captured at seven, a slave of wealthy, doting whites who instilled in her the "savagery" of the Africa they "rescued" her from . . . one wonders if she was even able to remember her homeland as she had known it, or as it really was.

Yet, because she did try to use her gift for poetry in a world that made her a slave, she was "so thwarted and hindered by . . . contrary instincts, that she . . . lost her health. . . ." In the last years of her brief life, burdened not only with the need to express her gift but also with a penniless, friendless "freedom" and several small children for whom she was forced to do strenuous work to feed, she lost her health, certainly. Suffering from malnutrition and neglect and who knows what mental agonies, Phillis Wheatley died.

So torn by "contrary instincts" was black, kidnapped, enslaved Phillis that her description of "the Goddess"—as she poetically called the Liberty she did not have—is ironically, cruelly humorous. And, in fact, has held Phillis up to ridicule for more than a century. It is usually read prior to hanging Phillis's memory as that of a fool. She wrote:

The Goddess comes, she moves divinely fair,
Olive and laurel binds her *golden* hair.
Wherever shines this native of the skies,
Unnumber'd charms and recent graces rise. [My italics]

It is obvious that Phillis, the slave, combed the "Goddess's" hair every morning; prior, perhaps, to bringing in the milk, or fixing her mistress's lunch. She took her imagery from the one thing she saw elevated above all others.

With the benefit of hindsight we ask, "How could she?"

But at last, Phillis, we understand. No more snickering when your stiff, struggling, ambivalent lines are forced on us. We know now that you were not an idiot or a traitor; only a sickly little black girl, snatched from your home and country and made a slave; a woman who still struggled to sing the song that was your gift, although in a land of barbarians who praised you for your bewildered tongue. It is not so much what you sang, as that you kept alive, in so many of our ancestors, *the notion of song.*

Black women are called, in the folklore that so aptly identifies one's status in society, "the *mule* of the world," because we have been handed the burdens that everyone else—*everyone* else—refused to carry. We have also been called "Matriarchs," "Superwomen," and "Mean and Evil Bitches." Not to mention "Castraters" and "Sapphire's Mama." When we have pleaded for understanding, our character has been distorted; when we have asked for simple caring, we have been handed empty inspirational appellations, then stuck in the farthest corner. When we have asked for love, we have been given children. In short, even our plainer gifts, our labors of fidelity and love, have been knocked down our throats. To be an artist and a black woman, even today, lowers our status in many respects, rather than raises it: and yet, artists we will be.

Therefore we must fearlessly pull out of ourselves and look at and identify with our lives the living creativity some of our great-grandmothers were not allowed to know. I stress *some* of them because it is well known that the majority of our great-grandmothers knew, even without "knowing" it, the reality of their spirituality, even if they didn't recognize it beyond what happened in the singing at church—and they never had any intention of giving it up.

How they did it—those millions of black women who were not Phillis Wheatley, or Lucy Terry or Frances Harper or Zora Hurston or Nella Larsen or Bessie Smith; or Elizabeth Catlett, or Katherine Dunham, either—brings me to the title of this essay, "In Search of Our Mothers' Gardens," which is a personal account that is yet shared, in its theme and its meaning, by all of us. I found, while thinking about the far-reaching world of the creative black woman, that often the truest answer to a question that really matters can be found very close.

In the late 1920s my mother ran away from home to marry my father. Marriage, if not running away, was expected of seventeen-year-old girls. By the time she was twenty, she had two children and was pregnant with a third. Five children later, I was born. And this is how I came to know my mother: she seemed a large, soft, loving-eyed woman who was rarely impatient in our home. Her quick, violent temper was on view only a few times a year, when she battled with the white landlord who had the misfortune to suggest to her that her children did not need to go to school.

She made all the clothes we wore, even my brothers' overalls. She made all the towels and sheets we used. She spent the summers canning vegetables and fruits. She spent the winter evenings making quilts enough to cover all our beds.

During the "working" day, she labored beside—not behind—my father in the fields. Her day began before sunup, and did not end until late at night. There was never a moment for her to sit down, undisturbed, to unravel her own private thoughts; never a time free from interruption—by work or the noisy inquiries of her many children. And yet, it is to my mother—and all our mothers who were not famous—that I went in search of the secret of what has fed the muzzled and often mutilated, but vibrant, creative spirit that the black woman has inherited, and that pops out in wild and unlikely places to this day.

But when, you will ask, did my overworked mother have time to know or care about feeding the creative spirit?

The answer is so simple that many of us have spent years discovering it. We have constantly looked high, when we should have looked high—and low.

For example: in the Smithsonian Institution in Washington, D.C., there hangs a quilt unlike any other in the world. In fanciful, inspired, and yet simple and identifiable figures, it portrays the story of the Crucifixion. It is considered rare, beyond price. Though it follows no known pattern of quilt-making, and though it is made of bits and pieces of worthless rags, it is obviously the work of a person of powerful imagination and deep spiritual feeling. Below this quilt I saw a note that says it was made by "an anonymous Black woman in Alabama, a hundred years ago."

If we could locate the "anonymous" black woman from Alabama, she would turn out to be one of our grandmothers—an artist who left her mark in the only materials she could afford, and in the only medium her position in society allowed her to use.

As Virginia Woolf wrote further, in *A Room of One's Own:*

> Yet genius of a sort must have existed among women as it must have existed among the working class. [Change this to "slaves" and "the wives and daughters of sharecroppers."] Now and again an Emily Brontë or a Robert Bums [change this to "a Zora Hurston or a Richard Wright"] blazes out and proves its presence. But certainly it never got itself on to paper. When, however, one reads of a witch being ducked, of a woman possessed by devils [or "Saint-hood"], of a wise woman selling herbs [our root workers], or even a very remarkable man who had a mother, then I think we are on the track of a lost novelist, a suppressed poet, of some mute and inglorious Jane Austen. . . . Indeed, I would venture to guess that Anon, who wrote so many poems without signing them, was often a woman. . . .

And so our mothers and grandmothers have, more often than not anonymously, handed on the creative spark, the seed of the flower they themselves never hoped to see: or like a sealed letter they could not plainly read.

And so it is, certainly, with my own mother. Unlike "Ma" Rainey's songs, which retained their creator's name even while blasting forth from Bessie Smith's mouth, no song or poem will bear my mother's name. Yet so many of the stories that I write, that we all write, are my mother's stories. Only recently did I fully realize this: that through years of listening to my mother's stories of her life, I have absorbed not only the stories themselves, but something of the manner in which she spoke, something of the urgency that involves the knowledge that her stories—like her life—must be recorded. It is probably for this reason that so much of what I have written is about characters whose counterparts in real life are so much older than I am.

But the telling of these stories, which came from my mother's lips as naturally as breathing, was not the only way my mother showed herself as an artist. For stories, too, were subject to being distracted, to dying without conclusion. Dinners must be started, and cotton must be gathered before the big rains. The artist that was and is my mother showed itself to me only after many years. This is what I finally noticed:

Like Mem, a character in *The Third Life of Grange Copeland*, my mother adorned with flowers whatever shabby house we were forced to live in. And not just your typical straggly country stand of zinnias, either. She planted ambitious gardens—and still does—with over fifty different varieties of plants that bloom profusely from early March until late November. Before she left home for the fields, she watered her flowers, chopped up the grass, and laid out new beds. When she returned from the fields she might divide clumps of bulbs, dig a cold pit, uproot and replant roses, or prune branches from her taller bushes or trees—until night came and it was too dark to see.

Whatever she planted grew as if by magic, and her fame as a grower of flowers spread over three counties. Because of her creativity with her flowers, even my memories of poverty are seen through a screen of blooms—sunflowers, petunias, roses, dahlias, forsythia, spirea, delphiniums, verbena . . . and on and on.

And I remember people coming to my mother's yard to be given cuttings from her flowers; I hear again the praise showered on her because whatever rocky soil she landed on, she turned into a garden. A garden so brilliant with colors, so original in its design, so magnificent with life and creativity, that to this day people drive by our house in Georgia—perfect strangers and imperfect strangers—and ask to stand or walk among my mother's art.

I notice that it is only when my mother is working in her flowers that she is radiant, almost to the point of being invisible—except as Creator: hand and eye. She is involved in work her soul must have. Ordering the universe in the image of her personal conception of Beauty.

Her face, as she prepares the Art that is her gift, is a legacy of respect she leaves to me, for all that illuminates and cherishes life. She has handed down respect for the possibilities—and the will to grasp them.

For her, so hindered and intruded upon in so many ways, being an artist has still been a daily part of her life. This ability to hold on, even in very simple ways, is work black women have done for a very long time.

This poem is not enough, but it is something, for the woman who literally covered the holes in our walls with sunflowers:

They were women then
My mama's generation
Husky of voice—Stout of
Step
With fists as well as

Hands
How they battered down
Doors
And ironed
Starched white
Shirts
How they led
Armies
Headragged Generals
Across mined
Fields
Booby-trapped
Kitchens
To discover books
Desks
A place for us
How they knew what we
Must know
Without knowing a page
Of it
Themselves.

Guided by my heritage of a love of beauty and a respect for strength—in search of my mother's garden, I found my own.

And perhaps in Africa over two hundred years ago, there was just such a mother; perhaps she painted vivid and daring decorations in oranges and yellows and greens on the walls of her hut; perhaps she sang—in a voice like Roberta Flack's—*sweetly* over the compounds of her village; perhaps she wove the most stunning mats or told the most ingenious stories of all the village storytellers. Perhaps she was herself a poet—though only her daughter's name is signed to the poems that we know.

Perhaps Phillis Wheatley's mother was also an artist.

Perhaps in more than Phillis Wheatley's biological life is her mother's signature made clear.

Poetry

To Be, or Not To Be . . .

(Soliloquy from Hamlet)

William Shakespeare

To be, or not to be, that is the question:
Whether 'tis nobler in the mind to suffer
The slings and arrows of outrageous fortune,
Or to take arms against a sea of troubles,
And by opposing end them. To die, to sleep—
No more; and by a sleep to say we end
The heartache, and the thousand natural shocks
That flesh is heir to. 'Tis a consummation
Devoutly to be wished—to die, to sleep—
To sleep, perchance to dream, ay there's the rub;
For in that sleep of death what dreams may come
When we have shuffled off this mortal coil
Must give us pause—there's the respect
That makes calamity of so long life.
For those who would bear the whips and scorns of time
Th' oppressor's wrong, the proud man's contumely,
The pangs of despised love, the law's delay,
The insolence of office, and the spurns
That patient merit of th' unworthy takes,
When he himself might his quietus make
With a bare bodkin? Who would fardels bear,
To grunt and sweat under a weary life,
But that the dread of something after death,
The undiscovered country, from whose bourn
No traveller returns, puzzles the will,
And makes us rather bear those ills we have
Than fly to others that we know not of?
Thus conscience does make cowards of us all;
And thus the native hue of resolution
Is sicklied o'er with the pale cast of thought,
And enterprises of great pitch and moment
With this regard their currents turn awry
And lose the name of action.

Sonnet 29

William Shakespeare

When, in disgrace with Fortune and men's eyes,
I all alone beweep my outcast state,
And trouble deaf heaven with my bootless cries,
And look upon myself and curse my fate,
Wishing me like to one more rich in hope,
Featured like him, like him with friends possessed,
Desiring this man's art and that man's scope,
With what I most enjoy contented least;
Yet in these thoughts myself almost despising,
Haply I think on thee, and then my state
(Like to the lark at break of day arising
From sullen earth) sings hymns at heaven's gate;
For thy sweet love remembered such wealth brings
That then I scorn to change my state with kings.

Porphyria's Lover

Robert Browning

The rain set early in to-night,
    The sullen wind was soon awake,
It tore the elm-tops down for spite,
    And did its worst to vex the lake:
    I listened with heart fit to break.
When glided in Porphyria; straight
    She shut the cold out and the storm,
And kneeled and made the cheerless grate
    Blaze up, and all the cottage warm;
    Which done, she rose, and from her form
Withdrew the dripping cloak and shawl,
    And laid her soiled gloves by, untied
Her hat and let the damp hair fall,
    And, last, she sat down by my side
    And called me. When no voice replied,
She put my arm around her waist,
    And made her smooth white shoulders bare,
And all her yellow hair displaced,
    And, stooping, made my cheek lie there,
    And spread, o'er all, her yellow hair,
Murmuring how she loved me—she
    Too weak, for all her heart's endeavour,
To set its struggling passion free
    From pride, and vainer ties dissever,
    And give herself to me for ever.
But passion sometimes would prevail,
    Nor could to-night's gay feast restrain
A sudden thought of one so pale
    For love of her, and all in vain:
    So, she was come through the wind and rain.
Be sure I looked up at her eyes
    Happy and proud; at last I knew
Porphyria worshipped me; surprise
    Made my heart swell, and still it grew
    While I debated what to do.

That moment she was mine, mine, fair,
    Perfectly pure and good: I found
A thing to do, and all her hair
    In one long yellow string I wound
    Three times her little throat around,
And strangled her. No pain felt she;
    I am quite sure she felt no pain.
As a shut bud that holds a bee,
    I warily oped her lids: again
    Laughed the blue eyes without a stain.
And I untightened next the tress
    About her neck; her cheek once more
Blushed bright beneath my burning kiss:
    I propped her head up as before,
    Only, this time my shoulder bore
Her head, which droops upon it still:
    The smiling rosy little head,
So glad it has its utmost will,
    That all it scorned at once is fled,
    And I, its love, am gained instead!
Porphyria's love: she guessed not how
    Her darling one wish would be heard.
And thus we sit together now,
    And all night long we have not stirred,
    And yet God has not said a word!

Ozymandias

Percy Bysshe Shelley

I met a traveler from an antique land
Who said: "Two vast and trunkless legs of stone
Stand in the desert . . . Near them on the sand,
Half-sunk, a shattered visage lies, whose frown,
And wrinkled lip, and sneer of cold command,
Tell that its sculptor well those passions read
Which yet survive, stamped on these lifeless things,
The hand that mocked them, and the heart that fed:
And on the pedestal these words appear:
'My name is Ozymandias, king of kings:
Look on my works, ye Mighty, and despair!'
Nothing beside remains. Round the decay
Of that colossal wreck, boundless and bare
The lone and level sands stretch far away."

Ulysses

Alfred, Lord Tennyson

It little profits that an idle king,
By this still hearth, among these barren crags,
Matched with an agèd wife, I mete and dole
Unequal laws unto a savage race,
That hoard, and sleep, and feed, and know not me.
I cannot rest from travel; I will drain
Life to the lees. All times I have enjoyed
Greatly, have suffered greatly, both with those
That love me, and alone; on shore, and when
Through scudding drifts the rainy Hyades
Vexed the dim sea. I am become a name;
For always roaming with a hungry heart
Much have I seen and known—cities of men
And manners, climates, councils, governments,
Myself not least, but honored of them all—
And drunk delight of battle with my peers,
Far on the ringing plains of windy Troy.
I am a part of all that I have met;
Yet all experience is an arch wherethrough
Gleams that untraveled world whose margin fades
For ever and for ever when I move.
How dull it is to pause, to make an end,
To rust unburnished, not to shine in use!
As though to breathe were life! Life piled on life
Were all too little, and of one to me
Little remains; but every hour is saved
From that eternal silence, something more,
A bringer of new things; and vile it were
For some three suns to store and hoard myself,
And this gray spirit yearning in desire
To follow knowledge like a sinking star,
Beyond the utmost bound of human thought.
    This is my son, mine own Telemachus,
To whom I leave the scepter and the isle,—
Well-loved of me, discerning to fulfill

This labor, by slow prudence to make mild
A rugged people, and through soft degrees
Subdue them to the useful and the good.
Most blameless is he, centered in the sphere
Of common duties, decent not to fail
In offices of tenderness, and pay
Meet adoration to my household gods,
When I am gone. He works his work, I mine.
    There lies the port; the vessel puffs her sail;
There gloom the dark, broad seas. My mariners,
Souls that have toiled, and wrought, and thought with me,—
That ever with a frolic welcome took
The thunder and the sunshine, and opposed
Free hearts, free foreheads—you and I are old;
Old age hath yet his honor and his toil.
Death closes all; but something ere the end,
Some work of noble note, may yet be done,
Not unbecoming men that strove with Gods.
The lights begin to twinkle from the rocks;
The long day wanes; the slow moon climbs; the deep
Moans round with many voices. Come, friends,
'Tis not too late to seek a newer world.
Push off, and sitting well in order smite
The sounding furrows; for my purpose holds
To sail beyond the sunset, and the baths
Of all the western stars, until I die.
It may be that the gulfs will wash us down;
It may be we shall touch the Happy Isles,
And see the great Achilles, whom we knew.
Though much is taken, much abides; and though
We are not now that strength which in old days
Moved earth and heaven, that which we are, we are,—
One equal temper of heroic hearts,
Made weak by time and fate, but strong in will
To strive, to seek, to find, and not to yield.

To Autumn

John Keats

I

Season of mists and mellow fruitfulness,
Close bosom-friend of the maturing sun;
Conspiring with him how to load and bless
With fruit the vines that round the thatch-eaves run;
To bend with apples the mossed cottage-trees,
And fill all fruit with ripeness to the core;
To swell the gourd, and plump the hazel shells
With a sweet kernel; to set budding more,
And still more, later flowers for the bees,
Until they think warm days will never cease.
For summer has o'er-brimmed their clammy cells.

II

Who hath not seen thee oft amid thy store?
Sometimes who ever seeks abroad may find
Thee sitting careless on a granary floor,
Thy hair soft-lifted by the winnowing wind;
Or on a half-reaped furrow sound asleep,
Drowsed with the fume of poppies, while thy hook
Spares the next swath and all its twinèd flowers:
And sometimes like a gleaner thou dost keep
Steady thy laden head across a brook;
Or by a cider-press, with patient look,
Thou watchest the last oozings hours by hours.

III

Where are the songs of spring? Aye, where are they?
Think not of them, thou hast thy music too,—
While barrèd clouds bloom the soft-dying day,
And touch the stubble-plains with rosy hue;
Then in a wailful choir the small gnats mourn
Among the river sallows, borne aloft
Or sinking as the light wind lives or dies;

And full-grown lambs loud bleat from hilly bourn;
Hedge-crickets sing; and now with treble soft
The redbreast whistles from a garden-croft;
And gathering swallows twitter in the skies.

To an Athlete Dying Young

A. E. Housman

The time you won your town the race
We chaired you through the market-place;
Man and boy stood cheering by,
And home we brought you shoulder-high.

Today, the road all runners come,
Shoulder-high we bring you home,
And set you at your threshold down,
Townsman of a stiller town.

Smart lad, to slip betimes away
From fields where glory does not stay
And early though the laurel grows
It withers quicker than the rose.

Eyes the shady night has shut
Cannot see the record cut,
And silence sounds no worse than cheers
After earth has stopped the ears:

Now you will not swell the rout
Of lads that wore their honors out,
Runners whom renown outran
And the name died before the man.

So set, before its echoes fade,
The fleet foot on the sill of shade,
And hold to the low lintel up
The still-defended challenge-cup.

And round that early-laureled head
Will flock to gaze the strengthless dead,
And find unwithered on its curls
The garland briefer than a girl's.

To His Coy Mistress

Andrew Marvell

Had we but world enough, and time,
This coyness, Lady, were no crime.
We would sit down and think which way
To walk and pass our long love's day.
Thou by the Indian Ganges' side
Shouldst rubies find: I by the tide
Of Humber would complain. I would
Love you ten years before the Flood,
And you should, if you please, refuse
Till the conversion of the Jews.
My vegetable love should grow
Vaster than empires, and more slow;
An hundred years should go to praise
Thine eyes and on thy forehead gaze;
Two hundred to adore each breast;
But thirty thousand to the rest;
An age at least to every part,
And the last age should show your heart;
For, Lady, you deserve this state,
Nor would I love at lower rate.
    But at my back I always hear
Time's wingèd chariot hurrying near;
And yonder all before us lie
Deserts of vast eternity.
Thy beauty shall no more be found,
Nor, in thy marble vault, shall sound
My echoing song: then worms shall try
That long preserved virginity,
And your quaint honour turn to dust,
And into ashes all my lust:
The grave's a fine and private place,
But none, I think, do there embrace.
    Now therefore, while the youthful hue
Sits on thy skin like morning dew,
And while thy willing soul transpires

At every pore with instant fires,
Now let us sport us while we may,
And now, like amorous birds of prey,
Rather at once our time devour
Than languish in his slow-chapt power.
Let us roll all our strength and all
Our sweetness up into one ball,
And tear our pleasures with rough strife
Thorough the iron gates of life:
Thus, though we cannot make our sun
Stand still, yet we will make him run.

Annabel Lee

Edgar Allan Poe

It was many and many a year ago,
    In a kingdom by the sea,
That a maiden there lived whom you may know
    By the name of ANNABEL LEE
And this maiden she lived with no other thought
    Than to love and be loved by me.

I was a child and *she* was a child,
    In this kingdom by the sea,
But we loved with a love that was more than love—
    I and my ANNABEL LEE—
With a love that the wingèd seraphs of heaven
    Coveted her and me.

And this was the reason that, long ago,
    In this kingdom by the sea,
A wind blew out of a cloud, chilling
    My beautiful ANNABEL LEE;
So that her high-born kinsmen came
    And bore her away from me,
To shut her up in a sepulchre
    In this kingdom by the sea.

The angels, not half so happy in heaven,
    Went envying her and me—
Yes!—that was the reason (as all men know,
    In this kingdom by the sea)
That the wind came out of the cloud by night,
    Chilling and killing my ANNABEL LEE.

But our love it was stronger by far than the love
    Of those who were older than we—
    Of many far wiser than we—
And neither the angels in heaven above,
    Nor the demons down under the sea,
Can ever dissever my soul from the soul
    Of the beautiful ANNABEL LEE:

For the moon never beams, without bringing me dreams
    Of the beautiful ANNABEL LEE;
And the stars never rise, but I feel the bright eyes
    Of the beautiful ANNABEL LEE:
And so, all the night-tide, I lie down by the side
Of my darling—my darling—my life and my bride,
    In the sepulchre there by the sea—
    In her tomb by the sounding sea.

The Colored Soldiers

Paul Laurence Dunbar

If the muse were mine to tempt it
    And my feeble voice were strong,
If my tongue were trained to measures,
    I would sing a stirring song.
Or would sing a song heroic
    Of those noble sons of Ham,
Of the gallant colored soldiers
    Who fought for Uncle Sam!

In the early days you scorned them,
    And with many a flip and flout
Said "These battles are the white man's,
    And the whites will fight them out."
Up the hills you fought and faltered,
    In the vales you strove and bled,
While your ears still heard the thunder
    Of the foes' advancing tread.

Then distress fell on the nation,
    And the flag was drooping low;
Should the dust pollute your banner?
    No! the nation shouted, No!
So when War, in savage triumph,
    Spread abroad his funeral pall—
Then you called the colored soldiers,
    And they answered to your call.

And like hounds unleashed and eager
    For the life blood of the prey,
Sprung they forth and bore them bravely
    In the thickest of the fray.
And where'er the fight was hottest,
    Where the bullets fastest fell,
There they pressed unblanched and fearless
    At the very mouth of hell.

Ah, they rallied to the standard
    To uphold it by their might;
None were stronger in the labors,
    None were braver in the fight.
From the blazing breach of Wagner
    To the plains of Olustee,
They were foremost in the fight
    Of the battles of the free.

And at Pillow! God have mercy
    On the deeds committed there,
And the souls of those poor victims
    Sent to Thee without a prayer.
Let the fulness of Thy pity
    O'er the hot wrought spirits sway
Of the gallant colored soldiers
    Who fell fighting on that day!

Yes, the Blacks enjoy their freedom,
    And they won it dearly, too;
For the life blood of their thousands
    Did the southern fields bedew.
In the darkness of their bondage,
    In the depths of slavery's night,
Their muskets flashed the dawning,
    And they fought their way to light.

They were comrades then and brothers,
    Are they more or less to-day?
They were good to stop a bullet
    And to front the fearful fray.
They were citizens and soldiers,
    When rebellion raised its head;
And the traits that made them worthy,—
    Ah! those virtues are not dead.

They have shared your nightly vigils,
    They have shared your daily toil;
And their blood with yours commingling
    Has enriched the Southern soil.
They have slept and marched and suffered
    'Neath the same dark skies as you,
They have met as fierce a foeman,
    And have been as brave and true.

And their deeds shall find a record
    In the registry of Fame;
For their blood has cleansed completely
    Every blot of Slavery's shame.
So all honor and all glory
    To those noble sons of Ham—
The gallant colored soldiers
    Who fought for Uncle Sam!

We Wear the Mask

Paul Laurence Dunbar

We wear the mask that grins and lies,
It hides our cheeks and shades our eyes,—
This debt we pay to human guile;
With torn and bleeding hearts we smile,
And mouth with myriad subtleties.

Why should the world be overwise,
In counting all our tears and sighs?
Nay, let them only see us, while
    We wear the mask.

We smile, but, O great Christ, our cries
To thee from tortured souls arise.
We sing, but oh the clay is vile
Beneath our feet, and long the mile;
But let the world dream otherwise,
    We wear the mask!

Elegy for Jane

My Student, Thrown by a Horse

Theodore Roethke

I remember the neckcurls, limp and damp as tendrils;
And her quick look, a sidelong pickerel smile;
And how, once startled into talk, the light syllables leaped for her,
And she balanced in the delight of her thought,
A wren, happy, tail into the wind,
Her song trembling the twigs and small branches.
The shade sang with her;
The leaves, their whispers turned to kissing;
And the mold sang in the bleached valleys under the rose.

Oh, when she was sad, she cast herself down into such a pure depth,
Even a father could not find her:
Scraping her cheek against straw;
Stirring the clearest water.

My sparrow, you are not here,
Waiting like a fern, making a spiny shadow.
The sides of wet stones cannot console me,
Nor the moss, wound with the last light.

If only I could nudge you from this sleep,
My maimed darling, my skittery pigeon.
Over this damp grave I speak the words of my love:
I, with no rights in this matter,
Neither father nor lover.

"Out, Out—"

Robert Frost

The buzz-saw snarled and rattled in the yard
And made dust and dropped stove-length sticks of wood,
Sweet-scented stuff when the breeze drew across it.
And from there those that lifted eyes could count
Five mountain ranges one behind the other
Under the sunset far into Vermont.
And the saw snarled and raffled, snarled and raffled,
As it ran light, or had to bear a load.
And nothing happened: day was all but done.
Call it a day, I wish they might have said
To please the boy by giving him the half hour
That a boy counts so much when saved from work.
His sister stood beside them in her apron
To tell them "Supper." At the word, the saw,
As if to prove saws knew what supper meant,
Leaped out at the boy's hand, or seemed to leap—
He must have given the hand. However it was,
Neither refused the meeting. But the hand!
The boy's first outcry was a rueful laugh,
As he swung toward them holding up the hand
Half in appeal, but half as if to keep
The life from spilling. Then the boy saw all—
Since he was old enough to know, big boy
Doing a man's work, though a child at heart—
He saw all spoiled "Don't let him cut my hand off—
The doctor, when he comes. Don't let him, sister!"
So. But the hand was gone already.
The doctor put him in the dark of ether.
He lay and puffed his lips out with his breath.
And then—the watcher at his pulse took fright.
No one believed. They listened at his heart,
Little—less—nothing!—and that ended it.
No more to build on there. And they, since they
Were not the one dead, turned to their affairs.

Mending Wall

Robert Frost

Something there is that doesn't love a wall.
That sends the frozen-ground-swell under it
And spills the upper boulders in the sun.
And makes gaps even two can pass abreast.
The work of hunters is another thing:
I have come after them and made repair
Where they have left not one stone on a stone,
But they would have the rabbit out of hiding,
To please the yelping dogs. The gaps I mean,
No one has seen them made or heard them made,
But at spring mending-time we find them there.
I let my neighbor know beyond the hill;
And on a day we meet to walk the line
And set the wall between us once again.
We keep the wall between us as we go.
To each the boulders that have fallen to each.
And some are loaves and some so nearly balls
We have to use a spell to make them balance:
"Stay where you are until our backs are turned!"
We wear our fingers rough with handling them.
Oh, just another kind of outdoor game,
One on a side. It comes to little more:
There where it is we do not need the wall:
He is all pine and I am apple orchard.
My apple trees will never get across
And eat the cones under his pines, I tell him.
He only says "Good fences make good neighbors."
Spring is the mischief in me, and I wonder
If I could put a notion in his head:
"Why do they make good neighbors? Isn't it
Where there are cows? But here there are no cows.
Before I built a wall I'd ask to know
What I was walling in or walling out.
And to whom I was like to give offense.
Something there is that doesn't love a wall,

That wants it down." I could say "Elves" to him.
But it's not elves exactly, and I'd rather
He said it for himself. I see him there.
Bringing a stone grasped firmly by the top
In each hand, like an old-stone savage armed.
He moves in darkness as it seems to me,
Not of woods only and the shade of trees.
He will not go behind his father's saying,
And he likes having thought of it so well
He says again, "Good fences make good neighbors."

The Creation

A Negro Sermon

James Weldon Johnson

And God stepped out on space,
And He looked around and said,
"I'm lonely—.
I'll make me a world."

And far as the eye of God could see
Darkness covered everything,
Blacker than a hundred midnights
Down in a cypress swamp.

Then God smiled.
And the light broke,
And the darkness rolled up on one side,
And the light stood shining on the other,
And God said, *"That's good!"*

Then God reached out and took the light in His hands,
And God rolled the light around in His hands,
Until He made the sun;
And He set that sun a-blazing in the heavens.
And the light that was left from making the sun
God gathered up in a shining ball
And flung against the darkness,
Spangling the night with the moon and stars.

Then down between
The darkness and the light
He hurled the world;
And God said, *"That's good!"*

Then God himself stepped down—
And the sun was on His right hand,
And the moon was on His left;
The stars were clustered about His head,
And the earth was under His feet.
And God walked, and where He trod

His footsteps hollowed the valleys out
And bulged the mountains up.

Then He stopped and looked and saw
That the earth was hot and barren.
So God stepped over to the edge of the world
And He spat out the seven seas;
He batted His eyes, and the lightnings flashed;
He clapped His hands, and the thunders rolled;
And the waters above the earth came down,
The cooling waters came down.

Then the green grass sprouted,
And the little red flowers blossomed,
The pine-tree pointed his finger to the sky,
And the oak spread out his arms;
The lakes cuddled down in the hollows of the ground,
And the rivers ran down to the sea;
And God smiled again,
And the rainbow appeared,
And curled itself around His shoulder.

Then God raised His arm and He waved His hand
Over the sea and over the land.
And He said, *"Bring forth! Bring forth!"*
And quicker than God could drop his hand,
Fishes and fowls
And beast and birds
Swam the rivers and the seas.
Roamed the forests and the woods,
And split the air with their wings,
And God said, *"That's good!"*

Then God walked around
And God looked around
On all that He had made.
He looked at His sun,
And He looked at His moon,
And He looked at His little stars;
He looked on His world
With all its living things,
And God said, *"I'm lonely still."*

Then God sat down
On the side of a hill where He could think;
By a deep, wide river He sat down;
With His head in His hands,
God thought and thought,
Till He thought, *"I'll make me a man!"*

Up from the bed of the river
God scooped the clay;
And by the bank of the river
He kneeled Him down;

And there the great God Almighty,
Who lit the sun and fixed it in the sky,
Who flung the stars to the most far corner of the night,
Who rounded the earth in the middle of His hand—
This Great God,
Like a mammy bending over her baby,
Kneeled down in the dust
Toiling over a lump of clay
Till He shaped it in His own image;

Then into it He blew the breath of life,
And man became a living soul.
Amen. Amen.

The Negro Speaks of Rivers

Langston Hughes

I've known rivers:
I've known rivers ancient as the world and
older than the flow of human blood in human
veins.

My soul has grown deep like the rivers.

I bathed in the Euphrates when dawns were
young.
I built my hut near the Congo and it lulled me
to sleep.

I looked upon the Nile and raised the pyramids
above it.
I heard the singing of the Mississippi when Abe
Lincoln went down to New Orleans, and I've
seen its muddy bosom turn all golden in the
sunset.

I've known rivers:
Ancient, dusky rivers.

My soul has grown deep like the rivers.

Mother to Son

Langston Hughes

Well, son, I'll tell you:
Life for me ain't been no crystal stair.
It's had tacks in it,
And splinters,
And boards torn up,
And places with no carpet on the floor—
Bare.
But all the time
I'se been a-climbin' on,
And reachin' landin's,
And turnin' corners,
And sometimes goin' in the dark
Where there ain't been no light.
So boy, don't you turn back.
Don't you set down on the steps
'Cause you finds it's kinder hard.
Don't you fall now—
For I'se still goin', honey,
I'se still climbin',
And life for me ain't been no crystal stair.

The Harlem Dancer

Claude McKay

Applauding youths laughed with young
prostitutes
And watched her perfect, half-clothed body
sway;
Her voice was like the sound of blended flutes
Blown by black players upon a picnic day.
She sang and danced on gracefully and calm,
The light gauze hanging loose about her form;
To me she seemed a proudly-swaying palm
Grown lovelier for passing through a storm.
Upon her swarthy neck black shiny curls
Luxuriant fell; and tossing coins in praise,
The wine-flushed, bold-eyed boys, and even the
girls,
Devoured her shape with eager, passionate
gaze;
But looking at her falsely-smiling face,
I knew her self was not in that strange place.

Those Winter Sundays

Robert Hayden

Sundays too my father got up early,
and put his clothes on in the blueblack cold,
then with cracked hands that ached
from labor in the weekday weather made
banked fires blaze. No one ever thanked him.

I'd wake and hear the cold splintering, breaking.
When the rooms were warm, he'd call,
and slowly I would rise and dress,
fearing the chronic angers of that house.

Speaking indifferently to him,
who had driven out the cold
and polished my good shoes as well.
What did I know, what did I know
of love's austere and lonely offices?

A Letter from Phillis Wheatley

London, 1773

Robert Hayden

Dear Obour
        Our crossing was without
event. I could not help, at times,
reflecting on that first—my Destined—
voyage long ago (I yet
have some remembrance of its Horrors)
and marveling at God's Ways.
    Last evening, her Ladyship presented me
to her illustrious Friends.
I scarce could tell them anything
of Africa, though much of Boston
and my hope of Heaven. I read
my latest Elegies to them.
"O Sable Muse!" the Countess cried,
embracing me, when I had done.
I held back tears, as is my wont,
and there were tears in Dear
Nathaniel's eyes.
    At supper—I dined apart
like captive Royalty—
the Countess and her Guests promised
signatures affirming me
True Poetess, albeit once a slave.
Indeed, they were most kind, and spoke,
moreover, of presenting me
at Court (I thought of Pocahontas)—
an Honor, to be sure, but one,
I should, no doubt, as Patriot decline.
    My health is much improved;
I feel I may, if God so Wills,
entirely recover here.
Idyllic England! Alas, there is
no Eden without its Serpent. Under
the chiming Complaisance I hear him Hiss;
I see his flickering tongue

when foppish would-be Wits
murmur of the Yankee Pedlar
and his Cannibal Mockingbird.
    Sister, forgive th'intrusion of
my Sombreness—Nocturnal Mood
I would not share with any save
your trusted Self. Let me disperse,
in closing, such unseemly Gloom
by mention of an Incident
you may, as I, consider Droll:
Today, a little Chimney Sweep,
his face and hands with soot quite Black,
staring hard at me, politely asked:
"Does you, M'lady, sweep chimneys too?"
I was amused, but dear Nathaniel
(ever Solicitous) was not.
    I pray the Blessings of our Lord
and Saviour Jesus Christ be yours
Abundantly. In His Name,

Phillis

Ballad of Birmingham

(On the bombing of a church in Birmingham, Alabama, 1963)

Dudley Randall

"Mother dear, may I go downtown
Instead of out to play,
And march the streets of Birmingham
In a Freedom March today?"

"No, baby, no, you may not go,
For the dogs are fierce and wild,
And clubs and hoses, guns and jails
Aren't good for a little child."

"But, mother, I won't be alone,
Other children will go with me,
And march the streets of Birmingham
To make our country free."

"No, baby, no, you may not go,
For I fear those guns will fire.
But you may go to church instead
And sing in the children's choir."

She has combed and brushed her night-dark hair,
And bathed rose petal sweet.
And drawn white gloves on her small brown hands,
And white shoes on her feet.

The mother smiled to know her child
Was in the sacred place,
But that smile was the last smile
To come upon her face.

For when she heard the explosion,
Her eyes grew wet and wild.
She raced through the streets of Birmingham
Calling for her child.

She clawed through bits of glass and brick,
Then lifted out a shoe.
"Oh, here's the shoe my baby wore,
But, baby, where are you?"

Daddy

Sylvia Plath

You do not do, you do not do
Any more, black shoe
In which I have lived like a foot
For thirty years, poor and white,
Barely daring to breathe or Achoo.

Daddy, I have had to kill you.
You died before I had time—
Marble-heavy, a bag full of God,
Ghastly statue with one grey toe
Big as a Frisco seal

And a head in the freakish Atlantic
Where it pours bean green over blue
In the waters off beautiful Nauset.
I used to pray to recover you.
Ach, du.

In the German tongue, in the Polish town
Scraped flat by the roller
Of wars, wars, wars.
But the name of the town is common.
My Polack friend

Says there are a dozen or two.
So I never could tell where you
Put your foot, your root,
I never could talk to you.
The tongue stuck in my jaw.

It stuck in a barb wire snare.
Ich, ich, ich, ich,
I could hardly speak.
I thought every German was you.
And the language obscene

An engine, an engine
Chuffing me off like a Jew.
A Jew to Dachau, Auschwitz, Belsen.
I began to talk like a Jew.
I think I may well be a Jew.

The snows of the Tyrol, the clear beer of Vienna
Are not very pure or true.
With my gipsy ancestress and my weird luck
And my Taroc pack and my Taroc pack
I may be a bit of a Jew.

I have always been scared of *you*,
With your Luftwaffe, your gobbledygoo.
And your neat moustache
And your Aryan eye, bright blue.
Panzer-man, panzer-man, O You—

Not God but a swastika
So black no sky could squeak through.
Every woman adores a Fascist,
The boot in the face, the brute
Brute heart of a brute like you.

You stand at the blackboard, daddy,
In the picture I have of you,
A cleft in your chin instead of your foot
But no less a devil for that, no not
Any less the black man who

Bit my pretty red heart in two.
I was ten when they buried you.
At twenty I tried to die
And get back, back, back to you.
I thought even the bones would do.

But they pulled me out of the sack,
And they stuck me together with glue.
And then I knew what to do.
I made a model of you,
A man in black with a Meinkampf look

And a love of the rack and the screw.
And I said I do, I do.
So daddy, I'm finally through.
The black telephone's off at the root,
The voices just can't worm through.

If I've killed one man, I've killed two—
The vampire who said he was you
And drank my blood for a year,
Seven years, if you want to know.
Daddy, you can lie back now.

There's a stake in your fat black heart
And the villagers never liked you.
They are dancing and stamping on you.
They always *knew* it was you.
Daddy, daddy, you bastard, I'm through.

One Art

Elizabeth Bishop

The art of losing isn't hard to master;
so many things seem filled with the intent
to be lost that their loss is no disaster.

Lose something every day. Accept the fluster
of lost door keys, the hour badly spent.
The art of losing isn't hard to master.

Then practice losing farther, losing faster:
places, and names, and where it was you meant
to travel. None of these will bring disaster.

I lost my mother's watch. And look! my last, or
next-to-last, of three loved houses went.
The art of losing isn't hard to master.

I lost two cities, lovely ones. And, vaster,
some realms I owned, two rivers, a continent.
I miss them, but it wasn't a disaster.

—Even losing you (the joking voice, a gesture
I love) I shan't have lied. It's evident
the art of losing's not too hard to master
though it may look like (Write it!) like disaster.

The Mother

Gwendolyn Brooks

Abortions will not let you forget.
You remember the children you got that you did not get,
The damp small pulps with a little or with no hair,
The singers and workers that never handled the air.
You will never neglect or beat
Them, or silence or buy with a sweet.
You will never wind up the sucking-thumb
Or scuttle off ghosts that come.
You will never leave them, controlling your luscious sigh,
Return for a snack of them, with gobbling mother-eye.

I have heard in the voices of the wind the voices of my dim killed children.
I have contracted. I have eased
My dim dears at the breasts they could never suck.
I have said, Sweets, if I sinned, if I seized
Your luck
And your lives from your unfinished reach,
If I stole your births and your names,
Your straight baby tears and your games,
Your stilted or lovely loves, your tumults, your marriages, aches,
    and your deaths,
If I poisoned the beginnings of your breaths,
Believe that even in my deliberateness I was not deliberate.
Though why should I whine,
Whine that the crime was other than mine?—
Since anyhow you are dead.
Or rather, or instead,
You were never made.

But that too, I am afraid,
Is faulty: oh, what shall I say, how is the truth to be said?
You were born, you had body, you died.
It is just that you never giggled or planned or cried.

Believe me, I loved you all.
Believe me, I knew you, though faintly, and I loved, I loved you
All.

Beware: Do Not Read This Poem

Ishmael Reed

tonite, *thriller* was
abt ol woman, so vain she
surrounded her self w/
    many mirrors

It got so bad that finally she
locked herself indoors & her
whole life became the
    mirrors

one day the villagers broke
into her house, but she was too
swift for them, she disappeared
        into a mirror
each tenant who bought the house
after that, lost a loved one to
        the ol woman in the mirror:
        first a little girl
        then a young woman
        then the young woman/s husband

the hunger of this poem is legendary
it has taken in many victims
back off from this poem
it has drawn in yr feet
back off from this poem
it has drawn in yr legs
back off from this poem
it is a greedy mirror
you are into this poem. from
    the waist down
nobody can hear you can they?

this poem has had you up to here
        belch
this poem aint got no manners
you cant call out frm this poem

relax now & go w/ this poem
move & roll on to this poem

do not resist this poem
this poem has yr eyes
this poem has his head
this poem has his arms
this poem has his fingers
this poem has his fingertips

this poem is the reader & the
reader this poem

statistic: the us bureau of missing persons reports
that in 1968 over 100,000 people disappeared
leaving no solid clues
nor trace only
a space in the lives of their friends

The Self-Hatred of Don L. Lee

Don L. Lee

i,
at one time,
loved
my
color—
it
opened sMall
doors of
tokenism
&
acceptance.

(doors called, "the only one" & "our negro")

after painfully
struggling
thru Du Bois,
Rogers, Locke,
Wright & others,
my blindness
was vanquished
by pitchblack
paragraphs of
"us, we, me, i"
awareness.

i
began
to love
only a
part of
me—
my inner
self which
is all
black—
&

developed a
vehement
hatred of
my light
brown
outer.

Big Momma

Don L. Lee

finally retired   pensionless
from cleaning somebody else's house
she remained home to clean
the one she didn't own.

in her kitchen where we often talked
the *chicago tribune* served as a tablecloth
for the two cups of tomato soup that went
along with my weekly visit & talkingto.

she was in a seriously-funny mood
& from the get-go she was down, realdown:

> roaches around here are like
> letters on a newspaper
> or
> u gonta be a writer, hunh
> when u gone write me some writen
> or
> the way niggers act around here
> if talk cd kill we'd all be dead.

she's somewhat confused about all this *blackness*
but said that it's good when negroes start putting themselves
first and added: we've always shopped at the colored
stores.
& the way niggers cut each other up
round here every weekend that white-
man don't haveta
worry bout no revolution specially
when he's gonta haveta pay for it too,
anyhow all he's gotta do is drop a
truck load of *dope* out there
on 43rd st. & all the niggers & yr
revolutionaries
be too busy getten high & then they'll
turn round

and fight each other over who got the
mostest.

we finished our soup and i moved to excuse myself,
as we walked to the front door she made a last comment:
now *luther* i knows you done changed a lots but if
you can think back, we never did eat too much pork
round here anyways, it was bad for the belly.
i shared her smile and agreed.

touching the snow lightly i headed for 43rd st.
at the corner i saw a brother crying while
trying to hold up a lamp post,
thru his watery eyes i cd see big momma's words.

at sixty-eight
she moves freely, is often right
and when there is food
eats joyously with her own
real teeth.

Self-Portrait

Asalean Springfield

The sunshine of childhood seemed never to end;
Wintry winds could only fan the glow within.
While sunbeams quenched the chill of snow,
And icicles ornamented every door,
Great falls and broken crowns upstaged no family show.

The heart was young, the mind, a sponge;
A nude Mona blushed at budding bod,
But into the jaws of pleasure was afraid to plunge.
A scarlet stamp foretold the wrath of God;
Read and tell, a daily rule and rod.

After the sock hops, the proms,
And rumbles receded from distant drums,
Wonder searched heart and soul throughout,
For an anchor to hold without fancy's doubt.

In Palm Garden on a warm summer's night,
The band played on; the wine was light.
With sun tan cheeks and a world of a stride,
The heart embarked upon a world untried;
Where tall trees open wider their open arms,
Little blossoms glow endlessly in the sun,
And Cu-birds coo a life long song.

Joy clasps sorrow in mutual sway;
Time makes demands that all must pay.
Hard tree rings and a three bar seal
Scored the rounds from Ypsi to Capitol Hill.
With the abyss-dim future expanded with a song,
Sharp peaks and pitfalls are but skips and stepping stones.
The struggle defied between man and man,
But God helps the child to defy time on the run.

Eternity in a span as the curtain folds back,
When faith joins reason in the search outright.
The spirit burning bright the heart once cold,
Refining the crude, purifying like gold.
As time sweeps over the horizontal expanse,
Spirit rises on wings to a mountain stance.

The map only defines an inverted cross,
Unless the mount cuts clear the Adamic cords.

Branching Out

Asalean Springfield

The Olive Tree, branching out under the sun;
Pruned in season to engrace and adorn:
Sufficient to stand yet destined to sway,
All but seven thousand will fall by the way.
An offer of peace grafted branches extend,
To a war torn world blinded by sin.

Dark earth proclaims the light of mind,
That mirrors the stars in false sublime;
In dull pursuit of a common sun,
That a higher light might once be formed.
Yet mundane nature is powerless to create,
Though prunes the olive dead laden with weight.

The cluster of years around faith recedes,
Infusing its substance through strange new reeds,
Or branches now grafted on the tattered tree,
Replacing lost limbs of inheritance free;
But the promise ever holds for the chosen ones,
Elected by grace when the wandering is done.

The Olive Tree, the infused lot, seven thousand manifold;
The law within, universal kin, the created oversoul.

Serpent Knowledge

Robert Pinsky

In something you have written in school, you say[1]
That snakes are born (or hatched) already knowing
Everything they will ever need to know—
Weazened and prematurely shrewd, like Merlin;
Something you read somewhere, I think, some textbook
Coy on the subject of the reptile brain.
(Perhaps the author half-remembered reading
About the Serpent of Experience
That changes manna to gall.) I don't believe it;
Even a snake's horizon must expand,
Inwardly, when an instinct is confirmed
By some new stage of life: to mate, kill, die.

Like angels, who have no genitals or place
Of national origin, however, snakes
Are not historical creatures; unlike chickens,
Who teach their chicks to scratch the dust for food—
Or people, who teach ours how to spell their names:
Not born already knowing all we need,
One generation differing from the next
In what it needs, and knows.
            So what I know,
What you know, what your sister knows (approaching
The age you were when I began this poem)
All differ, like different overlapping stretches
Of the same highway: with different lacks, and visions.
The words— "*Vietnam*"—that I can't use in poems
Without the one word threatening to gape
And swallow and enclose the poem, for you
May grow more finite; able to be touched.

The actual highway—snake's-back where it seems
That any strange thing may be happening, now,
Somewhere along its endless length—once twisted
And straightened, and took us past a vivid place:
Brave in the isolation of its profile,

"Ten miles from nowhere" on the rolling range,
A family graveyard on an Indian mound
Or little elevation above the grassland. . . .
Fenced in against the sky's huge vault at dusk
By a waist-high iron fence with spear-head tips,
The grass around and over the mound like surf.

A mile more down the flat fast road, the homestead:
Regretted, vertical, and unadorned
As its white gravestones on their lonely mound—
Abandoned now, the paneless windows breathing
Easily in the wind, and no more need
For courage to survive the open range
With just the graveyard for a nearest neighbor;
The sones of Limit—comforting and depriving.

Elsewhere along the highway, other limits—
Hanging in shades of neon from dusk to dusk,
The signs of people who know how to take
Pleasure in places where it seems unlikely:
New kinds of places, the "overdeveloped" strips
With their arousing, vacant-minded jumble;
Or garbegey lake-towns, and the tourist-pits
Where crimes unspeakably bizarre come true
To astonish countries older, or more savage . . .
As though the rapes and murders of the French
Or Indonesians were less inventive than ours,
Less goofy than those happenings that grow
Like air-plants—out of nothing, and alone.

They make us parents want to keep our children
Locked up, safe even from the daily papers
That keep the grisly record of that frontier
Where things unspeakable happen along the highways.

In today's paper, you see the teen-aged girl
From down the street; camping in Oregon
At the far point of a trip across the country,
Together with another girl her age,
They suffered and survived a random evil,
An unidentified, youngish man in jeans
Aimed his car off the highway, into the park
And at their tent (apparently at random)
And drove it over them once, and then again;
And then got out, and struck at them with a hatchet
Over and over, while they struggled; until
From fear, or for some other reason, or none,
He stopped; and got back into his car again
And drove off down the night-time highway. No rape,
Or robbery, no "motive." Not even words,
Or any sound from him that they remember.
The girl still conscious, by crawling, reached the road
And even some way down it; where some people

Drove by and saw her, and brought them both to help,
So doctors could save them—barely marked.
                                        You see
Our neighbor's picture in the paper: smiling,
A pretty child with a kerchief on her head
Covering where the surgeons had to shave it.
You read the story, and in a peculiar tone—
Factual, not unfeeling, like two policemen—
Discuss it with your sister. You seem to feel
Comforted that it happened far away,
As in a crazy place, in *Oregon*:
For me, a place of wholesome reputation;
For you, a highway where strangers go amok,
As in the universal provincial myth
That sees, in every stranger, a mad attacker . . .
(And in one's victims, it may be, a stranger).

Strangers: the Foreign who, coupling with their cousins
Or with their livestock, or even with wild beasts,
Spawn children with tails, or claws and spotted fur,
Ugly—and though their daughters are beautiful
seen dancing from the front, behind their backs
Or underneath their garments are the tails
Of reptiles, or teeth of bears.
                    So one might feel—
Thinking about the people who cross the mountains
And oceans of the earth with separate legends,
To die inside the squalor of sod huts,
Shanties, or tenements; and leave behind
Their legends, or the legend of themselves,
Broken and mended by the generations;
Their alien, orphaned, and disconsolate spooks.
Earth-trolls or Kallikaks or Snopes or golems,
Descended of Hessians, runaway slaves and Indians,
Legends confused and loose on the roads at night . . .
The Alien or Creature of the movies.
As people die, their monsters grow more tame;
So that the people who survived Saguntum,
Or in the towns that saw the Thirty Years' War,
Must have felt that the wash of blood and horror
Changed something, inside. Perhaps they came to see
The state or empire as a kind of Whale
Or Serpent, in whose body they must live—
Not that mere suffering could make us wiser,
Or nobler, but only older, and more ourselves. . . .

On television, I used to see, each week,
Americans descending in machines
With wasted bravery and blood; to spread
Pain and vast fires amid a foreign place,
Among the strangers to whom we were new—
Americans: a spook or golem, there.

I think it made our country older, forever.
I don't mean better or not better, but merely
As though a person should come to a certain place
And have his hair turn gray, that very night.

Someday, the War in Southeast Asia, somewhere—
Perhaps for you and your people younger than you—
Will be the kind of history and pain
Saguntum is for me; but never tamed
Or "history" for me, I think. I think
That I may always feel as if I lived
In a time when the country aged itself:
More lonely together in our common strangeness . . .
As if we were a family, and some members
Had done an awful thing on a road at night,
And all of us had grown white hair, or tails:
And though the tails or white hair would afflict
Only that generation then alive
And of a certain age, regardless whether
They were the ones that did or; planned the thing—
Or even heard about it—nevertheless
The members of that family ever after
Would bear some consequence or demarcation,
Forgotten maybe, taken for granted, a trait,
A new syllable buried in their name.

1. The poet is addressing his daughter.

What's Going On

Marvin Gaye, A. Cleveland, and R. Benson

Mother, mother
There's too many of you crying
Brother, brother, brother
There's too many of you dying.
You know we've got to find a way
To bring some lovin' here today—Ya
Father, father, father we don't need to escalate
You see, war is not the answer
For only love can conquer hate
You know we've got to find the way
To bring some lovin' here today.
Picket lines and picket signs
Don't punish me with brutality
Talk to me, so you can see
Oh, what's going on
What's going on
Ya, what's going on
Ah, what's going on
In the meantime
Right on baby
Right on
Right on
Father, father, everybody thinks we're wrong
Oh, but who are they to judge us
Simply because our hair is long
Oh, you know we've got to find a way
To bring some understanding here today
Oh
Picket lines and picket signs
Don't punish me with brutality
Come on talk to me
So you can see
What's going on

Ya, what's going on
Tell me, what's going on
I'll tell you what's going on—Uh
Right on baby
Right on baby

The Revolution Will Not Be Televised[1]

Gil Scott-Heron

You will not be able to stay home, brother.
You will not be able to plug in, turn on and cop out.
You will not be able to lose yourself on scag and
skip out for beer during commercials because
The revolution will not be televised.

The revolution will not be televised.
The revolution will not be brought to you by Xerox in four parts
without commercial interruption.
The revolution will not show you pictures of Nixon blowing a bugle
and leading a charge by John Mitchell, General Abramson and
Spiro Agnew to eat hog maws confiscated from a Harlem
sanctuary.
The revolution will not be televised.

The revolution will not be brought to you by
The Schaeffer Award Theatre and will not star
Natalie Wood and Steve McQueen or Bullwinkle and Julia.
The revolution will not give your mouth sex appeal.
The revolution will not get rid of the nubs.
The revolution will not make you look five pounds thinner.
The revolution will not be televised, brother.

There will be no pictures of you and Willie Mae
pushing that shopping cart down the block on the dead run
or trying to slide that color t.v. in a stolen ambulance.
NBC will not be able to predict the winner at 8:32 on reports from
twenty-nine districts.
The revolution will not be televised.

There will be no pictures of pigs shooting down brothers
on the instant replay.
There will be no pictures of pigs shooting down brothers
on the instant replay.
There will be no slow motion or still lifes of Roy Wilkins strolling
through Watts in a red, black and green liberation jumpsuit
that he has been saving for just the proper occasion.

Green Acres, Beverly Hillbillies and Hooterville Junction
will no longer be so damned relevant
and women will not care if Dick finally got down with Jane
on Search for Tomorrow
because black people will be in the streets looking for
A Brighter Day.
The revolution will not be televised.

There will be no highlights on the Eleven O'Clock News
and no pictures of hairy armed women liberationists
and Jackie Onassis blowing her nose.
The theme song will not be written by Jim Webb or Francis Scott
  Key
nor sung by Glen Campbell, Tom Jones, Johnny Cash,
Englebert Humperdink or Rare Earth.
The revolution will not be televised.

The revolution will not be right back after a
message about a white tornado, white lightning or white people.
You will not have to worry about a dove in your bedroom,
the tiger in your tank or the giant in your toilet bowl.
The revolution will not go better with coke.
The revolution will not fight germs that may cause bad breath.
The revolution *will* put you in the driver's seat.
The revolution will not be televised
        will not be televised
          not be televised
            be televised
The revolution will be no re-run, brothers.
The revolution will be LIVE.

Note

1. This recorded poem of 1970 by Gil Scott-Heron is not rap music *per se* but it had a vital influence on rap music's forms and themes.

Reflection on the Vietnam War Memorial

Jeffrey Harrison

Here it is, the back porch of the dead.
You can see them milling around in there,
    screened in by their own names,
    looking at us in the same
vague and serious way we look at them.

An underground house, a roof of grass—
one version of the underworld. It's all
    we know of death, a world
    like our own (but darker, blurred)
inhabited by beings like ourselves.

The location of the name you're looking for
can be looked up in a book whose resemblance
    to a phone book seems to claim
    some contact can be made
through the simple act of finding a name.

As we touch the name the stone absorbs our grief.
It takes us in—we see ourselves inside it.
    And yet we feel it as a wall
    and realize the dead are all
just names now, the separation final.

Digging

Seamus Heaney

Between my finger and my thumb
The squat pen rests; snug as a gun.

Under my window, a clean rasping sound
When the spade sinks into gravelly ground:
My father, digging. I look down

Till his straining rump among the flowerbeds
Bends low, comes up twenty years away
Stooping in rhythm through potato drills
Where he was digging.

The coarse boot nestled on the lug, the shaft
Against the inside knee was levered firmly.
He rooted out tall tops, buried the bright edge deep
To scatter new potatoes that we picked
Loving their cool hardness in our hands.
By God, the old man could handle a spade.
Just like his old man.

My grandfather cut more turf in a day
Than any other man on Toner's bog.
Once I carried him milk in a bottle
Corked sloppily with paper. He straightened up
To drink it, then fell to right away

Nicking and slicing neatly, heaving sods
Over his shoulder, going down and down
For the good turf. Digging.

The cold smell of potato mould, the squelch and slap
Of soggy peat, the curt cuts of an edge
Through living roots awaken in my head.
But I've no spade to follow men like them.

Between my finger and my thumb
The squat pen rests.
I'll dig with it.

Sunlight

Seamus Heaney

There was a sunlit absence.
The helmeted pump in the yard
heated its iron,
water honeyed

in the slung bucket
and the sun stood
like a griddle cooling
against the wall

of each long afternoon.
So, her hands scuffled
over the bakeboard,
the reddening stove

sent its plaque of heat
against her where she stood
in a floury apron
by the window.

Now she dusts the board
with a goose's wing,
now sits, broad-lapped,
with whitened nails

and measling shins:
here is a space
again, the scone rising
to the tick of two clocks.

And here is love
like a tinsmith's scoop
sunk past its gleam
in the meal-bin.

Black Woman

Toni Cade Bambara

Well, I said come here, Black Woman,
Ah-hmmm, don you hear me cryin, Lawd, Lawd!
I say run heah, Black Woman,
Sit on yo Black Daddy's knee, Lawd!
Mmmmm, I know yo house feel lonesome,
Ah, don you heah me whoopin, Lawd, Lawd,
Don yo house feel lonesome,
When yo biscuit roller gon,
Lawd, help my cryin time—
Don yo house feel lonesome, Mama,
When yo biscuit roller gon.

I say my house feel lonesome—
I know you heah me crying, oh Baby,
Ah-hmmm, ah, when I looked in my kitchen, Mama,
An I wen all thoo my dinin room
Ah-mmmm, when I woke up this mornin
I foun my biscuit roller done gone.

Goin to Texas, Mama,
Jus to heah the wild ox moan—
Lawd help mah cryin time—
Goin to Texas, Mama,
Jus to heah the wild ox moan,
An if they moan to suit me,
I'm goin to bring a wild ox home.
Ah-hmmm, I say I'm got to go to Texas, Black Mama—

I know you heah me cryin, Lawd, Lawd—
Ah-hmmm, I'm got to go to Texas, Black Mama,
Ahm-jus to heah the white cow, I say, moan!
Ah-hmmm, ah, if they moan to suit me, Lawd, Lawd,
I bleeve I'll bring a white cow back home.

Say, I feel superstitious, Mama,
'Bout my hoggin bread, Lawd help my hungry time,
I feel superstitious, Baby, 'bout my hoggin bread!

Ah-hmmm, Baby, I feel superstitious,
I say 'stitious, Black Woman!
Ah-hmmmm, ah you heah me cryin
Bout I don got hungry, Lawd, Lawd
Oh, Mama, I feel superstitious
Bout my hog, Lawd Gawd, it's mah bread.

I want you to tell me, Mama,
Ah-hmmm, I heah me cryin, oh Mama!
Ah-hmmm, I want you to tell me, Black Woman,
O wheah did you stay las night?
I love you, Black woman,
I tell the whole worl I do.
Ah-hmmm, I love you, Black Woman,
I know you heah me whoopin, Black Baby!
Ah-hmmm, I love you Black Woman
An I'll tell yo Daddy, I do, Lawd.

Short Stories

The Queen of Spades

Alexander Pushkin

Chapter I.

There was a card party at the rooms of Naroumoff of the Horse Guards. The long winter night passed away imperceptibly, and it was five o'clock in the morning before the company sat down to supper. Those who had won, ate with a good appetite; the others sat staring absently at their empty plates. When the champagne appeared, however, the conversation became more animated, and all took a part in it.

"And how did you fare, Sourin?" asked the host.

"Oh, I lost, as usual. I must confess that I am unlucky: I play mirandole, I always keep cool, I never allow anything to put me out, and yet I always lose!"

"And you did not once allow yourself to be tempted to back the red? . . . Your firmness astonishes me."

"But what do you think of Hermann?" said one of the guests, pointing to a young Engineer: "he has never had a card in his hand in his life, he has never in his life laid a wager, and yet he sits here till five o'clock in the morning watching our play."

"Play interests me very much," said Hermann: "but I am not in the position to sacrifice the necessary in the hope of winning the superfluous."

"Hermann is a German: he is economical—that is all!" observed Tomsky. "But if there is one person that I cannot understand, it is my grandmother, the Countess Anna Fedorovna."

"How so?" inquired the guests.

"I cannot understand," continued Tomsky, "how it is that my grandmother does not punt."

"What is there remarkable about an old lady of eighty not punting?" said Naroumoff.

"Then you do not know the reason why?"

"No, really; haven't the faintest idea."

"Oh! then listen. You must know that, about sixty years ago, my grandmother went to Paris, where she created quite a sensation. People used to run after her to catch a glimpse of the 'Muscovite Venus.' Richelieu made love to her, and my grandmother maintains that he almost blew out his brains in consequence of her cruelty. At that time ladies used to play at faro. On one occasion at the Court, she lost a very considerable sum to the Duke of Orleans. On returning home, my grandmother removed the patches from her face, took off her hoops, informed my grandfather of her loss at the gaming-table, and ordered him to pay the money. My deceased grandfather, as far as I remember, was a sort of house-steward to my grandmother. He dreaded her like fire; but, on hearing of such a heavy loss, he almost went out of his mind; he calculated the various sums she had lost, and pointed out to her that in six months she had spent half a million of francs, that neither their Moscow nor Saratoff estates were in Paris, and finally refused point blank to pay the debt. My grandmother gave him a box on the ear and slept by herself as a sign of her displeasure. The next day she sent for her husband, hoping that this domestic punishment had produced an effect upon him, but she found him inflexible. For the first time in her life, she entered into reasonings and explanations with him, thinking to be able to convince him by pointing out to him that there are debts and debts, and that there is a great difference between a Prince and a coachmaker. But it was all in vain, my grandfather still remained obdurate. But the matter did not rest there. My grandmother did not know what to do. She had shortly before become acquainted with a very remarkable man. You have heard of Count St. Germain, about whom so many marvellous stories are told. You know that he represented himself as the Wandering Jew, as the discoverer of the elixir of life, of the philosopher's stone, and so forth. Some laughed at him as a charlatan; but Casanova, in his memoirs, says that he was a spy. But be that as it may, St. Germain, in spite of the mystery surrounding him, was a very fascinating person, and was much sought after in the best circles of society. Even to this day my grandmother retains an affectionate recollection of him, and becomes quite angry if anyone speaks disrespectfully of him. My grandmother knew that St. Germain had large sums of money at his disposal. She resolved to have recourse to him, and she wrote a letter to him asking him to come to her without delay. The queer old man immediately waited upon her and found her overwhelmed with grief. She described to him in the blackest colours the barbarity of her husband, and ended by declaring that her whole hope depended upon his friendship and amiability.

"St. Germain reflected.

"'I could advance you the sum you want,' said he; 'but I know that you would not rest easy until you had paid me back, and I should not like to bring fresh troubles upon you. But there is another way of getting out of your difficulty: you can win back your money."

"'But, my dear Count,' replied my grandmother, 'I tell you that I haven't any money left.'"

"'Money is not necessary,' replied St. Germain: 'be pleased to listen to me.'"

"Then he revealed to her a secret, for which each of us would give a good deal . . ."

The young officers listened with increased attention. Tomsky lit his pipe, puffed away for a moment and then continued:

"That same evening my grandmother went to Versailles to the *jeu de la reine.* The Duke of Orleans kept the bank; my grandmother excused herself in an off-handed manner for not having yet paid her debt, by inventing some little story, and then began to play against him. She chose three cards and played them one after the other: all three won *sonika,* and my grandmother recovered every farthing that she had lost."

"Mere chance!" said one of the guests.

"A tale!" observed Hermann.

"Perhaps they were marked cards!" said a third.

"I do not think so," replied Tomsky gravely.

"What!" said Naroumoff, "you have a grandmother who knows how to hit upon three lucky cards in succession, and you have never yet succeeded in getting the secret of it out of her?"

"That's the deuce of it!" replied Tomsky: "she had four sons, one of whom was my father; all four were determined gamblers, and yet not to one of them did she ever reveal her secret, although it would not have been a bad thing either for them or for me. But this is what I heard from my uncle, Count Ivan Ilitch, and he assured me, on his honour, that it was true. The late Chaplitsky—the same who died in poverty after having squandered millions—once lost, in his youth, about three hundred thousand roubles—to Zoritch, if I remember rightly. He was in despair. My grandmother, who was always very severe upon the extravagance of young men, took pity, however, upon Chaplitsky. She gave him three cards, telling him to play them one after the other, at the same time exacting from him a solemn promise that he would never play at cards again as long as he lived. Chaplitsky then went to his victorious opponent, and they began a fresh game. On the first card he staked fifty thousand roubles and won *sonika;* he doubled the stake and won again, till at last, by pursuing the same tactics, he won back more than he had lost . . .

"But it is time to go to bed: it is a quarter to six already."

And indeed it was already beginning to dawn: the young men emptied their glasses and then took leave of each other.

Chapter II.

The old countess A— was seated in her dressing-room in front of her looking-glass. Three waiting-maids stood around her. One held a small pot of rouge, another a box of hair-pins, and the third a tall cap with bright red ribbons. The Countess had no longer the slightest pretensions to beauty, but she still preserved the habits of her youth, dressed in strict accordance with the fashion of seventy years before, and made as long and as careful a toilette as she would have done sixty years previously. Near the window, at an embroidery frame, sat a young lady, her ward.

"Good morning, grandmamma" said a young officer, entering the room. "*Bonjour, Mademoiselle Lise.* Grandmamma, I want to ask you something."

"What is it, Paul?"

"I want you to let me introduce one of my friends to you, and to allow me to bring him to the ball on Friday."

"Bring him direct to the ball and introduce him to me there. Were you at B—'s yesterday?"

"Yes; everything went off very pleasantly, and dancing was kept up until five o'clock. How charming Eletskaia was!"

"But, my dear, what is there charming about her? Isn't she like her grandmother, the Princess Daria Petrovna? By the way, she must be very old, the Princess Daria Petrovna."

"How do you mean, old?" cried Tomsky thoughtlessly; "she died seven years ago."

The young lady raised her head and made a sign to the young officer. He then remembered that the old Countess was never to be informed of the death of any of her contemporaries, and he bit his lips. But the old Countess heard the news with the greatest indifference.

"Dead!" said she; "and I did not know it. We were appointed maids of honour at the same time, and when we were presented to the Empress . . ."

And the Countess for the hundredth time related to her grandson one of her anecdotes.

"Come, Paul," said she, when she had finished her story, "help me to get up. Lizanka, where is my snuff-box?"

And the Countess with her three maids went behind a screen to finish her toilette. Tomsky was left alone with the young lady.

"Who is the gentleman you wish to introduce to the Countess?" asked Lizaveta Ivanovna in a whisper.

"Naroumoff. Do you know him?"

"No. Is he a soldier or a civilian?"

"A soldier."

"Is he in the Engineers?"

"No, in the Cavalry. What made you think that he was in the Engineers?"

The young lady smiled, but made no reply.

"Paul," cried the Countess from behind the screen, "send me some new novel, only pray don't let it be one of the present day style."

"What do you mean, grandmother?"

"That is, a novel, in which the hero strangles neither his father nor his mother, and in which there are no drowned bodies. I have a great horror of drowned persons."

"There are no such novels nowadays. Would you like a Russian one?"

"Are there any Russian novels? Send me one, my dear, pray send me one!"

"Good-bye, grandmother: I am in a hurry. . . . Good-bye, Lizaveta Ivanovna. What Made you think that Naroumoff was in the Engineers?"

And Tomsky left the boudoir.

Lizaveta Ivanovna was left alone: she laid aside her work and began to look out of the window. A few moments afterwards, at a corner house on the other side of the street, a young officer appeared. A deep blush covered her cheeks; she took up her work again and bent her head down over the frame. At the same moment the Countess returned completely dressed.

"Order the carriage, Lizaveta," said she; "we will go out for a drive."

Lizaveta arose from the frame and began to arrange her work.

"What is the matter with you, my child, are you deaf?" cried the Countess. "Order the carriage to be got ready at once."

"I will do so this moment," replied the young lady, hastening into the ante-room.

A servant entered and gave the Countess some books from Prince Paul Alexandrovitch.

"Tell him that I am much obliged to him," said the Countess. "Lizaveta! Lizaveta! where are you running to?"

"I am going to dress."

"There is plenty of time, my dear. Sit down here. Open the first volume and read to me aloud."

Her companion took the book and read a few lines.

"Louder," said the Countess. "What is the matter with you, my child? Have you lost your voice? Wait—give me that footstool—a little nearer—that will do!"

Lizaveta read two more pages. The Countess yawned.

"Put the book down," said she: "what a lot of nonsense! Send it back to Prince Paul with my thanks. . . . But where is the carriage?"

"The carriage is ready," said Lizaveta, looking out into the street.

"How is it that you are not dressed?" said the Countess: "I must always wait for you. It is intolerable, my dear!"

Liza hastened to her room. She had not been there two minutes, before the Countess began to ring with all her might. The three waiting-maids came running in at one door and the valet at another.

"How is it that you cannot hear me when I ring for you?" said the Countess. "Tell Lizaveta Ivanovna that I am waiting for her."

Lizaveta returned with her hat and cloak on.

"At last you are here!" said the Countess. "But why such an elaborate toilette? Whom do you intend to captivate? What sort of weather is it? It seems rather windy."

"No, Your Ladyship, it is very calm," replied the valet.

"You never think of what you are talking about. Open the window. So it is: windy and bitterly cold. Unharness the horses. Lizaveta, we won't go out—there was no need for you to deck yourself like that."

"What a life is mine!" thought Lizaveta Ivanovna.

And, in truth, Lizaveta Ivanovna was a very unfortunate creature. "The bread of the stranger is bitter," says Dante, "and his staircase hard to climb." But who can know what the bitterness of dependence is so well as the poor companion of an old lady of quality? The Countess A— had by no means a bad heart, but she was capricious, like a woman who had been spoilt by the world, as well as being avaricious and egotistical, like all old people who have seen their best days, and whose thoughts are with the past and not the present. She participated in all the vanities of the great world, went to balls, where she sat in a corner, painted and dressed in old-fashioned style, like a deformed but indispensable ornament of the ball-room; all the guests on entering approached her and made a profound bow, as if in accordance with a set ceremony, but after that nobody took any further notice of her. She received the whole town at her house, and observed the strictest etiquette, although she could no longer recognize the faces of people. Her numerous domestics, growing fat and old in her ante-chamber and servants' hall, did just as they liked, and vied with each other in robbing the aged Countess in the most bare-faced manner. Lizaveta Ivanovna was the martyr of the household. She made tea, and was reproached with using too much sugar; she read novels aloud to the Countess, and the faults of the author were visited upon her head; she accompanied the Countess in her walks, and was held answerable for the weather or the state of the pavement. A salary was attached to the post, but she very rarely received it although she was expected to dress like everybody else, that is to say, like very few indeed. In society she played the most pitiable rôle. Everybody knew her, and nobody paid her any attention. At balls she danced only when a partner was wanted, and ladies would only take hold of her arm when it was necessary to lead her out of the room to attend to their dresses. She was very self-conscious, and felt her position keenly, and she looked about her with impatience for a deliverer to come to her rescue; but the young men, calculating in their giddiness, honoured her with but very little attention, although Lizaveta Ivanovna was a hundred times prettier than the bare-faced and cold-hearted marriageable girls around whom they hovered. Many a time did she quietly slink away from the glittering but wearisome drawing-room, to go and cry in her own poor little room, in which stood a screen, a chest of drawers, a looking-glass and a painted bedstead, and where a tallow candle burnt feebly in a copper candle-stick.

One morning—this was about two days after the evening party described at the beginning of this story, and a week previous to the scene at which we have just assisted—Lizaveta Ivanovna was seated near the window at her embroidery frame, when, happening to look out into the street, she caught sight of a young Engineer officer, standing motionless with his eyes fixed upon her window. She lowered her head and went on again with her work. About five minutes afterwards she looked out again—the young officer was still standing in the same place. Not being in the habit of coquetting with passing officers, she did not continue to gaze out into the street, but went on sewing for a couple of hours, without raising her head. Dinner was announced. She rose up and began to put her embroidery away, but glancing casually out of the window, she perceived the officer again. This seemed to her very strange. After dinner she went to the window with a certain feeling of uneasiness, but the officer was no longer there—and she thought no more about him.

A couple of days afterwards, just as she was stepping into the carriage with the Countess, she saw him again. He was standing close behind the door, with his face half-concealed by his fur collar, but his dark eyes sparkled beneath his cap. Lizaveta felt alarmed, though she knew not why, and she trembled as she seated herself in the carriage.

On returning home, she hastened to the window—the officer was standing in his accustomed place, with his eyes fixed upon her. She drew back, a prey to curiosity and agitated by a feeling which was quite new to her.

From that time forward not a day passed without the young officer making his appearance under the window at the customary hour, and between him and her there was established a sort of mute acquaintance. Sitting in her place at work, she used to feel his approach; and raising her head, she would look at him longer and longer each day. The young man seemed to be very grateful to her: she saw with the sharp eye of youth, how a sudden flush covered his pale cheeks each time that their glances met. After about a week she commenced to smile at him. . . .

When Tomsky asked permission of his grandmother the Countess to present one of his friends to her, the young girl's heart beat violently. But hearing that Naroumoff was not an Engineer, she regretted that by her thoughtless question, she had betrayed her secret to the volatile Tomsky.

Hermann was the son of a German who had become a naturalized Russian, and from whom he had inherited a small capital. Being firmly convinced of the necessity of preserving his independence, Hermann did not touch his private income, but lived on his pay, without allowing himself the slightest luxury. Moreover, he was reserved and ambitious, and his companions rarely had an opportunity of making merry at the expense of his extreme parsimony. He had strong passions and an ardent imagination, but his firmness of disposition preserved him from the ordinary errors of young men. Thus, though a gamester at heart, he never touched a card, for he considered his position did not allow him—as he said—"to risk the necessary in the hope of winning the superfluous," yet he would sit for nights together at the card table and follow with feverish anxiety the different turns of the game.

The story of the three cards had produced a powerful impression upon his imagination, and all night long he could think of nothing else. "If," he thought to himself the following evening, as he walked along the streets of St. Petersburg, "if the old Countess would but reveal her secret to me! if she would only tell me the names of the three winning cards. Why should I not try my fortune? I must get introduced to her and win her favour—become her lover. . . . But all that will take time, and she is eighty-seven years old: she might be dead in a week, in a couple of days even! . . . But the story itself can it really be true? . . . No! Economy, temperance and industry: those are my three winning cards; by means of them I shall be able to double my capital—increase it sevenfold, and procure for myself ease and independence."

Musing in this manner, he walked on until he found himself in one of the principal streets of St. Petersburg, in front of a house of antiquated architecture. The street was blocked with equipages; carriages one after the other drew up in front of the brilliantly illuminated doorway. At one moment there stepped out on to the pavement the well-shaped little foot of some young beauty, at another the heavy boot of a cavalry officer, and then the silk stockings and shoes of a member of the diplomatic world. Furs and cloaks passed in rapid succession before the gigantic porter at the entrance.

Hermann stopped. "Whose house is this?" he asked of the watchman at the corner.

"The Countess A—'s," replied the watchman.

Hermann started. The strange story of the three cards again presented itself to his imagination. He began walking up and down before the house, thinking of its owner and her strange secret. Returning late to his modest lodging, he could not go to sleep for a long time, and when at last he did doze off, he could dream of nothing but cards, green tables, piles of banknotes and heaps of ducats. He played one card after the other, winning uninterruptedly, and then he gathered up the gold and filled his pockets with the notes. When he woke up late the next morning, he sighed over the loss of his imaginary wealth, and then sallying out into the town, he found himself once more in front of the Countess's residence. Some unknown power seemed to have attracted him thither. He stopped and looked up at the windows. At one of these he saw a head with luxuriant black hair, which was bent down probably over some book or an embroidery frame. The head was raised. Hermann saw a fresh complexion and a pair of dark eyes. That moment decided his fate.

Chapter III.

Lizaveta Ivanovna had scarcely taken off her hat and cloak, when the Countess sent for her and again ordered her to get the carriage ready. The vehicle drew up before the door, and they prepared to take their seats. Just at the moment when two footmen were assisting the old lady to enter the carriage, Lizaveta saw her Engineer standing close beside the wheel; he grasped her hand; alarm caused her to lose her presence of mind, and the young man disappeared—but not before he had left a letter between her fingers. She concealed it in her glove, and during the whole of the drive she neither saw nor heard anything. It was the custom of the Countess, when out for an airing in her carriage, to be constantly asking such questions as: "Who was that person that met us just now? What is the name of this bridge? What is written on that signboard?" On this occasion, however, Lizaveta returned such vague and absurd answers, that the Countess became angry with her.

"What is the matter with you, my dear?" she exclaimed. "Have you taken leave of your senses, or what is it? Do you not hear me or understand what I say? . . . Heaven be thanked, I am still in my right mind and speak plainly enough!"

Lizaveta Ivanovna did not hear her. On returning home she ran to her room, and drew the letter out of her glove: it was not sealed. Lizaveta read it. The letter contained a declaration of love; it was tender, respectful, and copied word for word from a German novel. But Lizaveta did not know anything of the German language, and she was quite delighted.

For all that, the letter caused her to feel exceedingly uneasy. For the first time in her life she was entering into secret and confidential relations with a young man. His boldness alarmed her. She reproached herself for her imprudent behaviour, and knew not what to do. Should she cease to sit at the window and, by assuming an appearance of indifference towards him, put a check upon the young officer's desire for further acquaintance with her? Should she send his letter back to him, or should she answer him in a cold and decided manner? There was nobody to whom she could turn in her perplexity, for she had neither female friend nor adviser. . . . At length she resolved to reply to him.

She sat down at her little writing-table, took pen and paper, and began to think. Several times she began her letter, and then tore it up: the way she had expressed herself seemed to her either too inviting or too cold and decisive. At last she succeeded in writing a few lines with which she felt satisfied.

"I am convinced," she wrote, "that your intentions are honourable, and that you do not wish to offend me by any imprudent behaviour, but our acquaintance must not begin in such a manner. I return you your letter, and I hope that I shall never have any cause to complain of this undeserved slight."

The next day, as soon as Hermann made his appearance, Lizaveta rose from her embroidery, went into the drawing-room, opened the ventilator and threw the letter into the street, trusting that the young officer would have the perception to pick it up.

Hermann hastened forward, picked it up and then repaired to a confectioner's shop. Breaking the seal of the envelope, he found inside it his own letter and Lizaveta's reply. He had expected this, and he returned home, his mind deeply occupied with his intrigue.

Three days afterwards, a bright-eyed young girl from a milliner's establishment brought Lizaveta a letter. Lizaveta opened it with great uneasiness, fearing that it was a demand for money, when suddenly she recognized Hermann's handwriting.

"You have made a mistake, my dear," she said: "this letter is not for me."

"Oh, yes, it is for you," replied the girl, smiling very knowingly. "Have the goodness to read it."

Lizaveta glanced at the letter. Hermann requested an interview.

"It cannot be," she cried, alamed at the audacious request, and the manner in which it was made. "This letter is certainly not for me."

And she tore it into fragments.

"If the letter was not for you, why have you torn it up?" said the girl. "I should have given it back to the person who sent it."

"Be good enough, my dear," said Lizaveta, disconcerted by this remark, "not to bring me any more letters for the future, and tell the person who sent you that he ought to be ashamed. . . ."

But Hermann was not the man to be thus put off. Every day Lizaveta received from him a letter, sent now in this way, now in that. They were no longer translated from the German. Hermann wrote them under the inspiration of passion, and spoke in his own language, and they bore full testimony to the inflexibility of his desire and the disordered condition of his uncontrollable imagination. Lizaveta no longer thought of sending them back to him: she became intoxicated with them and began to reply to them, and little by little her answers became longer and more affectionate. At last she threw out of the window to him the following letter:

"This evening there is going to be a ball at the Embassy. The Countess will be there. We shall remain until two o'clock. You have now an opportunity of seeing me alone. As soon as the Countess is gone, the servants will very probably go out, and there will be nobody left but the Swiss, but he usually goes to sleep in his lodge. Come about half-past eleven. Walk straight upstairs. If you meet anybody in the ante-room, ask if the Countess is at home. You will be told "No," in which case there will be nothing left for you to do but to go away again. But it is most probable that you will meet nobody. The maidservants will all be together in one room. On leaving the ante-room, turn to the left, and walk straight on until you reach the Countess's bedroom. In the bedroom, behind a screen, you will find two doors: the one on the right leads to a cabinet, which the Countess never enters; the one on the left leads to a corridor, at the end of which is a little winding staircase; this leads to my room."

Hermann trembled like a tiger, as he waited for the appointed time to arrive. At ten o'clock in the evening he was already in front of the Countess's house. The weather was terrible; the wind blew with great violence; the sleety snow fell in large flakes; the lamps emitted a feeble light, the streets were deserted; from time to time a sledge, drawn by a sorry-looking hack, passed by, on the look-out for a belated passenger. Hermann was enveloped in a thick overcoat, and felt neither wind nor snow.

At last the Countess's carriage drew up. Hermann saw two footmen carry out in their arms the bent form of the old lady, wrapped in sable fur, and immediately behind her, clad in a warm mantle, and with her head ornamented with a wreath of fresh flowers, followed Lizaveta. The door was closed. The carriage rolled away heavily through the yielding snow. The porter shut the street-door; the windows became dark.

Hermann began walking up and down near the deserted house; at length he stopped under a lamp, and glanced at his watch: it was twenty minutes past eleven. He remained standing under a lamp, his eyes fixed upon the watch, impatiently waiting for the remaining minutes to pass. At half-past eleven precisely, Hermann ascended the steps of the house, and made his way into the brightly-illuminated vestibule. The porter was not there. Hermann hastily ascended the staircase, opened the door of the ante-room and saw a footman sitting asleep in an antique chair by the side of a lamp. With a light firm step Hermann passed by him. The drawing-room and dining-room were in darkness, but a feeble reflection penetrated thither from the lamp in the ante-room.

Hermann reached the Countess's bedroom. Before a shrine, which was full of old images, a golden lamp was burning. Faded stuffed chairs and divans with soft cushions stood in melancholy symmetry around the room, the walls of which were hung with China silk. On one side of the room hung two portraits painted in Paris by Madame Lebrun. One of these represented a stout, red-faced man of about forty years of age in a bright-green uniform and with a star upon his breast; the other—a beautiful young woman, with an aquiline nose, forehead curls and a rose in her powdered hair. In the corners stood porcelain shepherds and shepherdesses, dining-room clocks from the workshop of the celebrated Leroy, bandboxes, roulettes, fans and the various playthings for the amusement of ladies that were in vogue at the end of the last century, when Montgolfier's balloons and Mesmer's magnetism were the rage. Hermann stepped behind the screen. At the back of it stood a little iron bedstead; on the right was the door which led to the cabinet; on the left—the other which led to the corridor. He opened the latter, and saw the little winding staircase which led to the room of the poor companion. . . . But he retraced his steps and entered the dark cabinet.

The time passed slowly. All was still. The clock in the drawing-room struck twelve; the strokes echoed through the room one after the other, and everything was quiet again. Hermann stood leaning against the cold stove. He was calm; his heart beat regularly, like that of a man resolved upon a dangerous but inevitable undertaking. One o'clock in the morning struck; then two; and he heard the distant noise of carriage-wheels. An involuntary agitation took possession of him. The carriage drew near and stopped. He heard the sound of the carriage-steps being let down. All was bustle within the house. The servants were running hither and thither, there was a confusion of voices, and the rooms were lit up. Three antiquated chambermaids entered the bedroom, and they were shortly afterwards followed by the Countess who, more dead than alive, sank into a Voltaire armchair. Hermann peeped through a chink. Lizaveta Ivanovna passed close by him, and he heard her hurried steps as she hastened up the little spiral staircase. For a moment his heart was assailed by something like a pricking of conscience, but the emotion was only transitory, and his heart became petrified as before.

The Countess began to undress before her looking-glass. Her rose-bedecked cap was taken off, and then her powdered wig was removed from off her white and closely-cut hair. Hairpins fell in showers around her. Her yellow satin dress, brocaded with silver, fell down at her swollen feet.

Hermann was a witness of the repugnant mysteries of her toilette; at last the Countess was in her night-cap and dressing-gown and in this costume, more suitable to her age, she appeared less hideous and deformed.

Like all old people in general, the Countess suffered from sleeplessness. Having undressed, she seated herself at the window in a Voltaire armchair and dismissed her maids. The candles were taken away, and once more the room was left with only one lamp burning in it. The Countess sat there looking quite yellow, mumbling with her flaccid lips and swaying to and fro. Her dull eyes expressed complete vacancy of mind, and looking at her, one would have though that the rocking of her body was not a voluntary action of her own, but was produced by the action of some concealed galvanic mechanism.

Suddenly the death-like face assumed an inexplicable expression. The lips ceased to tremble, the eyes became animated: before the Countess stood an unknown man.

"Do not be alarmed, for Heaven's sake, do not be alarmed!" said he in a low but distinct voice. "I have no intention of doing you any harm, I have only come to ask a favour of you."

The old woman looked at him in silence, as if she had not heard what he had said. Hermann thought that she was deaf, and, bending down towards her ear, he repeated what he had said. The aged Countess remained silent as before.

"You can insure the happiness of my life," continued Hermann, "and it will cost you nothing. I know that you can name three cards in order—"

Hermann stopped. The Countess appeared now to understand what he wanted; she seemed as if seeking for words to reply.

"It was a joke." she replied at last: "I assure you it was only a joke."

"There is no joking about the matter," replied Hermann angrily. "Remember Chaplitsky, whom you helped to win."

The Countess became visibly uneasy. Her features expressed strong emotion, but they quickly resumed their former immobility.

"Can you not name me these three winning cards?" continued Hermann.

The Countess remained silent; Hermann continued:

"For whom are you preserving your secret? For your grandsons? They are rich enough without it; they do not know the worth of money. Your cards would be of no use to a spendthrift. He who cannot preserve his paternal inheritance, will die in want, even though he had a demon at his service. I am not a man of that sort; I know the value of money. Your three cards will not be thrown away upon me. Come!" . . .

He paused and tremblingly awaited her reply. The Countess remained silent; Hermann fell upon his knees.

"If your heart has ever known the feeling of love," said he, "if you remember its rapture, if you have ever smiled at the cry of your newborn child, if any human feeling has ever entered into your breast, I entreat you by the feelings of a wife, a lover, a mother, by all that is most sacred in life, not to reject my prayer. Reveal to me your secret. Of what use is it to you? . . . May be it is connected with some terrible sin, with the loss of eternal salvation, with some bargain with the devil. . . . Reflect,—you are old; you have not long to live—I am ready to take your sins upon my soul. Only reveal to me your secret. Remember that the happiness of a man is in your hands, that not only I, but my children, and grandchildren will bless your memory and reverence you as a saint. . . ."

The old Countess answered not a word. Hermann rose to his feet.

"You old hag!" he exclaimed, grinding his teeth, "then I will make you answer!"

With these words he drew a pistol from his pocket.

At the sight of the pistol, the Countess for the second time exhibited strong emotion. She shook her head and raised her hands as if to protect herself from the shot . . . then she fell backwards and remained motionless.

"Come, an end to this childish nonsense!" said Hermann, taking hold of her hand. "I ask you for the last time: will you tell me the names of your three cards, or will you not?"

The Countess made no reply. Hermann perceived that she was dead!

Chapter IV.

Lizaveta Ivanovna was as sitting in her room, still in her ball dress, lost in deep thought. On returning home, she had hastily dismissed the chambermaid who very reluctantly came forward to assist her, saying that she would undress herself, and with a trembling heart had gone up to her own room, expecting to find Hermann there, but yet hoping not to find him. At the first glance she convinced herself that he was not there, and she thanked her fate for having prevented him keeping the appointment. She sat down without undressing, and began to recall to mind all the circumstances which in so short a time had carried her so far. It was not three weeks since the time when she first saw the young officer from the window—and yet she was already in correspondence with him, and he had succeeded in inducing her to grant him a nocturnal interview! She knew his name only through his having written it at the bottom of some of his letters; she had never spoken to him, had never heard his voice, and had never heard him spoken of until that evening. But, strange to say, that very evening at the ball, Tomsky, being piqued with the young Princess Pauline N—, who, contrary to her usual custom, did not flirt with him, wished to revenge himself by assuming an air of indifference: he therefore engaged Lizaveta Ivanovna and danced an endless mazurka with her. During the whole of the time he kept teasing her about her partiality for Engineer officers; he assured her that he knew far more than she imagined, and some of his jests were so happily aimed, that Lizaveta thought several times that her secret was known to him.

"From whom have you learnt all this?" she asked, smiling.

"From a friend of a person very well known to you," replied Tomsky, "from a very distinguished man."

"And who is this distinguished man?"

"His name is Hermann."

Lizaveta made no reply; but her hands and feet lost all sense of feeling.

"This Hermann," continued Tomsky, "is a man of romantic personality. He has the profile of a Napoleon, and the soul of a Mephistopheles. I believe that he has at least three crimes upon his conscience . . . How pale you have become!"

"I have a headache . . . But what did this Hermann—or whatever his name is—tell you?"

"Hermann is very much dissatisfied with his friend: he says that in his place he would act very differently . . . I even think that Hermann himself has designs upon you; at least, he listens very attentively to all that his friend has to say about you."

"And where has he seen me?"

"In church, perhaps; or on the parade—God alone knows where. It may have been in your room, while you were asleep, for there is nothing that he—"

Three ladies approaching him with the question: *"oubli ou regret?"* interrupted the conversation, which had become so tantalizingly interesting to Lizaveta.

The lady chosen by Tomsky was the Princess Pauline herself. She succeeded in effecting a reconciliation with him during the numerous turns of the dance, after which he conducted her to her chair. On returning to his place, Tomsky thought no more either of Hermann or Lizaveta. She longed to renew the interrupted conversation, but the mazurka came to an end, and shortly afterwards the old Countess took her departure.

Tomsky's words were nothing more than the customary small talk of the dance, but they sank deep into the soul of the young dreamer. The portrait, sketched by Tomsky, coincided with the picture she had formed within her own mind, and thanks to the latest romances, the ordinary countenance of her admirer became invested with attributes capable of alarming her and fascinating her imagination at the same time. She was now sitting with her bare arms crossed and with her head, still adorned with flowers, sunk upon her uncovered bosom. Suddenly the door opened and Hermann entered. She shuddered.

"Where were you?" she asked in a terrified whisper.

"In the old Countess's bedroom," replied Hermann: "I have just left her. The Countess is dead."

"My God! What do you say?"

"And I am afraid," added Hermann, "that I am the cause of her death."

Lizaveta looked at him, and Tomsky's words found an echo in her soul: "This man has at least three crimes upon his conscience!" Hermann sat down by the window near her, and related all that had happened.

Lizaveta listened to him in terror. So all those passionate letters, those ardent desires, this bold obstinate pursuit—all this was not love! Money—that was what his soul yearned for! She could not satisfy his desire and make him happy! The poor girl had been nothing but the blind tool of a robber, of the murderer of her aged benefactress! . . . She wept bitter tears of agonized repentance. Hermann gazed at her in silence: his heart, too, was a prey to violent emotion, but neither the tears of the poor girl, nor the wonderful charm of her beauty, enhanced by grief, could produce any impression upon his hardened soul. He felt no pricking of conscience at the thought of the dead old woman. One thing only grieved him: the irreparable loss of the secret from which he had expected to obtain great wealth.

"You are a monster!" said Lizaveta at last.

"I did not wish for her death," replied Hermann: "my pistol was not loaded."

Both remained silent.

The day began to dawn. Lizaveta extinguished her candle: a pale light illumined her room. She wiped her tear-stained eyes and raised them towards Hermann: he was sitting near the window, with his arms crossed and with a fierce frown upon his forehead. In this attitude he bore a striking resemblance to the portrait of Napoleon. This resemblance struck Lizaveta even.

"How shall I get you out of the house?" said she at last. "I thought of conducting you down the secret staircase, but in that case it would be necessary to go through the Countess's bedroom, and I am afraid."

"Tell me how to find this secret staircase—I will go alone."

Lizaveta arose, took from her drawer a key, handed it to Hermann and gave him the necessary instructions. Hermann pressed her cold, powerless hand, kissed her bowed head, and left the room.

He descended the winding staircase, and once more entered the Countess's bedroom. The dead old lady sat as if petrified; her face expressed profound tranquillity. Hermann stopped before her, and gazed long and earnestly at her, as if he wished to convince himself of the terrible reality; at last he entered the cabinet, felt behind the tapestry for the door, and then began to descend the dark staircase, filled with strange emotions. "Down this very staircase," thought he, "perhaps coming from the very same room, and at this very same hour six years ago, there may have glided, in an embroidered coat, with his hair dressed *a l'oiseau royal* and pressing to his heart his three-cornered hat, some young gallant who has long been mouldering in the grave, but the heart of his aged mistress has only to-day ceased to beat. . . ."

At the bottom of the staircase Hermann found a door, which he opened with a key, and then traversed a corridor which conducted him into the street.

Chapter V.

Three days after the fatal night, at nine o'clock in the morning, Hermann repaired to the Convent of —, where the last honours were to be paid to the mortal remains of the old Countess. Although feeling no remorse, he could not altogether stifle the voice of conscience, which said to him: "You are the murderer of the old woman!" In spite of his entertaining very little religious belief, he was exceedingly superstitious; and believing that the dead Countess might exercise an evil influence on his life, he resolved to be present at her obsequies in order to implore her pardon.

The church was full. It was with difficulty that Hermann made his way through the crowd of people. The coffin was placed upon a rich catafalque beneath a velvet baldachin. The deceased Countess lay within it, with her hands crossed upon her breast, with a lace cap upon her head and dressed in a white satin robe. Around the catafalque stood the members of her household: the servants in black *caftans,* with armorial ribbons upon their shoulders, and candles in their hands; the relatives—children, grandchildren, and great-grandchildren—in deep mourning.

Nobody wept; tears would have been *une affectation.* The Countess was so old, that her death could have surprised nobody, and her relatives had long looked upon her as being out of the world. A famous preacher pronounced the funeral sermon. In simple and touching words he described the peaceful passing away of the righteous, who had passed long years in calm preparation for a Christian end. "The angel of death found her," said the orator, "engaged in pious meditation and waiting for the midnight bridegroom."

The service concluded amidst profound silence. The relatives went forward first to take farewell of the corpse. Then followed the numerous guests, who had come to render the last homage to her who for so many years had been a participator in their frivolous amusements. After these followed the members of the Countess's household. The last of these was an old woman of the same age as the deceased. Two young women led her forward by the hand. She had not strength enough to bow down to the ground—she merely shed a few tears and kissed the cold hand of her mistress.

Hermann now resolved to approach the coffin. He knelt down upon the cold stones and remained in that position for some minutes; at last he arose, as pale as the deceased Countess herself; he ascended the steps of the catafalque and bent over the corpse. . . . At that moment it seemed to him that the dead woman darted a mocking look at him and winked with one eye. Hermann started back, took a false step and fell to the ground. Several persons hurried forward and raised him up. At the same moment Lizaveta Ivanovna was borne fainting into the porch of the church. This episode disturbed for some minutes the solemnity of the gloomy ceremony. Among the congregation arose a deep murmur, and a tall thin chamberlain, a near relative of the deceased, whispered in the ear of an Englishman who was standing near him, that the young officer was a natural son of the Countess, to which the Englishman coldly replied: "Oh!"

During the whole of that day, Hermann was strangely excited. Repairing to an out-of-the-way restaurant to dine, he drank a great deal of wine, contrary to his usual custom, in the hope of deadening his inward agitation. But the wine only served to excite his imagination still more. On returning home, he threw himself upon his bed without undressing, and fell into a deep sleep.

When he woke up it was already night, and the moon was shining into the room. He looked at his watch: it was a quarter to three. Sleep had left him; he sat down upon his bed and thought of the funeral of the old Countess.

At that moment somebody in the street looked in at his window, and immediately passed on again. Hermann paid no attention to this incident. A few moments afterwards he heard the door of his ante-room open. Hermann thought that it was his orderly, drunk as usual, returning from some nocturnal expedition, but presently he heard footsteps that were unknown to him: somebody was walking softly over the floor in slippers. The door opened, and a woman dressed in white, entered the room. Hermann mistook her for his old nurse, and wondered what could bring her there at that hour of the night. But the white woman glided rapidly across the room and stood before him—and Hermann recognized the Countess!

"I have come to you against my wish," she said in a firm voice: "but I have been ordered to grant your request. Three, seven, ace, will win for you if played in succession, but only on these conditions: that you do not play more than one card in twenty-four hours, and that you never play again during the rest of your life. I forgive you my death, on condition that you marry my companion, Lizaveta Ivanovna."

With these words she turned round very quietly, walked with a shuffling gait towards the door and disappeared. Hermann heard the street-door open and shut, and again he saw someone look in at him through the window.

For a long time Hermann could not recover himself. He then rose up and entered the next room. His orderly was lying asleep upon the floor, and he had much difficulty in waking him. The orderly was drunk as usual, and no information could be obtained from him. The street-door was locked. Hermann returned to his room, lit his candle, and wrote down all the details of his vision.

Chapter VI.

Two fixed ideas can no more exist together in the moral world than two bodies can occupy one and the same place in the physical world. "Three, seven, ace" soon drove out of Hermann's mind the thought of the dead Countess. "Three, seven, ace" were perpetually running through his head and continually being repeated by his lips. If he saw a young girl, he would say: "How slender she is! quite like the three of hearts." If anybody asked: "What is the time?" he would reply: "Five minutes to seven." Every stout man that he saw reminded him of the ace. "Three, seven, ace" haunted him in his sleep, and assumed all possible shapes. The threes bloomed before him in the forms of magnificent flowers, the sevens were represented by Gothic portals, and the aces became transformed into gigantic spiders. One thought alone occupied his whole mind—to make profitable use of the secret which he had purchased so dearly. He thought of applying for a furlough so as to travel abroad. He wanted to go to Paris and tempt fortune in some of the public gambling-houses that abounded there. Chance spared him all this trouble.

There was in Moscow a society of rich gamesters, presided over by the celebrated Chekalinsky who had passed all his life at the card-table and had amassed millions, accepting bills of exchange for his winnings and paying his losses in ready money. His long experience secured for him the confidence of his companions, and his open house, his famous cook, and his agreeable and fascinating manners gained for him the respect of the public. He came to St. Petersburg. The young men of the capital flocked to his rooms, forgetting balls for cards, and preferring the emotions of faro to the seductions of flirting. Naroumoff conducted Hermann to Chekalinsky's residence.

They passed through a suite of magnificent rooms, filled with attentive domestics. The place was crowded. Generals and Privy Counsellors were playing at whist; young men were lolling carelessly upon the velvet-covered sofas, eating ices and smoking pipes. In the drawing-room, at the head of a long table, around which were assembled about a score of players, sat the master of the house keeping the bank. He was a man of about sixty years of age, of a very dignified appearance; his head was covered with silvery-white hair; his full, florid countenance expressed good-nature, and his eyes twinkled with a perpetual smile. Naroumoff introduced Hermann to him. Chekatinsky shook him by the hand in a friendly manner, requested him not to stand on ceremony, and then went on dealing.

The game occupied some time. On the table lay more than thirty cards. Chekalinsky paused after each throw, in order to give the players time to arrange their cards and note down their losses, listened politely to their requests, and more politely still, put straight the corners of cards that some player's hand had chanced to bend. At last the game was finished. Chekalinsky shuffled the cards and prepared to deal again.

"Will you allow me to take a card?" said Hermann, stretching out his hand from behind a stout gentleman who was punting.

Chekalinsky smiled and bowed silently, as a sign of acquiescence. Naroumoff laughingly congratulated Hermann on his abjuration of that abstention from cards which he had practised for so long a period, and wished him a lucky beginning.

"Stake!" said Hermann, writing some figures with chalk on the back of his card.

"How much?" asked the banker, contracting the muscles of his eyes; excuse me, I cannot see quite clearly."

"Forty-seven thousand roubles," replied Hermann.

At these words every head in the room turned suddenly round, and all eyes were fixed upon Hermann.

"He has taken leave of his senses!" thought Naroumoff.

"Allow me to inform you," said Chekalinsky, with his eternal smile, "that you are playing very high; nobody here has ever staked more than two hundred and seventy-five roubles at once."

"Very well," replied Hermann; "but do you accept my card or not?"

Chekalinsky bowed in token of consent.

"I only wish to observe," said he, "that although I have the greatest confidence in my friends, I can only play against ready money. For my own part, I am quite convinced that your word is sufficient, but for the sake of the order of the game, and to facilitate the reckoning up, I must ask you to put the money on your card."

Hermann drew from his pocket a bank-note and handed it to Chekalinsky, who, after examining it in a cursory manner, placed it on Hermann's card.

He began to deal. On the right a nine turned up, and on the left a three.

"I have won!" said Hermann, showing his card.

A murmur of astonishment arose among the players. Chekalinsky frowned, but the smile quickly returned to his face.

"Do you wish me to settle with you?" he said to Hermann.

"If you please," replied the latter.

Chekalinsky drew from his pocket a number of bank-notes and paid at once. Hermann took up his money and left the table. Naroumoff could not recover from his astonishment. Hermann drank a glass of lemonade and returned home.

The next evening he again repaired to Chekalinsky's. The host was dealing. Hermann walked up to the table; the punters immediately made room for him. Chekalinsky greeted him with a gracious bow.

Hermann waited for the next deal, took a card and placed upon it his forty-seven thousand roubles, together with his winnings of the previous evening.

Chekalinsky began to deal. A knave turned up on the right, a seven on the left.

Hermann showed his seven.

There was a general exclamation. Chekalinsky was evidently ill at ease, but he counted out the ninety-four thousand roubles and handed them over to Hermann, who pocketed them in the coolest manner possible and immediately left the house.

The next evening Hermann appeared again at the table. Everyone was expecting him. The generals and Privy Counsellors left their whist in order to watch such extraordinary play. The young officers quitted their sofas, and even the servants crowded into the room. All pressed round Hermann. The other players left off punting, impatient to see how it would end. Hermann stood at the table and prepared to play alone against the pale, but still smiling Chekalinsky. Each opened a pack of cards. Chekalinsky shuffled. Hermann took a card and covered it with a pile of bank-notes. It was like a duel. Deep silence reigned around.

Chekalinsky began to deal; his hands trembled. On the right a queen turned up, and on the left an ace.

"Ace has won!" cried Hermann, showing his card.

"Your queen has lost," said Chekalinsky, politely.

Hermann started; instead of an ace, there lay before him the queen of spades! He could not believe his eyes, nor could he understand how he had made such a mistake.

At that moment it seemed to him that the queen of spades smiled ironically and winked her eye at him. He was struck by her remarkable resemblance. . . .

"The old Countess!" he exclaimed, seized with terror.

Chekalinsky gathered up his winnings. For some time, Hermann remained perfectly motionless. When at last he left the table, there was a general commotion in the room.

"Splendidly punted!" said the players. Chekalinsky shuffled the cards afresh, and the game went on as usual.

* * *

Hermann went out of his mind, and is now confined in room Number 17 of the Oboukhoff Hospital. He never answers any questions, but he constantly mutters with unusual rapidity: "Three, seven, ace! Three, seven, queen!"

Lizaveta Ivanovna has married a very amiable young man, a son of the former steward of the old Countess. He is in the service of the State somewhere, and is in receipt of a good income. Lizaveta is also supporting a poor relative.

Tomsky has been promoted to the rank of captain, and has become the husband of the Princess Pauline.

Young Goodman Brown

Nathaniel Hawthorne

> *A dreamer may dwell so long among fantasies that the things without him will seem as real as those within.*
>
> NATHANIEL HAWTHORNE

Young Goodman Brown came forth at sunset into the street at Salem village; but put his head back, after crossing the threshold, to exchange a parting kiss with his young wife. And Faith, as the wife was aptly named, thrust her own pretty head into the street, letting the wind play with the pink ribbons of her cap while she called to Goodman Brown.

"Dearest heart," whispered she, softly and rather sadly, when her lips were close to his ear, "prithee put off your journey until sunrise and sleep in your own bed tonight. A lone woman is troubled with such dreams and such thoughts that she's afeard of herself sometimes. Pray tarry with me this night, dear husband, of all nights in the year."

"My love and my Faith," replied Goodman Brown, "of all nights in the year, this one night must I tarry away from thee. My journey, as thou callest it, forth and back again, must needs be done 'twixt now and sunrise. What, my sweet, pretty wife, dost thou doubt me already, and we but three months married?"

"Then God bless you!" said Faith, with the pink ribbons; "and may you find all well when you come back."

"Amen!" cried Goodman Brown. "Say thy prayers, dear Faith, and go to bed at dusk, and no harm will come to thee."

So they parted; and the young man pursued his way until, being about to turn the corner by the meeting-house, he looked back and saw the head of Faith still peeping after him with a melancholy air, in spite of her pink ribbons.

"Poor little Faith!" thought he, for his heart smote him. "What a wretch am I to leave her on such an errand! She talks of dreams too. Methought as she spoke there was trouble in her face, as if a dream had warned her what work is to be done tonight. But no, no; 'twould kill her to think it. Well, she's a blessed angel on earth; and after this one night I'll cling to her skirts and follow her to heaven."

With this excellent resolve for the future, Goodman Brown felt himself justified in making more haste on his present evil purpose. He had taken a dreary road, darkened by all the gloomiest trees of the forest, which barely stood aside to let the narrow path creep through, and closed immediately behind. It was all as lonely as could be; and there is this peculiarity in such a solitude, that the traveller knows not who may be concealed by the innumerable trunks and the thick boughs overhead; so that with lonely footsteps he may yet be passing through an unseen multitude.

"There may be a devilish Indian behind every tree," said Goodman Brown to himself; and he glanced fearfully behind him as he added, "What if the devil himself should be at my very elbow!"

His head being turned back, he passed a crook of the road, and, looking forward again, beheld the figure of a man, in grave and decent attire, seated at the foot of an old tree. He arose at Goodman Brown's approach and walked onward side by side with him.

"You are late, Goodman Brown," said he. "The clock of the Old South was striking as I came through Boston, and that is full fifteen minutes agone."

"Faith kept me back a while," replied the young man, with a tremor in his voice, caused by the sudden appearance of his companion, though not wholly unexpected.

It was now deep dusk in the forest, and deepest in that part of it where these two were journeying. As nearly as could be discerned, the second traveller was about fifty years old, apparently in the same rank of life as Goodman Brown, and bearing a considerable resemblance to him, though perhaps more in expression than features. Still they might have been take for father and son. And yet, though the elder person was as simply clad as the younger, and as simple in manner too, he had an indescribable air of one who knew the world, and who would not have felt abashed at the governor's dinner table or in King William's court, were it possible that his affairs should call him thither. But the only thing about him that could be fixed upon as remarkable was his staff, which bore the likeness of a great black snake, so curiously wrought that it might almost be seen to twist and wriggle itself like a living serpent. This, of course, must have been an ocular deception, assisted by the uncertain light.

"Come, Goodman Brown," cried his fellow-traveller, "this is a dull place for the beginning of a journey. Take my staff, if you are so soon weary.

"Friend," said the other, exchanging his slow pace for a full stop, "having kept covenant by meeting thee here, it is my purpose now to return whence I came. I have scruples touching the matter thou wot'st of."

"Sayest thou so?" replied he of the serpent, smiling apart. "Let us walk on, nevertheless, reasoning as we go; and if I convince thee not thou shalt turn back. We are but a little way in the forest yet."

"Too far! too far!" exclaimed the goodman, unconsciously resuming his walk. "My father never went into the woods on such an errand, nor his father before him. We have been a race of honest men and good

Christians since the days of the martyrs; and shall I be the first of the name of Brown that ever took this path and kept—"

"Such company, thou wouldst say," observed the elder person, interpreting his pause. "Well said, Goodman Brown! I have been as well acquainted with your family as with ever a one among the Puritans; and that's no trifle to say. I helped your grandfather, the constable, when he lashed the Quaker woman so smartly through the streets of Salem; and it was I that brought your father a pitch-pine knot, kindled at my own hearth, to set fire to an Indian village, in King Philip's war. They were my good friends, both; and many a pleasant walk have we had along this path, and returned merrily after midnight. I would fain be friends with you for their sake."

"If it be as thou sayest," replied Goodman Brown, "I marvel they never spoke of these matters; or, verily, I marvel not, seeing that the least rumor of the sort would have driven them from New England. We are a people of prayer, and good works to boot, and abide no such wickedness."

"Wickedness or not," said the traveller with the twisted staff, "I have a very general acquaintance here in New England. The deacons of many a church have drunk the communion wine with me; the selectmen of divers towns make me their chairman; and a majority of the Great and General Court are firm supporters of my interest. The governor and I, too—But these are state secrets."

"Can this be so?" cried Goodman Brown, with a stare of amazement at his undisturbed companion. "Howbeit, I have nothing to do with the governor and council; they have their own ways, and are no rule for a simple husbandman like me. But, were I to go on with thee, how should I meet the eye of that good old man, our minister, at Salem village? Oh, his voice would make me tremble both Sabbath day and lecture day."

Thus far the elder traveller had listened with due gravity; but now burst into a fit of irrepressible mirth, shaking himself so violently that his snake-like staff actually seemed to wriggle in sympathy.

"Ha! ha! ha!" shouted he again and again; then composing himself, "Well, go on, Goodman Brown, go on; but, prithee, don't kill me with laughing."

"Well, then, to end the matter at once," said Goodman Brown, considerably nettled, "there is my wife, Faith. It would break her dear little heart; and I'd rather break my own."

"Nay, if that be the case," answered the other, "e'en go thy ways, Goodman Brown. I would not for twenty old women like the one hobbling before us that Faith should come to any harm."

As he spoke he pointed his staff at a female figure on the path, in whom Goodman Brown recognized a very pious and exemplary dame, who had taught him his catechism in youth, and was still his moral and spiritual adviser, jointly with the minister and Deacon Gookin.

"A marvel, truly, that Goody Cloyse should be so far in the wilderness at nightfall," said he. "But with your leave, friend, I shall take a cut through the woods until we have left this Christian woman behind.

Being a stranger to you, she might ask whom I was consorting with and whither I was going."

"Be it so," said his fellow-traveller. "Betake you to the woods, and let me keep the path."

Accordingly the young man turned aside, but took care to watch his companion, who advanced softly along the road until he had come within a staff s length of the old dame. She, meanwhile, was making the best of her way, with singular speed for so aged a woman, and mumbling some indistinct words—a prayer, doubtless—as she went. The traveller put forth his staff and touched her withered neck with what seemed the serpent's tail.

"The devil!" screamed the pious old lady.

"Then Goody Cloyse knows her old friend?" observed the traveller, confronting her and leaning on his writhing stick.

"Ah, forsooth, and is it your worship indeed?" cried the good dame. "Yea, truly it is, and in the very image of my old gossip, Goodman Brown, the grandfather of the silly fellow that now is. But—would your worship believe it?—my broomstick hath strangely disappeared, stolen, as I suspect, by that unhanged witch, Goody Cory, and that, too, when I was all anointed with the juice of smallage, and cinquefoil, and wolf's bane—"

"Mingled with fine wheat and the fat of a new-born babe," said the shape of old Goodman Brown.

"Ah, your worship knows the recipe," cried the old lady, cackling aloud. "So, as I was saying, being all ready for the meeting, and no horse to ride on, I made up my mind to foot it; for they tell me there is a nice young man to be taken into communion tonight. But now your good worship will lend me your arm, and we shall be there in a twinkling."

"That can hardly be," answered her friend. "I may not spare you my arm, Goody Cloyse; but here is my staff, if you will."

So saying, he threw it down at her feet, where, perhaps, it assumed life, being one of the rods which its owner had formerly lent to the Egyptian magi. Of this fact, however, Goodman Brown could not take cognizance. He had cast up his eyes in astonishment, and, looking down again, beheld neither Goody Cloyse nor the serpentine staff, but his fellow-traveller alone, who waited for him as calmly as if nothing had happened.

"That old woman taught me my catechism," said the young man; and there was a world of meaning in this simple comment.

They continued to walk onward, while the elder traveller exhorted his companion to make good speed and persevere in the path, discoursing so aptly that his arguments seemed rather to spring up in the bosom of his auditor than to be suggested by himself. As they went, he plucked a branch of maple to serve for a walking stick, and began to strip it of the twigs and little boughs, which were wet with evening dew. The moment his fingers touched them they became strangely withered and dried up as with a week's sunshine. Thus the pair proceeded, at a good free pace, until suddenly, in a gloomy hollow of the road, Goodman Brown sat himself down on the stump of a tree and refused to go any farther.

"Friend," said he, stubbornly, "my mind is made up. Not another step will I budge on this errand. What if a wretched old woman do choose to go to the devil when I thought she was going to heaven: is that any reason why I should quit my dear Faith and go after her?"

"You will think better of this by and by," said his acquaintance, composedly. "Sit here and rest yourself a while; and when you feel like moving again, there is my staff to help you along."

Without more words, he threw his companion the maple stick, and was as speedily out of sight as if he had vanished into the deepening gloom. The young man sat a few moments by the roadside, applauding himself greatly, and thinking with how clear a conscience he should meet the minister in his morning walk, nor shrink from the eye of good old Deacon Gookin. And what calm sleep would be his that very night, which was to have been spent so wickedly, but so purely and sweetly now, in the arms of Faith! Amidst these pleasant and praiseworthy meditations, Goodman Brown heard the tramp of horses along the road, and deemed it advisable to conceal himself within the verge of the forest, conscious of the guilty purpose that had brought him thither, though now so happily turned from it.

On came the hoof tramps and the voices of the riders, two grave old voices, conversing soberly as they drew near. These mingled sounds appeared to pass along the road, within a few yards of the young man's hiding-place; but, owing doubtless to the depth of the gloom at that particular spot, neither the travellers nor their steeds were visible. Though their figures brushed the small boughs by the wayside, it could not be seen that they intercepted, even for a moment, the faint gleam from the strip of bright sky athwart which they must have passed. Goodman Brown alternately crouched and stood on tiptoe, pulling aside the branches and thrusting forth his head as far as he durst without discerning so much as a shadow. it vexed him the more, because he could have sworn, were such a thing possible, that he recognized the voices of the minister and Deacon Gookin, jogging along quietly, as they were wont to do, when bound to some ordination or ecclesiastical council. While yet within hearing, one of the riders stopped to pluck a switch.

"Of the two, reverend sir," said the voice like the deacon's, "I had rather miss an ordination dinner than tonight's meeting. They tell me that some of our community are to be here from Falmouth and beyond, and others from Connecticut and Rhode Island, besides several of the Indian powwows, who, after their fashion, know almost as much deviltry as the best of us. Moreover, there is a goodly young woman to be taken into communion."

"Mighty well, Deacon Gookin!" replied the solemn old tones of the minister. "Spur up, or we shall be late. Nothing can be done, you know, until I get on the ground."

The hoofs clattered again; and the voices, talking so strangely in the empty air, passed on through the forest, where no church had ever been gathered or solitary Christian prayed. Whither, then, could these holy men be journeying so deep into the heathen wilderness? Young Goodman Brown caught hold of a tree for support, being ready to sink down on the ground, faint and overburdened with the heavy sickness of his heart. He looked up to the sky, doubting whether there really was a heaven above him. Yet there was the blue arch, and the stars brightening in it.

"With heaven above and Faith below, I will yet stand firm against the devil!" cried Goodman Brown.

While he still gazed upward into the deep arch of the firmament and had lifted his hands to pray, a cloud, though no wind was stirring, hurried across the zenith and hid the brightening stars. The blue sky was still visible, except directly overhead, where this black mass of cloud was sweeping swiftly northward. Aloft in the air, as if from the depths of the cloud, came a confused and doubtful sound of voices. Once the listener fancied that he could distinguish the accents of townspeople of his own, men, and women, both pious and ungodly, many of whom he had met at the communion table, and had seen others rioting at the tavern. The next moment, so indistinct were the sounds, he doubted whether he had heard aught but the murmur of the old forest, whispering without a wind. Then came a stronger swell of those familiar tones, heard daily in the sunshine at Salem village, but never until now from a cloud of night. There was one voice, of a young woman, uttering lamentations, yet with an uncertain sorrow, and entreating for some favor, which, perhaps, it would grieve her to obtain; and all the unseen multitude, both saints and sinners, seemed to encourage her onward.

"Faith!" shouted Goodman Brown, in a voice of agony and desperation; and the eclioes of the forest mocked him, crying, "Faith! Faith!" as if bewildered wretches were seeking her all through the wilderness.

The cry of grief, rage, and terror was yet piercing the night, when the unhappy husband held his breath for a response. There was a scream, drowned immediately in a louder murmur of voices, fading into far-off laughter, as the dark cloud swept away, leaving the clear and silent sky above Goodman Brown. But something fluttered lightly down through the air and caught on the branch of a tree. The young man seized it, and beheld a pink ribbon.

"My Faith is gone!" cried he, after one stupefied moment. "There is no good on earth; and sin is but a name. Come, devil; for to thee is this world given."

And, maddened with despair, so that he laughed loud and long, did Goodman Brown grasp his staff and set forth again, at such a rate that he seemed to fly along the forest path rather than to walk or run. The road grew wilder and drearier and more faintly traced, and vanished at length, leaving him in the heart of the dark wilderness, still rushing onward with the instinct that guides mortal man to evil. The whole forest was peopled with frightful sounds—the creaking of the trees, the howling of wild beasts, and the yell of Indians; while sometimes the wind tolled like a distant church bell, and sometimes gave a broad roar around the traveller, as if all Nature were laughing him to scorn. But he was himself the chief horror of the scene, and shrank not from its other horrors.

"Ha! ha! ha!" roared Goodman Brown when the wind laughed at him. "Let us hear which will laugh loudest. Think not to frighten me with your deviltry. Come witch, come wizard, come Indian powwow, come devil himself, and here comes Goodman Brown. You may as well fear him as he fear you."

In truth, all through the haunted forest there could be nothing more frightful than the figure of Goodman Brown. On he flew among the black pines, brandishing his staff with frenzied gestures, now giving vent to an inspiration of horrid blasphemy, and now shouting forth such laughter as set all the echoes of the forest laughing like demons around him. The fiend in his own shape is less hideous than when he rages in the breast of man. Thus sped the demoniac on his course, until, quivering among the trees, he saw a red light before him, as when the felled trunks and branches of a clearing have been set on fire, and throw up their lurid blaze against the sky, at the hour of midnight. He paused, in a lull of the tempest that had driven him onward, and heard the swell of what seemed a hymn, rolling solemnly from a distance with the weight of many voices. He knew the tune; it was a familiar one in the choir of the village meeting-house. The verse died heavily away, and was lengthened by a chorus, not of human voices, but of all the sounds of the benighted wilderness pealing in awful harmony together. Goodman Brown cried out, and his cry was lost to his own ear by its unison with the cry of the desert.

In the interval of silence he stole forward until the light glared full upon his eyes. At one extremity of an open space, hemmed in by the dark wall of the forest, arose a rock, bearing some rude, natural resemblance either to an altar or a pulpit, and surrounded by four blazing pines, their tops aflame, their stems untouched, like candles at an evening meeting. The mass of foliage that had overgrown the summit of the rock was all on fire, blazing high into the night and fitfully illuminating the whole field. Each pendent twig and leafy festoon was in a blaze. As the red light arose and fell, a numerous congregation alternately shone forth, then disappeared in shadow, and again grew, as it were, out of the darkness, peopling the heart of the solitary woods at once.

"A grave and dark-clad company," quoth Goodman Brown.

In truth they were such. Among them, quivering to and fro between gloom and splendor, appeared faces that would be seen next day at the council board of the province, and others which, Sabbath after Sabbath, looked devoutly heavenward, and benignantly over the crowded pews, from the holiest pulpits in the land. Some affirm that the lady of the governor was there. At least there were high dames well known to her, and wives of honored husbands, and widows, a great multitude, and ancient maidens, all of excellent repute, and fair young girls, who trembled lest their mothers should espy them. Either the sudden gleams of light flashing over the obscure field bedazzled Goodman Brown, or he recognized a score of the Church members of Salem village famous for their especial sanctity. Good old Deacon Gookin had arrived, and waited at the skirts of that venerable saint, his revered pastor. But, irreverently consorting with these grave, reputable, and pious people, these elders of the church, these chaste dames and dewy virgins, there were men of dissolute lives and women of spotted fame, wretches given over to all mean and filthy vice, and suspected even of horrid crimes. It was strange to see that the good shrank not from the wicked, nor were the sinners abashed by the saints. Scattered also among their pale-faced enemies were the Indian priests, or powwows, who had often scared their native forest with more hideous incantations than any known to English witchcraft.

"But where is Faith?" thought Goodman Brown; and, as hope came into his heart, he trembled.

Another verse of the hymn arose, a slow and mournful strain, such as the pious love, but joined to words which expressed all that our nature can conceive of sin, and darkly hinted at far more. Unfathomable to mere mortals is the lore of fiends. Verse after verse was sung; and still the chorus of the desert swelled between like the deepest tone of a mighty organ; and with the final peal of that dreadful anthem there came a sound, as if the roaring wind, the rushing streams, the howling beasts, and every other voice of the unconverted wilderness were mingling and according with the voice of guilty man in homage to the prince of all. The four blazing pines threw up a loftier flame, and obscurely discovered shapes and visages of horror on the smoke wreaths above the impious assembly. At the same moment the fire on the rock shot redly forth and formed a glowing arch above its base, where now appeared a figure. With reverence be it spoken, the figure bore no slight similitude, both in garb and manner, to some grave divine of the New England churches.

"Bring forth the converts!" cried a voice that echoed through the field and rolled into the forest.

At the word, Goodman Brown stepped forth from the shadow of the trees and approached the congregation, with whom he felt a loathful brotherhood by the sympathy of all that was wicked in his heart. He could have well-nigh sworn that the shape of his own dead father beckoned him to advance, looking downward from a smoke wreath, while a woman, with dim features of despair, threw out her hand to warn him back. Was it his mother? But he had no power to retreat one step, nor to resist, even in thought, when the minister and good old Deacon Gookin seized his arms and led him to the blazing rock. Thither came also the slender form of a veiled female, led between Goody Cloyse, that pious teacher of the catechism, and Martha Carrier, who had received the devil's promise to be queen of hell. A rampant hag was she. And there stood the proselytes beneath the canopy of fire.

"Welcome, my children," said the dark figure, "to the communion of your race. Ye have found thus young your nature and your destiny. My children, look behind you!"

They turned; and flashing forth, as it were, in a sheet of flame, the fiend worshippers were seen; the smile of welcome gleamed darkly on every visage.

"There," resumed the sable form, "are all whom ye have reverenced from youth. Ye deemed them holier than yourselves, and shrank from your own sin, contrasting it with their lives of righteousness and prayerful aspirations heavenward. Yet here are they all in my worshipping assembly. This night it shall be granted you to know their secret deeds: how hoary-bearded elders of the church have whispered wanton words to the young maids of their households; how many a woman, eager for widows' weeds, has given her husband a drink at bedtime and let him sleep his last sleep in her bosom; how beardless youths have made haste to inherit their fathers' wealth; and how fair damsels—blush not, sweet ones—have dug little graves in the garden, and bidden me, the sole guest, to an infant's funeral. By the sympathy of your human hearts for sin ye shall scent out all the places—whether in church, bedchamber, street, field, or forest—where crime has been committed, and shall exult to behold the whole earth one stain of guilt, one mighty blood spot. Far more than this. It shall be yours to penetrate, in every bosom, the deep mystery of sin, the fountain of all wicked arts, and which inexhaustibly supplies more evil impulses than human power—than my power at its utmost—can make manifest in deeds. And now, my children, look upon each other."

They did so; and, by the blaze of the hell-kindled torches, the wretched man beheld his Faith, and the wife her husband, trembling before that unhallowed altar.

"Lo, there ye stand, my children," said the figure, in a deep and solemn tone, almost sad with its despairing awfulness, as if his once angelic nature could yet mourn for our miserable race. "Depending upon one another's hearts, ye had still hoped that virtue were not all a dream. Now are ye undeceived. Evil is the nature of mankind. Evil must be your only happiness. Welcome again, my children, to the communion of your race."

"Welcome," repeated the fiend worshippers, in one cry of despair and triumph.

And there they stood, the only pair, as it seemed, who were yet hesitating on the verge of wickedness in this dark world. A basin was hollowed, naturally, in the rock. Did it contain water, reddened by the lurid light? or was it blood? or, perchance, a liquid flame? Herein did the shape of evil dip his hand and prepare to lay the mark of baptism upon their foreheads, that they might be partakers of the mystery of sin, more conscious of the secret guilt of others, both in deed and thought, than they could now be of their own. The husband cast one look at his pale wife, and Faith at him. What polluted wretches would the next glance show them to each other, shuddering alike at what they disclosed and what they saw!

"Faith! Faith!" cried the husband, "look up to heaven, and resist the wicked one."

Whether Faith obeyed he knew not. Hardly had he spoken when he found himself amid calm night and solitude, listening to a roar of the wind which died heavily away through the forest. He staggered against the rock, and felt it chill and damp; while a hanging twig, that had been all on fire, besprinkled his cheek with the coldest dew.

The next morning young Goodman Brown came slowly into the street of Salem village, staring around him like a bewildered man. The good old minister was taking a walk along the graveyard to get an appetite for breakfast and meditate his sermon, and bestowed a blessing, as he passed, on Goodman Brown. He shrank from the venerable saint as if to avoid an anathema. Old Deacon Gookin was at domestic worship, and the holy words of his prayer were heard through the open window. "What God doth the wizard pray to?" quoth Goodman Brown. Goody Cloyse, that excellent old Christian, stood in the early sunshine at her own lattice, catechizing a little girl who had brought her a pint of morning's milk. Goodman Brown snatched away the child as from the grasp of the fiend himself. Turning the corner by the meeting-house, he spied the head of Faith, with the pink ribbons, gazing anxiously forth, and bursting into such joy at the sight of him that she skipped along the street and almost kissed her husband before the whole village. But Goodman Brown looked sternly and sadly into her face, and passed on without a greeting.

Had Goodman Brown fallen asleep in the forest and only dreamed a wild dream of a witch-meeting?

Be it so if you will; but, alas! it was a dream of evil omen for young Goodman Brown. A stern, a sad, a darkly meditative, a distrustful, if not a desperate man did he become from the night of that fearful dream. On the Sabbath day, when the congregation were singing a holy psalm, he could not listen because an anthem of sin rushed loudly upon his ear and drowned all the blessed strain. When the minister spoke from the pulpit with power and fervid eloquence, and, with his hand on the open Bible, of the sacred truths of our religion, and of saint-like lives and triumphant deaths, and of future bliss or misery unutterable, then did Goodman Brown turn pale, dreading lest the roof should thunder down upon the gray blasphemer and his hearers. Often, waking suddenly at midnight, he shrank from the bosom of Faith; and at morning or eventide, when the family knelt down at prayer, he scowled and muttered to himself, and gazed sternly at his wife, and turned away. And when he had lived long, and was borne to his grave a hoary corpse, followed by Faith, an aged woman, and children and grandchildren, a goodly procession, besides neighbors not a few, they carved no hopeful verse upon his tombstone, for his dying hour was gloom.

A Rose for Emily

William Faulkner

I

When Miss Emily Grierson died, our whole town went to her funeral: the men through a sort of respectful affection for a fallen monument, the women mostly out of curiosity to see the inside of her house, which no one save an old manservant—a combined gardener and cook—had seen in at least ten years.

It was a big, squarish frame house that had once been white, decorated with cupolas and spires and scrolled balconies in the heavily lightsome style of the seventies, set on what had once been our most select street. But garages and cotton gins had encroached and obliterated even the august names of that neighborhood; only Miss Emily's house was left, lifting its stubborn and coquettish decay above the cotton wagons and the gasoline pumps—an eyesore among eyesores. And now Miss Emily had gone to join the representatives of those august names where they lay in the cedar-bemused cemetery among the ranked and anonymous graves of Union and Confederate soldiers who fell at the battle of Jefferson.

Alive, Miss Emily had been a tradition, a duty, and a care; a sort of hereditary obligation upon the town, dating from that day in 1894 when Colonel Sartoris, the mayor—he who fathered the edict that no Negro woman should appear on the streets without an apron—remitted her taxes, the dispensation dating from the death of her father on into perpetuity. Not that Miss Emily would have accepted charity. Colonel Sartoris invented an involved tale to the effect that Miss Emily's father had loaned money to the town, which the town, as a matter of business, preferred this way of repaying. Only a man of Colonel Sartoris' generation and thought could have invented it, and only a woman could have believed it.

When the next generation, with its more modern ideas, became mayors and aldermen, this arrangement created some little dissatisfaction. On the first of the year they mailed her a tax notice. February came, and there was no reply. They wrote her a formal letter, asking her to call at the sheriff's office at her convenience. A week later the mayor wrote her himself, offering to call or to send his car for her, and received in reply a note on paper of an archaic shape, in a thin, flowing calligraphy in faded ink, to the effect that she no longer went out at all. The tax notice was also enclosed without comment.

They called a special meeting of the Board of Aldermen. A deputation waited upon her, knocked at the door through which no visitor had passed since she ceased giving china-painting lessons eight or ten years earlier. They were admitted by the old Negro into a dim hall from which a stairway mounted into still more shadow. It smelled of dust and disuse—a close, dank smell. The Negro led them into the parlor. It was furnished in heavy, leather-covered furniture. When the Negro opened the blinds of one window, they could see that the leather was cracked; and when they sat down, a faint dust rose sluggishly about their thighs, spinning with slow motes in the single sun-ray. On a tarnished gilt easel before the fireplace stood a crayon portrait of Miss Emily's father.

They rose when she entered—a small, fat woman in black, with a thin gold chain descending to her waist and vanishing into her belt, leaning on an ebony cane with a tarnished gold head. Her skeleton was small and spare; perhaps that was why what would have been merely plumpness in another was obesity in her. She looked bloated, like a body long submerged in motionless water, and of that pallid hue. Her eyes, lost in the fatty ridges of her face, looked like two small pieces of coal pressed into a lump of dough as they moved from one face to another while the visitors stated their errand.

She did not ask them to sit. She just stood in the door and listened quietly until the spokesman came to a stumbling halt. Then they could hear the invisible watch ticking at the end of the gold chain.

Her voice was dry and cold. "I have no taxes in Jefferson. Colonel Sartoris explained it to me. Perhaps one of you can gain access to the city records and satisfy yourselves."

"But we have. We are the city authorities, Miss Emily. Didn't you get a notice from the sheriff, signed by him?"

"I received a paper, yes," Miss Emily said. "Perhaps he considers himself the sheriff. . . . I have no taxes in Jefferson."

"But there is nothing on the books to show that, you see. We must go by the—"

"See Colonel Sartoris. I have no taxes in Jefferson."

"But, Miss Emily—"

"See Colonel Sartoris." (Colonel Sartoris had been dead almost ten years.) "I have no taxes in Jefferson. Tobe!" The Negro appeared. "Show these gentlemen out."

II

So she vanquished them, horse and foot, just as she had vanquished their fathers thirty years before about the smell. That was two years after her father's death and a short time after her sweetheart—the one we believed would marry her—had deserted her. After her father's death she went out very little; after her sweetheart went away, people hardly saw her at all. A few of the ladies had the temerity to call, but were not received, and the only sign of life about the place was the Negro man—a young man then—going in and out with a market basket.

"Just as if a man—any man—could keep a kitchen properly," the ladies said; so they were not surprised when the smell developed. It was another link between the gross, teeming world and the high and mighty Griersons.

A neighbor, a woman, complained to the mayor, Judge Stevens, eighty years old.

"But what will you have me do about it, madam?" he said.

"Why, send her word to stop it," the woman said. "Isn't there a law?"

"I'm sure that won't be necessary," Judge Stevens said. "It's probably just a snake or a rat that nigger of hers killed in the yard. I'll speak to him about it."

The next day he received two more complaints, one from a man who came in diffident deprecation. "We really must do something about it, Judge. I'd be the last one in the world to bother Miss Emily, but we've got to do something." That night the Board of Aldermen met—three graybeards and one younger man, a member of the rising generation.

"It's simple enough," he said. "Send her word to have her place cleaned up. Give her a certain time to do it in, and if she don't . . ."

"Dammit, sir," Judge Stevens said, "will you accuse a lady to her face of smelling bad?"

So the next night, after midnight, four men crossed Miss Emily's lawn and slunk about the house like burglars, sniffing along the base of the brickwork and at the cellar openings while one of them performed a regular sowing motion with his hand out of a sack slung from his shoulder. They broke open the cellar door and sprinkled lime there, and in all the out-buildings. As they recrossed the lawn, a window that had been dark was lighted and Miss Emily sat in it, the light behind her, and her upright torso motionless as that of an idol. They crept quietly across the lawn and into the shadow of the locusts that lined the street. After a week or two the smell went away.

That was when people had begun to feel really sorry for her. People in our town, remembering how old lady Wyatt, her great-aunt, had gone completely crazy at last, believed that the Griersons held themselves a little too high for what they really were. None of the young men were quite good enough for Miss Emily and such. We had long thought of them as a tableau, Miss Emily a slender figure in white in the background, her father a spraddled silhouette in the foreground, his back to her and clutching a horsewhip, the two of them framed by the back-flung front door. So when she got to be thirty and was still single, we were not pleased exactly, but vindicated; even with insanity in the family she wouldn't have turned down all of her chances if they had really materialized.

When her father died, it got about that the house was all that was left to her; and in a way, people were glad. At last they could pity Miss Emily. Being left alone, and a pauper, she had become humanized. Now she too would know the old thrill and the old despair of a penny more or less.

The day after his death all the ladies prepared to call at the house and offer condolence and aid, as is our custom. Miss Emily met them at the door, dressed as usual and with no trace of grief on her face. She told them that her father was not dead. She did that for three days, with the ministers calling on her, and the doctors, trying to persuade her to let them dispose of the body. Just as they were about to resort to law and force, she broke down, and they buried her father quickly.

We did not say she was crazy then. We believed she had to do that. We remembered all the young men her father had driven away, and we knew that with nothing left, she would have to cling to that which had robbed her, as people will.

III

She was sick for a long time. When we saw her again, her hair was cut short, making her look like a girl, with a vague resemblance to those angels in colored church windows—sort of tragic and serene.

The town had just let the contracts for paving the sidewalks, and in the summer after her father's death they began the work. The construction company came with niggers and mules and machinery, and a foreman named Homer Barron, a Yankee—a big, dark, ready man, with a big voice and eyes lighter than his face. The little boys would follow in groups to hear him cuss the niggers, and the niggers singing in time to the rise and fall of picks. Pretty soon he knew everybody in town. Whenever you heard a lot of laughing anywhere about the square, Homer Barron would be in the center of the group. Presently we began to see him and Miss Emily on Sunday afternoons driving in the yellow-wheeled buggy and the matched team of bays from the livery stable.

At first we were glad that Miss Emily would have an interest, because the ladies all said, "Of course a Grierson would not think seriously of a Northerner, a day laborer." But there were still others, older people, who said that even grief could not cause a real lady to forget *noblesse oblige*—without calling it *noblesse oblige.* They just said, "Poor Emily. Her kinsfolk should come to her." She had some kin in Alabama; but years ago her father had fallen out with them over the estate of old lady Wyatt, the crazy woman, and there was no communication between the two families. They had not even been represented at the funeral.

And as soon as the old people said, "Poor Emily," the whispering began. "Do you suppose it's really so?" they said to one another. "Of course it is. What else could. . ." This behind their hands; rustling of craned silk and satin behind jalousies closed upon the sun of Sunday afternoon as the thin, swift clop-clop-clop of the matched team passed: "Poor Emily."

She carried her head high enough—even when we believed that she was fallen. It was as if she demanded more than ever the recognition of her dignity as the last Grierson; as if it had wanted that touch of earthiness to reaffirm her imperviousness. Like when she bought the rat poison, the arsenic. That was over a year after they had begun to say "Poor Emily," and while the two female cousins were visiting her.

"I want some poison," she said to the druggist. She was over thirty then, still a slight woman, though thinner than usual, with cold, haughty black eyes in a face the flesh of which was strained across the temples and about the eyesockets as you imagine a lighthouse-keeper's face ought to look. "I want some poison," she said.

"Yes, Miss Emily. What kind? For rats and such? I'd recom—"

"I want the best you have. I don't care what kind."

The druggist named several. "They'll kill anything up to an elephant. But what you want is—"

"Arsenic," Miss Emily said. "Is that a good one?"

"Is . . . arsenic? Yes, ma'am. But what you want—"

"I want arsenic."

The druggist looked down at her. She looked back at him, erect, her face like a strained flag. "Why, of course," the druggist said. "If that's what you want. But the law requires you to tell what you are going to use it for."

Miss Emily just stared at him, her head tilted back in order to look him eye for eye, until he looked away and went and got the arsenic and wrapped it up. The Negro delivery boy brought her the package; the druggist didn't come back. When she opened the package at home there was written on the box, under the skull and bones: "For rats."

IV

So the next day we all said, "She will kill herself"; and we said it would be the best thing. When she had first begun to be seen with Homer Barron, we had said, "She will marry him." Then we said, "She will persuade him yet," because Homer himself had remarked—he liked men, and it was known that he drank with the younger men in the Elks' Club—that he was not a marrying man. Later we said, "Poor Emily" behind the jalousies as they passed on Sunday afternoon in the glittering buggy, Miss Emily with her head high and Homer Barron with his hat cocked and a cigar in his teeth, reins and whip in a yellow glove.

Then some of the ladies began to say that it was a disgrace to the town and a bad example to the young people. The men did not want to interfere, but at last the ladies forced the Baptist minister—Miss Emily's people were Episcopal—to call upon her. He would never divulge what happened during that interview, but he refused to go back again. The next Sunday they again drove about the streets, and the following day the minister's wife wrote to Miss Emily's relations in Alabama.

So she had blood-kin under her roof again and we sat back to watch developments. At first nothing happened. Then we were sure that they were to be married. We learned that Miss Emily had been to the jeweler's and ordered a man's toilet set in silver, with the letters H.B. on each piece. Two days later we learned that she had bought a complete outfit of men's clothing, including a nightshirt, and we said, "They are married." We were really glad. We were glad because the two female cousins were even more Grierson than Miss Emily had ever been.

So we were not surprised when Homer Barron—the streets had been finished some time since—was gone. We were a little disappointed that there was not a public blowing-off, but we believed that he had gone on to prepare for Miss Emily's coming, or to give her a chance to get rid of the cousins. (By that time it was a cabal, and we were all Miss Emily's allies to help circumvent the cousins.) Sure enough, after another week they departed. And, as we had expected all along, within three days Homer Barron was back in town. A neighbor saw the Negro man admit him at the kitchen door at dusk one evening.

And that was the last we saw of Homer Barron. And of Miss Emily for some time. The Negro man went in and out with the market basket, but the front door remained closed. Now and then we would see her at a window for a moment, as the men did that night when they sprinkled the lime, but for almost six months she did not appear on the streets. Then we knew that this was to be expected too; as if that quality of her father which had thwarted her woman's life so many times had been too virulent and too furious to die.

When we next saw Miss Emily, she had grown fat and her hair was turning gray. During the next few years it grew grayer and grayer until it attained an even pepper-and-salt iron-gray, when it ceased turning. Up to the day of her death at seventy-four it was still that vigorous iron-gray, like the hair of an active man.

From that time on her front door remained closed, save for a period of six or seven years, when she was about forty, during which she gave lessons in china-painting. She fitted up a studio in one of the downstairs rooms, where the daughters and granddaughters of Colonel Sartoris' contemporaries were sent to her with the same regularity and in the same spirit that they were sent on Sundays with a twenty-five-cent piece for the collection plate. Meanwhile her taxes had been remitted.

The newer generation became the backbone and the spirit of the town, and the painting pupils grew up and fell away and did not send their children to her with boxes of color and tedious brushes and pictures cut from the ladies' magazines. The front door closed upon the last one and remained closed for good. When the town got free postal delivery, Miss Emily alone refused to let them fasten the metal numbers above her door and attach a mailbox to it. She would not listen to them.

Daily, monthly, yearly we watched the Negro grow grayer and more stooped, going in and out with the market basket. Each December we sent her a tax notice , which would be returned by the post office a week later, unclaimed. Now and then we would see her in one of the downstairs windows—she had evidently shut up the top floor of the house—like the carven torso of an idol in a niche, looking or not looking at us, we could never tell which. Thus she passed from generation to generation—dear, inescapable, impervious, tranquil, and perverse.

And so she died. Fell ill in the house filled with dust and shadows, with only a doddering Negro man to wait on her. We did not even know she was sick; we had long since given up trying to get any information from the Negro. He talked to no one, probably not even to her, for his voice had grown harsh and rusty, as if from disuse.

She died in one of the downstairs rooms, in a heavy walnut bed with a curtain, her gray head propped on a pillow yellow and moldy with age and lack of sunlight.

V.

The Negro met the first of the ladies at the front door and let them in, with their hushed, sibilant voices and their quick, curious glances, and then he disappeared. He walked right through the house and out the back and was not seen again.

The two female cousins came at once. They held the funeral on the second day, with the town coming to look at Miss Emily beneath a mass of bought flowers, with the crayon face of her father musing profoundly above the bier and the ladies sibilant and macabre; and the very old men—some in their brushed Confederate uniforms—on the porch and the lawn, talking of Miss Emily as if she had been a contemporary of theirs, believing that they had danced with her and courted her perhaps, confusing time with its mathematical progression, as the old do, to whom all the past is not a diminishing road, but, instead, a huge meadow which no winter ever quite touches, divided from them now by the narrow bottleneck of the most recent decade of years.

Already we knew that there was one room in that region above stairs which no one had seen in forty years, and which would have to be forced. They waited until Miss Emily was decently in the ground before they opened it.

The violence of breaking down the door seemed to fill this room with pervading dust. A thin, acrid pall as of the tomb seemed to lie everywhere upon this room decked and furnished as for a bridal: upon the valance curtains of faded rose color, upon the rose-shaded lights, upon the dressing table, upon the delicate array of crystal and the man's toilet things backed with tarnished silver, silver so tarnished that the monogram was obscured. Among them lay a collar and tie, as if they had just been removed, which, lifted, left upon the surface a pale crescent in the dust. Upon the chair hung the suit, carefully folded; beneath it the two mute shoes and the discarded socks.

The man himself lay in the bed.

For a long while we just stood there, looking down at the profound and fleshless grin. The body had apparently once lain in the attitude of an embrace, but now the long sleep that outlasts love, that conquers even the grimace of love, had cuckolded him. What was left of him, rotted beneath what was left of the nightshirt, had become inextricable from the bed in which he lay; and upon him and upon the pillow beside him lay that even coating of the patient and biding dust.

Then we noticed that in the second pillow was the indentation of a head. One of us lifted something from it, and leaning forward, that faint and invisible dust dry and acrid in the nostrils, we saw a long strand of iron-gray hair.

Rock, Church

Langston Hughes

Elder William Jones was one of them rock-church preachers who know how to make the spirit rise and the soul get right. Sometimes in the pulpit he used to start talking real slow, and you'd think his sermon warn't gonna be nothing; but by the time he got through, the walls of the temple would be almost rent, the doors busted open, and the benches turned over from pure shouting on the part of the brothers and sisters.

He were a great preacher, was Reverend William Jones. But he warn't satisfied—he wanted to be greater than he was. He wanted to be another Billy Graham or Elmer Gantry or a resurrected Daddy Grace. And that's what brought about his downfall—ambition!

Now, Reverend Jones had been for nearly a year the pastor of one of them little colored churches in the back alleys of St. Louis that are open every night in the week for preaching, singing, and praying, where sisters come to shake tambourines, shout, swing gospel songs, and get happy while the Reverend presents the Word.

Elder Jones always opened his part of the services with "In His Hand," his theme song, and he always closed his services with the same. Now, the rhythm of "In His Hand" was such that once it got to swinging, you couldn't help but move your arms or feet or both, and since the Reverend always took up collection at the beginning and ending of his sermons, the dancing movement of the crowd at such times was always toward the collection table—which was exactly where the Elder wanted it to be.

In His hand!
In His hand!
I'm safe and sound
I'll be bound—
Settin' in Jesus' hand!

"Come one! Come all! Come, my Lambs," Elder Jones would shout, "and put it down for Jesus!"

Poor old washer-ladies, big fat cooks, long lean truck drivers and heavyset roustabouts would come up and lay their money down, two times every evening for Elder Jones.

That minister was getting rich right there in that St. Louis alley.

In His hand!
In His hand!
I'll have you know
I'm white as snow—
Settin' in Jesus' hand!

With the piano just a-going, tambourines a-flying, and people shouting right on up to the altar.

"Rock, church, rock!" Elder Jones would cry at such intensely lucrative moments.

But he were too ambitious. He wouldn't let well enough alone. He wanted to be a big shot and panic Harlem, gas Detroit, sew up Chicago, then move on to Hollywood. He warn't satisfied with just St. Louis.

So he got to thinking, "Now, what can I do to get everybody excited, to get everybody talking about my church, to get the streets outside crowded and my name known all over, even unto the far reaches of the nation? Now, what can I do?"

Billy Sunday had a sawdust trail, so he had heard. Reverend Becton had two valets in the pulpit with him as he cast off garment after garment in the heat of preaching, and used up dozens of white handkerchiefs every evening wiping his brow while calling on the Lord to come. Meanwhile, the Angel of Angelus Temple had just kept on getting married and divorced and making the front pages of everybody's newspapers.

"I got to be news, too, in my day and time," mused Elder Jones. "This town's too small for me! I want the world to hear my name!"

Now, as I've said before, Elder Jones was a good preacher—and a good-looking preacher, too. He could cry real loud and moan real deep, and he could move the sisters as no other black preacher on this side of town had ever moved them before. Besides, in his youth, as a sinner, he had done a little light hustling around Memphis and Vicksburg—so he knew just how to appeal to the feminine nature.

Since his recent sojourn in St. Louis, Elder Jones had been looking for a special female Lamb to shelter in his private fold. Out of all the sisters in his church, he had finally chosen Sister Maggie Bradford. Not that Sister Maggie was pretty. No, far from it. But Sister Maggie was well fed, brownskin, good-natured, fat, and *prosperous.* She owned four two-family houses that she rented out, upstairs and down, so she made a good living. Besides, she had sweet and loving ways as well as the interest of her pastor at heart.

Elder Jones confided his personal ambitions to said Sister Bradford one morning when he woke up to find her by his side.

"I want to branch out, Maggie," he said. "I want to be a really big man! Now, what can I do to get the 'tention of the world on me? I mean, in a religious way?"

They thought and they thought. Since it was a Fourth of July morning, and Sister Maggie didn't have to go collect rents, they just lay there and thought.

Finally, Sister Maggie said, "Bill Jones, you know something I ain't never forgot that I seed as a child? There was a preacher down in Mississippi named old man Eubanks who one time got himself dead and buried and then rose from the dead. Now, I ain't never forgot that. Neither has nobody else in that part of the Delta. That's something memorable. Why don't you do something like that?"

"How did he do it, Sister Maggie?"

"He ain't never told nobody how he do it, Brother Bill. He say it were the Grace of God, that's all."

"It might a-been," said Elder Jones. "It might a-been."

He lay there and thought awhile longer. By and by he said, "But, honey, I'm gonna do something better'n that. I'm gonna be nailed on a cross."

"Do, Jesus!" said Sister Maggie Bradford. "Jones, you's a mess!"

Now, the Elder, in order to pull off his intended miracle, had, of necessity, to take somebody else into his confidence, so he picked out Brother Hicks, his chief deacon, one of the main pillars of the church long before Jones came as pastor.

It was too bad, though, that Jones never knew that Brother Hicks (more familiarly known as Bulldog) used to be in love with Sister Bradford. Sister Bradford neglected to tell the new reverend about any of her former sweethearts. So how was Elder Jones to know that some of them still coveted her, and were envious of him in their hearts?

"Hicks," whispered Elder Jones in telling his chief deacon of his plan to die on the cross and then come back to life, "that miracle will make me the greatest minister in the world. No doubt about it! When I get to be world-renowned, Bulldog, and go traveling about the firmament, I'll take you with me as my chief deacon. You shall be my right hand, and Sister Maggie Bradford shall be my left. Amen!"

"I hear you," said Brother Hicks. "I hope it comes true."

But if Elder Jones had looked closely, he would have seen an evil light in his deacon's eyes.

"It will come true," said Elder Jones, "if you keep your mouth shut and follow out my instructions—exactly as I lay 'em down to you. I trust you, so listen! You know and I know that I ain't gonna *really* die. Neither is I *really* gonna be nailed. That's why I wants you to help me. I wants you to have me a great big cross made, higher than the altar—so high I has to have a stepladder to get up to it to be nailed thereon, and you to nail me. The higher the better, so's they won't see the straps—'cause I'm gonna be tied on by straps, you hear. The light'll be rose-colored so they can't see the straps. Now, here you come and do the nailin'—nobody else but you. Put them nails between my fingers and toes, not through 'em—*between*—and don't nail too deep. Leave the heads kinder stickin' out. You get the jibe?"

"I get the jibe, said Brother Bulldog Hicks.

"Then you and me'll stay right on there in the church all night and all day till the next night when the people come back to see me rise. Ever so often, you can let me down to rest a little bit. But as long as I'm on the cross, I play off like I'm dead, particularly when reporters come around. On Monday night, hallelujah! I will rise, and take up collection!"

"Amen!" said Brother Hicks.

Well, you couldn't get a-near the church on the night that Reverend Jones had had it announced by press, by radio, and by word of mouth that he would be crucified *dead*, stay dead, and rise. Negroes came from all over St. Louis, East St. Louis, and mighty nigh everywhere else to be present at the witnessing of the miracle. Lots of 'em didn't believe in Reverend Jones, but lots of 'em *did*. Sometimes false prophets can bamboozle you so you can't tell yonder from whither—and that's the way Jones had the crowd.

The church was packed and jammed. Not a seat to be found, and tears were flowing (from sorrowing sisters' eyes) long before the Elder even approached the cross which, made out of new lumber right straight from the sawmill, loomed up behind the pulpit. In the rose-colored lights, with big paper lilies that Sister Bradford had made decorating its head and foot, the cross looked mighty pretty.

Elder Jones preached a mighty sermon that night, and hot as it was, there was plenty of leaping and jumping and shouting in that crowded church. It looked like the walls would fall. Then when he got through preaching, Elder Jones made a solemn announcement. As he termed it, for a night and a day, his last pronouncement.

"Church! Tonight, as I have told the world, I'm gonna die. I'm gonna be nailed to this cross and let the breath pass from me. But tomorrow, Monday night, August the twenty-first, at twelve p.m., I am coming back to life. Amen! After twenty-four hours on the cross, hallelujah! And all the city of St. Louis can be saved—if they will just come out to see me. Now, before I mounts the steps to the cross, let us sing for the last time "In His Hand"—'cause I tell you, that's where I am! As we sing, let everybody come forward to the collection table and help this church before I go. Give largely!"

The piano tinkled, the tambourines rang, hands clapped. Elder Jones and his children sang:

In His hand!
In His hand!
You'll never stray
Down the Devil's way—
Settin' in Jesus' hand!

Oh, in His hand!
In His hand!
Though I may die
I'll mount on high—
Settin' in Jesus' hand!

"Let us pray." And while every back was bowed in prayer, the Elder went up the stepladder to the cross. Brother Hicks followed with the hammer and nails. Sister Bradford wailed at the top of her voice. Woe filled the amen corner. Emotion rocked the church.

Folks outside was saying all up and down the street, "Lawd, I wish we could have got in. Listen yonder at that noise! I wonder what is going on!"

Elder Jones was about to make himself famous—that's what was going on. And all would have went well had it not been for Brother Hicks—a two-faced rascal. Somehow that night the Devil got into Bulldog Hicks and took full possession.

The truth of the matter is that Hicks got to thinking about Sister Maggie Bradford, and how Reverend Jones had worked up to be her Number-One Man. That made him mad. The old green snake of jealousy began to coil around his heart, right there in the meeting, right there on the steps of the cross. Lord, have mercy! At the very high point of the ceremonies!

Hicks had the hammer in one hand and his other hand was full of nails as he mounted the ladder behind his pastor. He was going up to nail Elder Jones on that sawmill cross.

"While I'm nailin', I might as well nail him right," Hicks thought. "A low-down klinker—comin' here out of Mississippi to take my woman away from me! He'll never know the pleasure of my help in none o' his schemes to out-Divine Father! No, sir!"

Elder Jones had himself all fixed up with a system of straps round his waist, round his shoulder blades, and round his wrists and ankles, hidden under his long black coat. These straps fastened in hooks on the back of the cross, out of sight of the audience, so he could just hang up there all sad and sorrowful-looking, and make out like he was being nailed. Brother Bulldog Hicks was to plant the nails between his fingers and toes. Hallelujah! Rock, church, rock!

Excitement was intense.

All went well until the nailing began. Elder Jones removed his shoes and socks and, in his bare black feet, bade farewell to his weeping congregation. As be leaned back against the cross and allowed Brother Hicks to compose him there, the crowd began to moan. But it was when Hicks placed the first nail between Elder Jones's toes that they became hysterical. Sister Bradford out yelled them all.

Hicks placed that first nail between the big toe and the next toe of the left foot and began to hammer. The foot was well strapped down, so the Elder couldn't move it. The closer the head of the nail got to his toes, the harder Hicks struck it. Finally the hammer collided with Elder Jones's foot, *bam* against his big toe.

"Aw-oh!" he moaned under his breath. "Go easy, man!"

"Have mercy, shouted the brothers and sisters of the church. "Have mercy on our Elder!"

Once more the hammer struck his toe. But the all too human sound of his surprised and agonized "Ouch!" was lost in the tumult of the shouting church.

"Bulldog, I say, go easy" hissed the Elder. "This *ain't* real."

Brother Hicks desisted, a grim smile on his face. Then he turned his attention to the right foot. There he placed another nail between the toes, and began to hammer. Again, as the nail went into the wood, he showed no signs of stopping when the hammer reached the foot. He just kept on landing cruel metallic blows on the Elder's bare toenails until the preacher howled with pain, no longer able to keep back a sudden hair-raising cry. The sweat popped out on his forehead and dripped down on his shirt.

At first the Elder thought, naturally, that it was just a slip of the hammer on the deacon's part. Then he thought the man must have gone crazy—like the rest of the audience. Then it hurt him so bad, he didn't know what he thought—so he just hollered, "Aw-ooo-oo-o!"

It was a good thing the church was full of noise, or they would have heard a strange dialogue.

"My God, Hicks, what are you doing?" the Elder cried, staring wildly at his deacon on the ladder.

"I'm nailin' you to the cross, Jones! And man, I'm *really* nailin'."

"Aw-oow-ow! Don't you know you're hurting me? I told you not to nail so hard!"

But the deacon was unruffled.

"Who'd you say's gonna be your right hand when you get down from here and start your travelings?" Hicks asked.

"You, brother," the sweating Elder cried.

"And who'd you say was gonna be your left hand?"

"Sister Maggie Bradford," moaned Elder Jones from the cross.

"Naw she ain't," said Brother Hicks, whereupon he struck the Reverend's toe a really righteous blow.

"Lord, help me!" cried the tortured minister. The weeping congregation echoed his cry. It was certainly real. The Elder was being crucified!

Brother Bulldog Hicks took two more steps up the ladder, preparing to nail the hands. With his evil face right in front of Elder Jones, he hissed: "I'll teach you nappy-headed jackleg ministers to come to St. Louis and think you-all can walk away with any woman you's a mind to. I'm gonna teach you to leave my women alone. Here—here's a nail!"

Brother Hicks placed a great big spike right in the palm of Elder Jones's left hand. He was just about to drive it in when the frightened Reverend let out a scream that could be heard two blocks away. At the same time he began to struggle to get down. Jones tried to bust the straps, but they was too strong for him.

If he could just get one foot loose to kick Brother Bulldog Hicks!

Hicks lifted the hammer to let go when the Reverend's second yell, this time, was loud enough to be heard in East St. Louis. It burst like a bomb above the shouts of the crowd—and it had its effect. Suddenly the congregation was quiet. Everybody knew that was no way for a dying man to yell.

Sister Bradford realized that something had gone wrong, so she began to chant the song her beloved pastor had told her to chant at the propitious moment after the nailing was done. Now, even though the nailing was not done, Sister Bradford thought she had better sing:

Elder Jones will rise again,
Elder Jones will rise again,
Rise again, rise again!
Elder Jones will rise again,
Yes, my Lawd!

But nobody took up the refrain to help her carry it on. Everybody was too interested in what was happening in front of them, so Sister Bradford's voice just died out.

Meanwhile Brother Hicks lifted the hammer again, but Elder Jones spat right in his face. He not only spat, but suddenly called his deacon a name unworthy of man or beast. Then he let out another frightful yell and, in mortal anguish, called, "Sister Maggie Bradford, lemme down from here! I say, come and get . . . me . . . down . . . *from here!*

Those in the church that had not already stopped moaning and shouting did so at once. You could have heard a pin drop. Folks were petrified.

Brother Hicks stood on the ladder, glaring with satisfaction at Reverend Jones, his hammer still raised. Under his breath the panting Elder dared him to nail another nail, and threatened to kill him stone-dead with a .44 if he did.

"Just lemme get loost from here, and I'll fight you like a natural man," he gasped, twisting and turning like a tree in a storm.

"Come down, then," yelled Hicks, right out loud from the ladder. "Come on down! As sure as water runs, Jones, I'll show you up for what you is—a woman-chasing no-good low-down faker! I'll beat you to a batter with my bare hands!"

"Lawd, have mercy!" cried the church.

Jones almost broke a blood vessel trying to get loose from his cross. "Sister Maggie, come and lemme down," he pleaded, sweat streaming from his face.

But Sister Bradford was covered with confusion. In fact, she was petrified. What could have gone wrong for the Elder to call on her like this in public in the very midst of the thing that was to bring him famous-glory and make them all rich, preaching throughout the land with her at his side? Sister Bradford's head was in a whirl, her heart was in her mouth.

"Elder Jones, you means you really wants to get down?" she asked weakly from her seat in the amen corner.

"Yes," cried the Elder, "can't you hear? I done called on you twenty times to let me down!"

At this point Brother Hicks gave the foot nails one more good hammering. The words that came from the cross were not to be found in the Bible.

In a twinkling, Sister Bradford was at Jones's side. Realizing at last that the Devil must've done got into Hicks (like it used to sometimes in the days when she knowed him), she went to the aid of her battered Elder, grabbed the foot of the ladder, and sent Hicks sprawling across the pulpit.

"You'll never crucify my Elder," she cried, "not for real." Energetically she began to cut the straps away that bound the Reverend. Soon poor Jones slid to the floor, his feet too sore from the hammer's blows to even stand on them without help.

"Just lemme get at Hicks," was all Reverend Jones could gasp. "He knowed I didn't want them nails that close." In the dead silence that took possession of the church, everybody heard him moan, "Lawd, lemme get at Hicks," as he bobbled away on the protecting arm of Sister Maggie.

"Stand back, Bulldog," Sister Maggie said to the deacon, "and let your pastor pass. Soon as he's able, he'll flatten you out like a shadow—but now, I'm in charge. Stand back, I say, and let him pass!"

Hicks stood back. The crowd murmured. The minister made his exit. Thus ended the ambitious career of Elder William Jones. He never did pastor in St. Louis any more. Neither did he fight Hicks. He just snuck away for parts unknown.

No Place to Make Love

Langston Hughes

We had no place to make love. We could kiss in doorways, or hold hands in the movies, or somebody might lend us a car. But most of the time we had no place at all to make love. I tell you all that, mister, so you'll understand how come we're here, me and Mary, and not looking for charity, either.

Poor and young. But old enough to work, old enough for our parents to make us bring home the bacon every week. But never old enough for anybody to worry about our love life, mister. Kids in our kind of families have to worry about that themselves. The old folks are all loved out, so they don't care how we get along—just so we don't get hitched too soon and take their paychecks away.

Me and Mary wasn't hardly making enough to get married. I had my mother to take care of and two kid sisters. Pa died on us last year. Both of us had to quit school in our teens. Mary had a big family, too, house full of sisters and brothers. Being the oldest, she had to bring home every cent she made in the hosiery mill. Even white like me and Mary, it's no fun growing up down South in a poor family, I tell you, mister.

Though poor, our parents were awful religious. They liked revivals, liked to put money in the church and help support the Bible school. If me and Mary went to the movies on Sunday, they'd kick because we didn't go to church, too. Her parents didn't like me worth a damn, nohow. They knew I wanted to marry her—and they thought they'd lose a paying boarder if Mary left.

Me, I wanted to take her away. Sure! Come up North somewhere. But there was Ma on my hands, sufferin' and complainin' all the time. Mister, you know, sometimes I think kids ought to be born without parents.

The old folks knew we didn't have no place even to make love. They didn't care. A small town ain't like a big one where you can find a place. When you're too poor to own even an old Ford, what can you do—with your house full of family all the time?

Well, last summer we went out in the woods. And we found a hill. And we pretended we was in the Garden of Eden—only we could always hear the cars and buses passing in the road below.

Then one night coming home about dusk-dark, with the sun gone down and the first stars coming out, Mary said, "Honey, I guess I ought to tell you. I drunk that stuff again twice last week, but it didn't do no good—and it's almost two months now."

I said, "Listen! I don't care if it don't work. Let's get married. I like kids, don't you?"

So I found out how much a license costs—and the next Sunday when we went out on our hill, we was man and wife. I took her home to sleep on the davenport in the parlor with me—and Ma raised hell. Said we'd starve to death getting married, young as we was—and she'd be glad of it. Said what right did we have getting married anyhow, and she not knowing about it? What business did we have birthing a child with no money? My mother got sick and went to bed. All our folks like to died. Yet and still, Mary's parents told everybody they was glad to have her out of the house, glad to get rid of her, pregnant as she was. But they did everything they could to break us up. For the first time in history her mother got friendly with my mother, and they would spend hours talking about how foolish we was, and how ungrateful for all their raising.

My mother was as nasty to Mary as she could be. So was my sisters. Before we'd get home from work, they'd have their dinner all fixed and et. Mary would have to go cook another dinner for ourselves—and me buying all the food for them, too. Relatives sure can be mean when they want to!

Ma kept saying she was going to tell the authorities how we'd put our ages up to get married without telling our parents, and she'd have the whole thing annulled. But then there was our baby on the road. She knew when it came, I'd have to take care of the kid anyhow—so she didn't do nothing but talk.

"Hardly dry behind the ears yet, and a baby coming," she'd say.

I'd grin, but it made me mad. She didn't mean it in fun. Old folks are fierce. Between the lot of 'em, they made us so tired, Mary and me, that we just got on the evening train one night and come to New York. We both want our kid to be born in a friendly place, that's why, away from relatives.

Of course, I didn't know it was so hard to get a job in New York, or so cold up here. You see, mister, I ain't never been away from home before. With the kid coming and all, I *got* to have something to do, recession or not. Mary ain't able to work, and they won't even let me shovel snow for the city. You got to be a registered voter, the man says.

You're the second welfare investigator what's been here. The first one, the white man, said he couldn't do a thing. We don't come under his jurisdiction. Maybe you can tell us what jurisdiction we do come under? What about young folks like us who want a decent place for our kid to be born in? I'm getting desperate, mister! Mary is, too. We got rent to pay. We don't want our kid to be born out in the cold, maybe growing up like we did—without even a place to make love. I don't want relief, mister, but I do want a job. I know you understand. You niggers have a hard time, too, don't you?

On The Road

Langston Hughes

He was not interested in the snow. When he got off the freight, one early evening during the depression, Sargeant never even noticed the snow. But he must have felt it seeping down his neck, cold, wet, sopping in his shoes. But if you had asked him, he wouldn't have known it was snowing. Sargeant didn't see the snow, not even under the bright lights of the main street, falling white and flaky against the night. He was too hungry, too sleepy, too tired.

The Reverend Mr. Dorset, however, saw the snow when he switched on his porch light, opened the front door of his parsonage, and found standing there before him a big black man with snow on his face, a human piece of night with snow on his face—obviously unemployed.

Said the Reverend Mr. Dorset before Sargeant even realized he'd opened his mouth: "I'm sorry. No! Go right on down this street four blocks and turn to your left, walk up seven and you'll see the Relief Shelter. I'm sorry. No!" He shut the door.

Sargeant wanted to tell the holy man that he had already been to the Relief Shelter, been to hundreds of relief shelters during the depression years, the beds were always gone and supper was over, the place was full, and they drew the color line anyhow. But the minister said, "No," and shut the door. Evidently he didn't want to hear about it. And he *had* a door to shut.

The big black man turned away. And even yet he didn't see the snow, walking right into it. Maybe be sensed it, cold, wet, sticking to his jaws, wet on his black hands, sopping in his shoes. He stopped and stood on the sidewalk hunched over—hungry, sleepy, cold—looking up and down. Then he looked right where he was—in front of a church. Of course! A church! Sure, right next to a parsonage, certainly a church.

It had *two* doors.

Broad white steps in the night all snowy white. Two high arched doors with slender stone pillars on either side. And way up, a round lacy window with a stone crucifix in the middle and Christ on the crucifix in stone. All this was pale in the streetlights, solid and stony pale in the snow.

Sargeant blinked. When he looked up, the snow fell into his eyes. For the first time that night he *saw* the snow. He shook his head. He shook the snow from his coat sleeves, felt hungry, felt lost, felt not lost, felt cold. He walked up the steps of the church. He knocked at the door. No answer. He tried the handle. Locked. He put his shoulder against the door and his long black body slanted like a ramrod. He pushed. With loud rhythmic grunts, like the grunts in a chain-gang song, he pushed against the door.

"I'm tired . . . Huh! . . . Hongry . . . Uh! . . . I'm sleepy . . . Huh! I'm cold . . . I got to sleep somewheres," Sargeant said. "This here is a church, ain't it? Well, uh!"

He pushed against the door.

Suddenly, with an undue cracking and screaking, the door began to give way to the tall black Negro who pushed ferociously against the door.

By now two or three white people had stopped in the street, and Sargeant was vaguely aware of some of them yelling at him concerning the door. Three or four more came running, yelling at him.

"Hey!" they said. "Hey!"

"Un-huh," answered the big tall Negro, "I know it's a white folks' church, but I got to sleep somewhere." He gave another lunge at the door. "Huh!"

And the door broke open.

But just when the door gave way, two white cops arrived in a car, ran up the steps with their clubs, and grabbed Sargeant. But Sargeant for once had no intention of being pulled or pushed away from the door.

Sargeant grabbed, but not for anything so weak as a broken door. He grabbed for one of the tall stone pillars beside the door, grabbed at it and caught it. And held it. The cops pulled and Sargeant pulled. Most of the people in the street got behind the cops and helped them pull.

"A big black unemployed Negro holding on to our church!" thought the people. "The idea!"

The cops began to beat Sargeant over the head, and nobody protested. But he held on.

And then the church fell down.

Gradually, the big stone front of the church fell down, the walls and the rafters, the crucifix and the Christ. Then the whole thing fell down, covering the cops and the people with bricks and stones and debris. The whole church fell down in the snow.

Sargeant got out from under the church and went walking on up the street with the stone pillar on his shoulder. He was under the impression that he had buried the parsonage and the Reverend Mr. Dorset who said, "No!" So he laughed, and threw the pillar six blocks up the street and went on.

Sargeant thought he was alone, but listening to the crunch, crunch, crunch on the snow of his own footsteps, he heard other footsteps, too, doubling his own. He looked around and there was Christ walking along beside him, the same Christ that had been on the cross on the church—still stone with a rough stone surface, walking along beside him just like he was broken off the cross when the church fell down.

"Well, I'll be dogged," said Sargeant. "This here's the first time I ever seed you off the cross."

"Yes," said Christ, crunching his feet in the snow. "You had to pull the church down to get me off the cross."

"You glad?" said Sargeant.

"I sure am," said Christ.

They both laughed.

"I'm a hell of a fellow, ain't I?" said Sargeant. "Done pulled the church down!"

"You did a good job," said Christ. "They have kept me nailed on a cross for nearly two thousand years."

"Whee-ee-e!" said Sargeant. "I know you are glad to get off."

"I sure am," said Christ.

They walked on in the snow. Sargeant looked at the man of stone.

"And you been up there two thousand years?"

"I sure have," Christ said.

"Well, if I had a little cash," said Sargeant, "I'd show you around a bit."

"I been around," said Christ.

"Yeah, but that was a long time ago."

"All the same," said Christ, "I've been around."

They walked on in the snow until they came to the railroad yards. Sargeant was tired, sweating and tired.

"Where you goin'?" Sargeant said, stopping by the tracks. He looked at Christ. Sargeant said, "I'm just a bum on the road. How about you? Where you goin'?"

"God knows," Christ said, "but I'm leavin' here."

They saw the red and green lights of the railroad yard half veiled by the snow that fell out of the night. Away down the track they saw a fire in a hobo jungle.

"I can go there and sleep," Sargeant said.

"You can?"

"Sure," said Sargeant. "That place ain't got no doors."

Outside the town, along the tracks, there were barren trees and bushes below the embankment, snow-gray in the dark. And down among the trees and bushes there were makeshift houses made out of boxes and tin and old pieces of wood and canvas. You couldn't see them in the dark, but you knew they were there if you'd ever been on the road, if you had ever lived with the homeless and hungry in a depression.

"I'm side-tracking," Sargeant said. "I'm tired."

"I'm gonna make it on to Kansas City," said Christ.

"O.K.," Sargeant said. "So long!"

He went down into the hobo jungle and found himself a place to sleep. He never did see Christ no more. About six a.m. a freight came by. Sargeant scrambled out of the jungle with a dozen or so more hoboes and ran along the track, grabbing at the freight. It was dawn, early dawn, cold and gray.

"Wonder where Christ is by now?" Sargeant thought. "He musta gone on way on down the road. He didn't sleep in this jungle."

Sargeant grabbed the train and started to pull himself up into a moving coal car, over the edge of a wheeling coal car. But strangely enough, the car was full of cops. The nearest cop rapped Sargeant soundly across the knuckles with his nightstick. Wham! Rapped his big black hands for clinging to the top of the car. Wham! But Sargeant did not turn loose. He clung on and tried to pull himself into the car. He hollered at the top of his voice, "Damn it, lemme in this car!"

"Shut up," barked the cop. "You crazy coon!" He rapped Sargeant across the knuckles and punched him in the stomach. "You ain't out in no jungle now. This ain't no train. You in jail."

Wham! across his bare black fingers clinging to the bars of his cell. Wham! between the steel bars low down against his shins.

Suddenly Sargeant realized that he really was in jail. He wasn't on no train. The blood of the night before had dried on his face, his head hurt terribly, and a cop outside in the corridor was hitting him across the knuckles for holding on to the door, yelling and shaking the cell door.

"They musta took me to jail for breaking down the door last night," Sargeant thought, "that church door."

Sargeant went over and sat on a wooden bench against the cold stone wall. He was emptier than ever. His clothes were wet, clammy cold wet, and shoes sloppy with snow water. It was just about dawn. There he was, locked up behind a cell door, nursing his bruised fingers.

The bruised fingers were his, but not the *door*.

Not the *club*, but the fingers.

"You wait," mumbled Sargeant, black against the jail wall. "I'm gonna break down this door, too."

"Shut up—or I'll paste you one," said the cop.

"I'm gonna break down this door," yelled Sargeant as he stood up in his cell.

Then he must have been talking to himself because he said, "I wonder where Christ's gone? I wonder if he's gone to Kansas City?"

Sonny's Blues

James Baldwin

I read about it in the paper, in the subway, on my way to work. I read it, and I couldn't believe it, and I read it again. Then perhaps I just stared at it, at the newsprint spelling out his name, spelling out the story. I stared at it in the swinging lights of the subway car, and in the faces and bodies of the people, and in my own face, trapped in the darkness which roared outside.

It was not to be believed and I kept telling myself that, as I walked from the subway station to the high school. And at the same time I couldn't doubt it. I was scared, scared for Sonny. He became real to me again. A great block of ice got settled in my belly and kept melting there slowly all day long, while I taught my classes algebra. It was a special kind of ice. It kept melting, sending trickles of ice water all up and down my veins, but it never got less. Sometimes it hardened and seemed to expand until I felt my guts were going to come spilling out or that I was going to choke or scream. This would always be at a moment when I was remembering some specific thing Sonny had once said or done.

When he was about as old as the boys in my classes his face had been bright and open, there was a lot of copper in it; and he'd had wonderfully direct brown eyes, and great gentleness and privacy. I wondered what he looked like now. He had been picked up, the evening before, in a raid on an apartment downtown, for peddling and using heroin.

I couldn't believe it: but what I mean by that is that I couldn't find any room for it anywhere inside me. I had kept it outside me for a long time. I hadn't wanted to know. I had had suspicions, but I didn't name them, I kept putting them away. I told myself that Sonny was wild, but he wasn't crazy. And he'd always been a good boy, he hadn't ever turned hard or evil or disrespectful, the way kids can, so quick, so quick, especially in Harlem. I didn't want to believe that I'd ever see my brother going down, coming to nothing, all that light in his face gone out, in the condition I'd already seen so many others. Yet it had happened and here I was, talking about algebra to a lot of boys who might, every one of them for all I knew, be popping off needles every time they went to the head. Maybe it did more for them than algebra could.

I was sure that the first time Sonny had ever had horse, he couldn't have been much older than these boys were now. These boys, now, were living as we'd been living then, they were growing up with a rush and their heads bumped abruptly against the low ceiling of their actual possibilities. They were filled with rage. All they really knew were two darknesses, the darkness of their lives, which was now closing in on them, and the darkness of the movies, which had blinded them to that other darkness, and in which they now, vindictively, dreamed, at once more together than they were at any other time, and more alone.

When the last bell rang, the last class ended, I let out my breath. It seemed I'd been holding it for all that time. My clothes were wet—I may have looked as though I'd been sitting in a steam bath, all dressed up, all afternoon. I sat alone in the classroom a long time. I listened to the boys outside, downstairs, shouting and cursing and laughing. Their laughter struck me for perhaps the first time. It was not the joyous laughter which—God knows why—one associates with children. It was mocking and insular, its intent was to denigrate. It was disenchanted, and in this, also, lay the authority of their curses. Perhaps I was listening to them because I was thinking about my brother and in them I heard my brother. And myself.

One boy was whistling a tune, at once very complicated and very simple, it seemed to be pouring out of him as though he were a bird, and it sounded very cool and moving through all that harsh, bright air, only just holding its own through all those other sounds.

I stood up and walked over to the window and looked down into the courtyard. It was the beginning of the spring and the sap was rising in the boys. A teacher passed through them every now and again, quickly, as though he or she couldn't wait to get out of that courtyard, to get those boys out of their sight and off their minds. I started collecting my stuff. I thought I'd better get home and talk to Isabel.

The courtyard was almost deserted by the time I got downstairs. I saw this boy standing in the shadow of a doorway, looking just like Sonny. I almost called his name. Then I saw that it wasn't Sonny, but somebody we used to know, a boy from around our block. He'd been Sonny's friend. He'd never been mine, having been too young for me, and, anyway, I'd never liked him. And now, even though he was a grown-up man, he still hung around that block, still spent hours on the street corners, was always high and raggy. I used to run into him from time to time and he'd often work around to asking me for a quarter or fifty cents. He always had some real good excuse, too, and I always gave it to him, I don't know why.

But now, abruptly, I hated him. I couldn't stand the way he looked at me, partly like a dog, partly like a cunning child. I wanted to ask him what the hell he was doing in the school courtyard.

He sort of shuffled over to me, and he said, "I see you got the papers. So you already know about it."

"You mean about Sonny? Yes, I already know about it. How come they didn't get you?"

He grinned. It made him repulsive and it also brought to mind what he'd looked like as a kid. "I wasn't there. I stay away from them people."

"Good for you." I offered him a cigarette and I watched him through the smoke. "You come all the way down here just to tell me about Sonny?"

"That's right." He was sort of shaking his head and his eyes looked strange, as though they were about to cross. The bright sun deadened his damp dark brown skin and it made his eyes look yellow and showed up the dirt in his kinked hair. He smelled funky. I moved a little away from him and I said, "Well, thanks. But I already know about it and I got to get home."

"I'll walk you a little ways," he said. We started walking. There were a couple of kids still loitering in the courtyard and one of them said goodnight to me and looked strangely at the boy beside me.

"What're you going to do?" he asked me. "I mean, about Sonny?"

"Look. I haven't seen Sonny for over a year, I'm not sure I'm going to do anything. Anyway, what the hell *can* I do?"

"That's right," he said quickly, "ain't nothing you can do. Can't much help old Sonny no more, I guess."

It was what I was thinking and so it seemed to me he had no right to say it.

"I'm surprised at Sonny, though," he went on—he had a funny way of talking, he looked straight ahead as though he were talking to himself—"I thought Sonny was a smart boy, I thought he was too smart to get hung."

"I guess he thought so too," I said sharply, "and that's how he got hung. And how about you? You're pretty goddamn smart, I bet."

Then he looked directly at me, just for a minute. "I ain't smart," he said. "If I was smart, I'd have reached for a pistol a long time ago."

"Look. Don't tell *me* your sad story, if it was up to me, I'd give you one." Then I felt guilty—guilty, probably, for never having supposed that the poor bastard *had* a story of his own, much less a sad one, and I asked, quickly, "What's going to happen to him now?"

He didn't answer this. He was off by himself some place. "Funny thing," he said, and from his tone we might have been discussing the quickest way to get to Brooklyn, "when I saw the papers this morning, the first thing I asked myself was if I had anything to do with it. I felt sort of responsible."

I began to listen more carefully. The subway station was on the corner, just before us, and I stopped. He stopped, too. We were in front of a bar and he ducked slightly, peering in, but whoever he was looking for didn't seem to be there. The juke box was blasting away with something black and bouncy and I half watched the barmaid as she danced her way from the juke box to her place behind the bar. And I watched her face as she laughingly responded to something someone said to her, still keeping time to the music. When she smiled one saw the little girl, one sensed the doomed, still-struggling woman beneath the battered face of the semi-whore.

"I never *give* Sonny nothing," the boy said finally, "but a long time ago I come to school high and Sonny asked me how it felt." He paused, I couldn't bear to watch him, I watched the barmaid, and I listened to the music which seemed to be causing the pavement to shake. "I told him it felt great." The music stopped, the barmaid paused and watched the juke box until the music began again. "It did."

All this was carrying me some place I didn't want to go. I certainly didn't want to know how it felt. It filled everything, the people, the houses, the music, the dark, quicksilver barmaid, with menace; and this menace was their reality.

"What's going to happen to him now?" I asked again.

"They'll send him away some place and they'll try to cure him." He shook his head. "Maybe he'll even think he's kicked the habit. Then they'll let him loose"—he gestured, throwing his cigarette into the gutter. "That's all."

"What do you mean, that's *all?*"

But I knew what he meant.

"I *mean,* that's *all.*" He turned his head and looked at me, pulling down the corners of his mouth. "Don't you know what I mean?" he asked, softly.

"How the hell *would* I know what you mean?" I almost whispered it, I don't know why.

"That's right," he said to the air, "how would *he* know what I mean?" He turned toward me again, patient and calm, and yet I somehow felt him shaking, shaking as though he were going to fall apart. I felt that ice in my guts again, the dread I'd felt all afternoon; and again I watched the barmaid, moving about the bar, washing glasses, and singing. "Listen. They'll let him out and then it'll just start all over again. That's what I mean."

"You mean—they'll let him out. And then he'll just start working his way back in again. You mean he'll never kick the habit. Is that what you mean?"

"That's right," he said, cheerfully. "*You* see what I mean."

"Tell me," I said it last, "why does he want to die? He must want to die, he's killing himself, why does he want to die?"

He looked at me in surprise. He licked his lips. "He don't want to die. He wants to live. Don't nobody want to die, ever."

Then I wanted to ask him—too many things. He could not have answered, or if he had, I could not have borne the answers. I started walking. "Well, I guess it's none of my business."

"It's going to be rough on old Sonny," he said. We reached the subway station. "This is your station?" he asked. I nodded. I took one step down. "Damn!" he said, suddenly. I looked up at him. He grinned again. "Damn it if I didn't leave all my money home. You ain't got a dollar on you, have you? Just for a couple of days, is all."

All at once something inside gave and threatened to come pouring out of me. I didn't hate him any more. I felt that in another moment I'd start crying like a child.

"Sure," I said. "Don't sweat." I looked in my wallet and didn't have a dollar, I only had a five. "Here," I said. "That hold you?"

He didn't look at it—he didn't want to look at it. A terrible, closed look came over his face, as though he were keeping the number on the bill a secret from him and me. "Thanks," he said, and now he was dying to see me go. "Don't worry about Sonny. Maybe I'll write him or something."

"Sure," I said. "You do that. So long."

"Be seeing you," he said. I went on down the steps.

And I didn't write Sonny or send him anything for a long time. When I finally did, it was just after my little girl died, he wrote me back a letter which made me feel like a bastard.

Here's what he said:

> Dear brother,
>
> You don't know how much I needed to hear from you. I wanted to write you many a time but I dug how much I must have hurt you and so I didn't write. But now I feel like a man who's been trying to climb up out of some deep, real deep and funky hole and just saw the sun up there, outside. I got to get outside.
>
> I can't tell you much about how I got here. I mean I don't know how to tell you. I guess I was afraid of something or I was trying to escape from something and you know I have never been very strong in the head (smile). I'm glad Mama and Daddy are dead and can't see what's happened to their son and I swear if I'd known what I was doing I would never have hurt you so, you and a lot of other fine people who were nice to me and who believed in me.
>
> I don't want you to think it had anything to do with me being a musician. It's more than that. Or maybe less than that. I can't get anything straight in my head down here and I try not to think about what's going to happen to me when I get outside again. Sometime I think I'm going to flip and *never* get outside and sometime I think I'll come straight back. I tell you one thing, though, I'd rather blow my brains out than go through this again. But that's what they all say, so they tell me. If I tell you when I'm coming to New York and if you could meet me, I sure would appreciate it. Give my love to Isabel and the kids and I was sure sorry to hear about little Gracie. I wish I could be like Mama and say the Lord's will be done, but I don't know it seems to me that trouble is the one thing that never does get stopped and I don't know what good it does to blame it on the Lord. But maybe it does some good if you believe it.
>
> Your brother,
> Sonny

Then I kept in constant touch with him and I sent him whatever I could and I went to meet him when he came back to New York. When I saw him many things I thought I had forgotten came flooding back to me. This was because I had begun, finally, to wonder about Sonny, about the life that Sonny lived inside. This life, whatever it was, had made him older and thinner and it had deepened the distant stillness in which he had always moved. He looked very unlike my baby brother. Yet, when he smiled, when we shook hands, the baby brother I'd never known looked out from the depths of his private life, like an animal waiting to be coaxed into the light.

"How you been keeping?" he asked me.

"All right. And you?"

"Just fine." He was smiling all over his face. "It's good to see you again."

"It's good to see you."

The seven years' difference in our ages lay between us like a chasm: I wondered if these years would ever operate between us as a bridge. I was remembering, and it made it hard to catch my breath, that I had been there when he was born; and I had heard the first words he had ever spoken. When he started to walk, he walked from our mother straight to me. I caught him just before he fell when he took the first steps he ever took in this world.

"How's Isabel?"

"Just fine. She's dying to see you."

"And the boys?"

"They're fine, too. They're anxious to see their uncle."

"Oh, come on. You know they don't remember me."

"Are you kidding? Of course they remember you."

He grinned again. We got into a taxi. We had a lot to say to each other, far too much to know how to begin.

As the taxi began to move, I asked, "You still want to go to India?"

He laughed. "You still remember that. Hell, no. This place is Indian enough for me."

"It used to belong to them," I said.

And he laughed again. "They damn sure knew what they were doing when they got rid of it."

Years ago, when he was around fourteen, he'd been all hipped on the idea of going to India. He read books about people sitting on rocks, naked, in all kinds of weather, but mostly bad, naturally, and walking barefoot through hot coals and arriving at wisdom. I used to say that it sounded to me as though they were getting away from wisdom as fast as they could. I think he sort of looked down on me for that.

"Do you mind," he asked, "if we have the driver drive alongside the park? On the west side—I haven't seen the city in so long."

"Of course not," I said. I was afraid that I might sound as though I were humoring him, but I hoped he wouldn't take it that way.

So we drove along, between the green of the park and the stony, lifeless elegance of hotels and apartment buildings, toward the vivid, killing streets of our childhood. These streets hadn't changed, though housing projects jutted up out of them now like rocks in the middle of a boiling sea. Most of the houses in which we had grown up had vanished, as had the stores from which we had stolen, the basements in which we had first tried sex, the rooftops from which we had hurled tin cans and bricks. But houses exactly like the houses of our past yet dominated the landscape, boys exactly like the boys we once had been found themselves smothering in these houses, came down into the streets for light and air and found themselves encircled by disaster. Some escaped the trap, most didn't. Those who got out always left something of themselves behind, as some animals amputate a leg and leave it in the trap. It might be said, perhaps, that I had escaped, after all, I was a school teacher; or that Sonny had, he hadn't lived in Harlem for years. Yet, as the cab moved uptown through streets which seemed, with a rush, to darken with dark people, and as I covertly studied Sonny's face, it came to me that what we both were seeking through our separate cab windows was that part of ourselves which had been left behind. It's always at the hour of trouble and confrontation that the missing member aches.

We hit 110th Street and started rolling up Lenox Avenue. And I'd known this avenue all my life, but it seemed to me again, as it had seemed on the day I'd first heard about Sonny's trouble, filled with a hidden menace which was its very breath of life.

"We almost there," said Sonny.

"Almost." We were both too nervous to say anything more.

We lived in a housing project. It hasn't been up long. A few days after it was up it seemed uninhabitably new, now, of course, it's already rundown. It looks like a parody of the good, clean, faceless life—God knows the people who live in it do their best to make it a parody. The beat-looking grass lying around isn't enough to make their lives green, the hedges will never hold out the streets, and they know it. The big windows fool no one, they aren't big enough to make space out of no space. They don't bother with the windows, they watch the TV screen instead. The playground is most popular with the children who don't play at jacks, or skip rope, or roller skate, or swing, and they can be found in it after dark. We moved in partly because it's not too far from where I teach, and partly for the kids; but it's really just like the houses in which Sonny and I grew up. The same things happen, they'll have the same things to remember. The moment Sonny and I started into the house I had the feeling that I was simply bringing him back into the danger he had almost died trying to escape.

Sonny has never been talkative. So I don't know why I was sure he'd be dying to talk to me when supper was over the first night. Everything went fine, the oldest boy remembered him, and the youngest boy liked him, and Sonny had remembered to bring something for each of them; and Isabel, who is really much nicer than I am, more open and giving, had gone to a lot of trouble about dinner and was genuinely glad to see him. And she's always been able to tease Sonny in a way that I haven't. It was nice to see her face so vivid again and to hear her laugh and watch her make Sonny laugh. She wasn't, or, anyway, she didn't seem to be, at all uneasy or embarrassed. She chatted as though there were no subject which had to be avoided and she got Sonny past his first, faint stiffness. And thank God she was there, for I was filled with that icy dread again. Everything I did seemed awkward to me, and everything I said sounded freighted with hidden meaning. I was trying to remember everything I'd heard about dope addiction and I couldn't help watching Sonny for signs. I wasn't doing it out of malice. I was trying to find out something about my brother. I was dying to hear him tell me he was safe.

"Safe!" my father grunted, whenever Mama suggested trying to move to a neighborhood which might be safer for children. "Safe, hell! Ain't no place safe for kids, nor nobody."

He always went on like this, but he wasn't, ever, really as bad as he sounded, not even on weekends, when he got drunk. As a matter of fact, he was always on the lookout for "something a little better," but he died before he found it. He died suddenly, during a drunken weekend in the middle of the war, when Sonny was fifteen. He and Sonny hadn't ever got on too well. And this was partly because Sonny was the apple of his father's eye. It was because he loved Sonny so much and was frightened for him, that he was always fighting with him. It doesn't do any good to fight with Sonny. Sonny just moves back, inside himself, where he can't be reached. But the principal reason that they never hit it off is that they were so much alike. Daddy was big and rough and loud-talking, just the opposite of Sonny, but they both had—that same privacy.

Mama tried to tell me something about this, just after Daddy died. I was home on leave from the army.

This was the last time I ever saw my mother alive. Just the same, this picture gets all mixed up in my mind with pictures I had of her when she was younger. The way I always see her is the way she used to be on a Sunday afternoon, say, when the old folks were talking after the big Sunday dinner. I always see her wearing pale blue. She'd be sitting on the sofa. And my father would be sitting in the easy chair, not far from her. And the living room would be full of church folks and relatives. There they sit, in chairs all around the living room, and the night is creeping up outside, but nobody knows it yet. You can see the darkness growing against the windowpanes and you hear the street noises every now and again, or maybe the jangling beat of a tambourine from one of the churches close by, but it's real quiet in the room. For a moment nobody's talking, but every face looks darkening, like the sky outside. And my mother rocks a little from the waist, and my father's eyes are closed. Everyone is looking at something a child can't see. For a minute they've forgotten the children. Maybe a kid is lying on the rug, half asleep. Maybe somebody's got a kid in his lap and is absent-mindedly stroking the kid's head. Maybe there's a kid, quiet and big-eyed, curled up in a big chair in the corner. The silence, the darkness coming, and the darkness in the faces frightens the child obscurely. He hopes that the hand which strokes his forehead will never stop—will never die. He hopes that there will never come a time when the old folks won't be sitting around the living room, talking about where they've come from, and what they've seen, and what's happened to them and their kinfolk.

But something deep and watchful in the child knows that this is bound to end, is already ending. In a moment someone will get up and turn on the light. Then the old folks will remember the children and they won't talk any more that day. And when the light fills the room, the child is filled with darkness. He knows that every time this happens he's moved just a little closer to that darkness outside. The darkness outside is what the old folks have been talking about. It's what they've come from. It's what they endure. The child knows that they won't talk any more because if he knows too much about what's happened to *them,* he'll know too much too soon, about what's going to happen to *him.*

The last time I talked to my mother, I remember I was restless. I wanted to get out and see Isabel. We weren't married then and we had a lot to straighten out between us.

There Mama sat, in black, by the window. She was humming an old church song, *Lord, you brought me from a long ways off.* Sonny was out somewhere. Mama kept watching the streets.

"I don't know," she said, "if I'll ever see you again, after you go off from here. But I hope you'll remember the things I tried to teach you."

"Don't talk like that," I said, and smiled. "You'll be here a long time yet."

She smiled, too, but she said nothing. She was quiet for a long time. And I said, "Mama, don't you worry about nothing. I'll be writing all the time, and you be getting the checks. . . ."

"I want to talk to you about your brother," she said, suddenly. "If anything happens to me he ain't going to have nobody to look out for him.

"Mama," I said, "ain't nothing going to happen to you *or* Sonny. Sonny's all right. He's a good boy and he's got good sense."

"It ain't a question of his being a good boy," Mama said, "nor of his having good sense. It ain't only the bad ones, nor yet the dumb ones that gets sucked under." She stopped, looking at me. "Your Daddy once had a brother," she said, and she smiled in a way that made me feel she was in pain. "You didn't never know that, did you?"

"No," I said, "I never knew that," and I watched her face.

"Oh, yes," she said, "your Daddy had a brother." She looked out of the window again. "I know you never saw your Daddy cry. But *I* did—many a time, through all these years."

I asked her, "What happened to his brother? How come nobody's ever talked about him?"

This was the first time I ever saw my mother look old.

"His brother got killed," she said, "when he was just a little younger than you are now. I knew him. He was a fine boy. He was maybe a little full of the devil, but he didn't mean nobody no harm."

Then she stopped and the room was silent, exactly as it had sometimes been on those Sunday afternoons. Mama kept looking out into the streets.

"He used to have a job in the mill," she said, "and, like all young folks, he just liked to perform on Saturday nights. Saturday nights, him and your father would drift around to different places, go to dances and things like that, or just sit around with people they knew, and your father's brother would sing, he had a fine voice, and play along with himself on his guitar. Well, this particular Saturday night, him and your father was coming home from some place, and they were both a little drunk and there was a moon that night, it was bright like day. Your father's brother was feeling kind of good, and he was whistling to himself, and he had his guitar slung over his shoulder. They was coming down a hill and beneath them was a road that turned off from the highway. Well, your father's brother, being always kind of frisky, decided to run down this hill, and he did, with that guitar banging and clanging behind him, and he ran across the road, and he was making water behind a tree. And your father was sort of amused at him and he was still coming down the hill, kind of slow. Then he heard a car motor and that same minute his brother stepped from behind the tree, into the road, in the moonlight. And he started to cross the road. And your father started to run down the hill, he says he don't know why. This car was full of white men. They was all drunk, and when they seen your father's brother they let out a great whoop and holler and they aimed the car straight at him. They was having fun, they just wanted to scare him, the way they do sometimes, you know. But they was drunk. And I guess the boy, being drunk, too, and scared, kind of lost his head. By the time he jumped it was too late. Your father says he heard his brother scream when the car rolled over him, and he heard the wood of that guitar when it give, and he heard them strings go flying, and he heard them white men shouting, and the car kept on a-going and it ain't stopped till this day. And, time your father got down the hill, his brother weren't nothing but blood and pulp."

Tears were gleaming on my mother's face. There wasn't anything I could say.

"He never mentioned it," she said, "because I never let him mention it before you children. Your Daddy was like a crazy man that night and for many a night thereafter. He says he never in his life seen anything as dark as that road after the lights of that car had gone away. Weren't nothing, weren't nobody on that road, just your Daddy and his brother and that busted guitar. Oh, yes. Your Daddy never did really get right again. Till the day he died he weren't sure but that every white man he saw was the man that killed his brother."

She stopped and took out her handkerchief and dried her eyes and looked at me.

"I ain't telling you all this," she said, "to make you scared or bitter or to make you hate nobody. I'm telling you this because you got a brother. And the world ain't changed."

I guess I didn't want to believe this. I guess she saw this in my face. She turned away from me, toward the window again, searching those streets.

"But I praise my Redeemer," she said at last, "that He called your Daddy home before me. I ain't saying it to throw no flowers at myself, but, I declare, it keeps me from feeling too cast down to know I helped your father get safely through this world. Your father always acted like he was the roughest, strongest man on earth. And everybody took him to be like that. But if he hadn't had *me* there—to see his tears!"

She was crying again. Still, I couldn't move. I said, "Lord, Lord, Mama, I didn't know it was like that."

"Oh, honey," she said, "there's a lot that you don't know. But you are going to find it out." She stood up from the window and came over to me. "You got to hold on to your brother," she said, "and don't let him fall, no matter what it looks like is happening to him and no matter how evil you gets with him. You going to be evil with him many a time. But don't you forget what I told you, you hear?"

"I won't forget," I said. "Don't you worry, I won't forget. I won't let nothing happen to Sonny."

My mother smiled as though she were amused at something she saw in my face. Then, "You may not be able to stop nothing from happening. But you got to let him know you's *there.*"

Two days later I was married, and then I was gone. And I had a lot of things on my mind and I pretty well forgot my promise to Mama until I got shipped home on a special furlough for her funeral.

And, after the funeral, with just Sonny and me alone in the empty kitchen, I tried to find out something about him.

"What do you want to do?" I asked him.

"I'm going to be a musician," he said.

For he had graduated, in the time I had been away, from dancing to the juke box to finding out who was playing what, and what they were doing with it, and he had bought himself a set of drums.

"You mean, you want to be a drummer?" I somehow had the feeling that being a drummer might be all right for other people but not for my brother Sonny.

"I don't think," he said, looking at me very gravely, "that I'll ever be a good drummer. But I think I can play a piano."

I frowned. I'd never played the role of the older brother quite so seriously before, had scarcely ever, in fact, *asked* Sonny a damn thing. I sensed myself in the presence of something I didn't really know how to handle, didn't understand. So I made my frown a little deeper as I asked: "What kind of musician do you want to be?"

He grinned, "How many kinds do you think there are?"

"Be *serious,*" I said.

He laughed, throwing his head back, and then looked at me. "I *am* serious."

"Well, then, for Christ's sake, stop kidding around and answer a serious question. I mean, do you want to be a concert pianist, you want to play classical music and all that, or—or what?" Long before I finished he was laughing again. "For Christ's *sake,* Sonny!"

He sobered, but with difficulty. "I'm sorry, But you sound so—*scared!*" and he was off again.

"Well, you may think it's funny now, baby, but it's not going to be so funny when you have to make your living at it, let me tell you *that.*" I was furious because I knew he was laughing at me and I didn't know why.

"No," he said, very sober now, and afraid, perhaps, that he'd hurt me, "I don't want to be a classical pianist. That isn't what interests me. I mean"—he paused, looking hard at me, as though his eyes would help me to understand, and then gestured helplessly, as though perhaps his hand would help—"I mean, I'll have a lot of studying to do, and I'll have to study *everything,* but, I mean, I want to play *with*—jazz musicians." He stopped. "I want to play jazz," he said.

Well, the word had never before sounded as heavy, as real, as it sounded that afternoon in Sonny's mouth. I just looked at him and I was probably frowning a real frown by this time. I simply couldn't see why on earth he'd want to spend his time hanging around nightclubs, clowning around on bandstands, while people pushed each other around a dance floor. It seemed—beneath him, somehow. I had never thought about it before, had never been forced to, but I suppose I had always put jazz musicians in a class with what Daddy called "good-time people."

"Are you *serious?*"

"Hell, *yes,* I'm serious."

He looked more helpless than ever, and annoyed, and deeply hurt. I suggested, helpfully: "You mean—like Louis Armstrong?"

His face closed as though I'd struck him. "No. I'm not talking about none of that old-time, down home crap."

"Well, look, Sonny, I'm sorry, don't get mad. I just don't altogether get it, that's all. Name somebody—you know, a jazz musician you admire."

"Bird."

"Who?"

"Bird! Charlie Parker! Don't they teach you nothing in the goddamn army?"

I lit a cigarette. I was surprised and then a little amused to discover that I was trembling. "I've been out of touch," I said. "You'll have to be patient with me. Now. Who's this Parker character?"

"He's just one of the greatest jazz musicians alive," said Sonny, sullenly, his hands in his pockets, his back to me. "Maybe *the* greatest," he added, bitterly, "that's probably why *you* never heard of him."

"All right," I said, "I'm ignorant. I'm sorry. I'll go out and buy all the cat's records right away, all right?"

"It don't," said Sonny, with dignity, "make any difference to me. I don't care what you listen to. Don't do me no favors."

I was beginning to realize that I'd never seen him so upset before. With another part of my mind I was thinking that this would probably turn out to be one of those things kids go through and that I shouldn't make it seem important by pushing it too hard. Still, I didn't think it would do any harm to ask: "Doesn't all this take a lot of time? Can you make a living at it?"

He turned back to me and half leaned, half sat, on the kitchen table. "Everything takes time," he said, "and—well, yes, sure, I can make a living at it. But what I don't seem to be able to make you understand is that it's the only thing I want to do."

"Well, Sonny," I said, gently, "you know people can't always do exactly what they *want* to do—"

"*No,* I don't know that," said Sonny, surprising me. "I think people *ought* to do what they want to do, what else are they alive for?"

"You getting to be a big boy," I said desperately, "it's time you started thinking about your future."

"I'm thinking about my future," said Sonny, grimly. "I think about it all the time."

I gave up. I decided, if he didn't change his mind, that we could always talk about it later. "In the meantime," I said, "you got to finish school." We had already decided that he'd have to move in with Isabel and her folks. I knew this wasn't the ideal arrangement because Isabel's folks are inclined to be dicty and they hadn't especially wanted Isabel to marry me. But I didn't know what else to do. "And we have to get you fixed up at Isabel's."

There was a long silence. He moved from the kitchen table to the window. "That's a terrible idea. You know it yourself."

"Do you have a *better* idea?"

He just walked up and down the kitchen for a minute. He was as tall as I was. He had started to shave. I suddenly had the feeling that I didn't know him at all.

He stopped at the kitchen table and picked up my cigarettes. Looking at me with a kind of mocking, amused defiance, he put one between his lips. "You mind?"

"You smoking already?"

He lit the cigarette and nodded, watching me through the smoke. "I just wanted to see if I'd have the courage to smoke in front of you." He grinned and blew a great cloud of smoke to the ceiling. "It was easy." He looked at my face. "Come on, now. I bet you was smoking at my age, tell the truth."

I didn't say anything but the truth was on my face, and he laughed. But now there was something very strained in his laugh. "Sure. And I bet that ain't all you was doing."

He was frightening me a little. "Cut the crap," I said. "We already decided that you was going to go and live at Isabel's. Now what's got into you all of a sudden?"

"*You* decided it," he pointed out. "*I* didn't decide nothing." He stopped in front of me, leaning against the stove, arms loosely folded. "Look, brother. I don't want to stay in Harlem no more, I really don't." He was very earnest. He looked at me, then over toward the kitchen window. There was something in his eyes I'd never seen before, some thoughtfulness, some worry all his own. He rubbed the muscle of one arm. "It's time I was getting out of here."

"Where do you want to *go,* Sonny?"

"I want to join the army. Or the navy, I don't care. If I say I'm old enough, they'll believe me."

Then I got mad. It was because I was so scared. "You must be crazy. You goddamn fool, what the hell do you want to go and join the *army* for?"

"I just told you. To get out of Harlem."

"Sonny, you haven't even finished *school.* And if you really want to be a musician, how do you expect to study if you're in the *army?*"

He looked at me, trapped, and in anguish. "There's ways. I might be able to work out some kind of deal. Anyway, I'll have the G.I. Bill when I come out."

"*If* you come out." We stared at each other. "Sonny, please. Be reasonable. I know the setup is far from perfect. But we got to do the best we can."

"I ain't learning nothing in school," he said. "Even when I go." He turned away from me and opened the window and threw his cigarette out into the narrow alley. I watched his back. "At least, I ain't learning nothing you'd want me to learn." He slammed the window so hard I thought the glass would fly out, and turned back to me. "And I'm sick of the stink of these garbage cans!"

"Sonny," I said, "I know how you feel. But if you don't finish school now, you're going to be sorry later that you didn't." I grabbed him by the shoulders. "And you only got another year. It ain't so bad. And I'll come back and I swear I'll help you do *whatever* you want to do. Just try to put up with it till I come back. Will you please do that? For me?"

He didn't answer and he wouldn't look at me.

"Sonny. You hear me?"

He pulled away. "I hear you. But you never hear anything *I* say."

I didn't know what to say to that. He looked out of the window and then back at me. "OK," he said, and sighed. "I'll try."

Then I said, trying to cheer him up a little, "They got a piano at Isabel's. You can practice on it."

And as a matter of fact, it did cheer him up for a minute. "That's right," he said to himself. "I forgot that." His face relaxed a little. But the worry, the thoughtfulness, played on it still, the way shadows play on a face which is staring into the fire.

But I thought I'd never hear the end of that piano. At first, Isabel would write me, saying how nice it was that Sonny was so serious about his music and how, as soon as he came in from school, or wherever he had been when he was supposed to be at school, he went straight to that piano and stayed there until suppertime. And, after supper, he went back to that piano and stayed there until everybody went to bed. He was at the piano all day Saturday and all day Sunday. Then he bought a record player and started playing records. He'd play one record over and over again, all day long sometimes, and he'd improvise along with it on the piano. Or he'd play one section of the record, one chord, one change, one progression, then he'd do it on the piano. Then back to the record. Then back to the piano.

Well, I really don't know how they stood it. Isabel finally confessed that it wasn't like living with a person at all, it was like living with sound. And the sound didn't make any sense to her, didn't make any sense to any of them—naturally. They began, in a way, to be afflicted by this presence that was living in their home. It was as though Sonny were some sort of god, or monster. He moved in an atmosphere which wasn't like theirs at all. They fed him and he ate, he washed himself, he walked in and out of their door; he certainly wasn't nasty or unpleasant or rude, Sonny isn't any of those things; but it was as though he were all wrapped up in some cloud, some fire, some vision all his own; and there wasn't any way to reach him.

At the same time, he wasn't really a man yet, he was still a child, and they had to watch out for him in all kinds of ways. They certainly couldn't throw him out. Neither did they dare to make a great scene about that piano because even they dimly sensed, as I sensed, from so many thousands of miles away, that Sonny was at that piano playing for his life.

But he hadn't been going to school. One day a letter came from the school board and Isabel's mother got it—there had, apparently, been other letters but Sonny had torn them up. This day, when Sonny came in, Isabel's mother showed him the letter and asked where he'd been spending his time. And she finally got it out of him that he'd been down in Greenwich Village, with musicians and other characters, in a white girl's apartment. And this scared her and she started to scream at him and what came up, once she began—though she denies it to this day—was what sacrifices they were making to give Sonny a decent home and how little he appreciated it.

Sonny didn't play the piano that day. By evening, Isabel's mother had calmed down but then there was the old man to deal with, and Isabel herself. Isabel says she did her best to be calm but she broke down and started crying. She says she just watched Sonny's face. She could tell, by watching him, what was happening with him. And what was happening was that they penetrated his cloud, they had reached him. Even if their fingers had been a thousand times more gentle than human fingers ever are, he could hardly help feeling that they had stripped him naked and were spitting on that nakedness. For he also had to see that his presence, that music, which was life or death to him, had been torture for them and that they had endured it, not at all for his sake, but only for mine. And Sonny couldn't take that. He can take it a little better today than he could then but he's still not very good at it and, frankly, I don't know anybody who is.

The silence of the next few days must have been louder than the sound of all the music ever played since time began. One morning, before she went to work, Isabel was in his room for something and she suddenly realized that all of his records were gone. And she knew for certain that he was gone. And he was. He went as far as the navy would carry him. He finally sent me a postcard from some place in Greece and that was the first I knew that Sonny was still alive. I didn't see him any more until we were both back in New York and the war had long been over.

He was a man by then, of course, but I wasn't willing to see it. He came by the house from time to time, but we fought almost every time we met. I didn't like the way he carried himself, loose and dreamlike all the time, and I didn't like his friends, and his music seemed to be merely an excuse for the life he led. It sounded just that weird and disordered.

Then we had a fight, a pretty awful fight, and I didn't see him for months. By and by I looked him up, where he was living, in a furnished room in the Village, and I tried to make it up. But there were lots of other people in the room and Sonny just lay on his bed, and he wouldn't come downstairs with me, and he treated these other people as though they were his family and I weren't. So I got mad and then he got mad, and then I told him that he might just as well be dead as live the way he was living.

Then he stood up and he told me not to worry about him any more in life, that he *was* dead as far as I was concerned. Then he pushed me to the door and the other people looked on as though nothing were happening, and he slammed the door behind me. I stood in the hallway, staring at the door. I heard somebody laugh in the room and then the tears came to my eyes. I started down the steps, whistling to keep from crying. I kept whistling to myself, *You going to need me, baby, one of these cold, rainy days.*

I read about Sonny's trouble in the spring. Little Grace died in the fall. She was a beautiful little girl. But she only lived a little over two years. She died of polio and she suffered. She had a slight fever for a couple of days, but it didn't seem like anything and we just kept her in bed. And we would certainly have called the doctor, but the fever dropped, she seemed to be all right. So we thought it had just been a cold. Then, one day, she was up, playing, Isabel was in the kitchen fixing lunch for the two boys when they'd come in from school, and she heard Grace fall down in the living room. When you have a lot of children you don't always start running when one of them falls, unless they start screaming or something. And, this time, Grace was quiet. Yet, Isabel says that when she heard that *thump* and then that silence, something happened in her to make her afraid. And she ran to the living room and there was little Grace on the floor, all twisted up, and the reason she hadn't screamed was that she couldn't get her breath. And when she did scream, it was the worst sound, Isabel says, that she'd ever heard in all her life, and she still hears it sometimes in her dreams. Isabel will sometimes wake me up with a low, moaning, strangled sound and I have to be quick to awaken her and hold her to me and where Isabel is weeping against me seems a mortal wound.

I think I may have written Sonny the very day that little Grace was buried. I was sitting in the living room in the dark, by myself, and I suddenly thought of Sonny. My trouble made his real.

One Saturday afternoon, when Sonny had been living with us, or, anyway, been in our house, for nearly two weeks, I found myself wandering aimlessly about the living room, drinking from a can of beer, and trying to work up the courage to search Sonny's room. He was out, he was usually out whenever I was home, and Isabel had taken the children to see their grandparents. Suddenly I was standing still in front of the living room window, watching Seventh Avenue. The idea of searching Sonny's room made me still. I scarcely dared to admit to myself what I'd be searching for. I didn't know what I'd do if I found it. Or if I didn't.

On the sidewalk across from me, near the entrance to a barbecue joint, some people were holding an old-fashioned revival meeting. The barbecue cook, wearing a dirty white apron, his conked hair reddish and metallic in the pale sun, and a cigarette between his lips, stood in the doorway, watching them. Kids and older people paused in their errands and stood there, along with some older men and a couple of very tough-looking women who watched everything that happened on the avenue, as though they owned it, or were maybe owned by it. Well, they were watching this, too. The revival was being carried on by three sisters in black, and a brother. All they had were their voices and their Bibles and a tambourine. The brother was testifying and while he testified two of the sisters stood together, seeming to say, amen, and the third sister walked around with the tambourine outstretched and a couple of people dropped coins into it. Then the brother's testimony ended and the sister who had been taking up the collection dumped the coins into her palm and transferred them to the pocket of her long black robe. Then she raised both hands, striking the tambourine against the air, and then against one hand, and she started to sing. And the two other sisters and the brother joined in.

It was strange, suddenly, to watch, though I had been seeing these street meetings all my life. So, of course, had everybody else down there. Yet, they paused and watched and listened and I stood still at the window. *"Tis the old ship of Zion,"* they sang, and the sister with the tambourine kept a steady, jangling beat, *"it has rescued many a thousand!"* Not a soul under the sound of their voices was hearing this song for the first time, not one of them had been rescued. Nor had they seen much in the way of rescue work being done around them. Neither did they especially believe in the holiness of the three sisters and the brother, they knew too much about them, knew where they lived, and how. The woman with the tambourine, whose voice dominated the air, whose face was bright with joy, was divided by very little from the woman who stood watching her, a cigarette between her heavy, chapped lips, her hair a cuckoo's nest, her face scarred and swollen from many beatings, and her black eyes glittering like coal. Perhaps they both knew this, which was why, when, as rarely, they addressed each other, they addressed each other as Sister. As the singing filled the air the watching, listening faces underwent a change, the eyes focusing on something within; the music seemed to soothe a poison out of them; and time seemed, nearly, to fall away from the sullen, belligerent, battered faces, as though they were fleeing back to their first condition, while dreaming of their last. The barbecue cook half shook his head and smiled, and dropped his cigarette and disappeared into his joint. A man fumbled in his pockets for change and stood holding it in his hand impatiently, as though he had just remembered a pressing appointment further up the avenue. He looked furious. Then I saw Sonny, standing on the edge of the crowd. He was carrying a wide, flat notebook with a green cover, and it made him look, from where I was standing, almost like a schoolboy. The coppery sun brought out the copper in his skin, he was very faintly smiling, standing very still. Then the singing stopped, the tambourine turned into a collection plate again. The furious man dropped in his coins and vanished, so did a couple of the women, and Sonny dropped some change in the plate, looking directly at the woman with a little smile. He started across the avenue, toward the house. He has a slow, loping walk, something like the way Harlem hipsters walk, only he's imposed on this his own half-beat. I had never really noticed it before.

I stayed at the window, both relieved and apprehensive. As Sonny disappeared from my sight, they began singing again. And they were still singing when his key turned in the lock.

"Hey," he said.

"Hey, yourself. You want some beer?"

"No. Well, maybe." But he came up to the window and stood beside me, looking out. "What a warm voice," he said.

They were singing *If I could only hear my mother pray again!*

"Yes," I said, "and she can sure beat that tambourine."

"But what a terrible song," he said, and laughed. He dropped his notebook on the sofa and disappeared into the kitchen. "Where's Isabel and the kids?"

"I think they went to see their grandparents. You hungry?"

"No." He came back into the living room with his can of beer. "You want to come some place with me tonight?"

I sensed, I don't know how, that I couldn't possibly say no. "Sure. Where?"

He sat down on the sofa and picked up his notebook and started leafing through it. "I'm going to sit in with some fellows in a joint in the Village."

"You mean, you're going to play, tonight?"

"That's right." He took a swallow of his beer and moved back to the window. He gave me a sidelong look. "If you can stand it."

"I'll try," I said.

He smiled to himself and we both watched as the meeting across the way broke up. The three sisters and the brother, heads bowed, were singing *God be with you till we meet again.* The faces around them were very quiet. Then the song ended. The small crowd dispersed. We watched the three women and the lone man walk slowly up the avenue.

"When she was singing before," said Sonny, abruptly, "her voice reminded me for a minute of what heroin feels like sometimes—when it's in your veins. It makes you feel sort of warm and cool at the same time. And distant. And—and sure." He sipped his beer, very deliberately not looking at me. I watched his face. "It makes you feel—in control. Sometimes you've got to have that feeling."

"Do you?" I sat down slowly in the easy chair.

"Sometimes." He went to the sofa and picked up his notebook again. "Some people do."

"In order," I asked, "to play?" And my voice was very ugly, full of contempt and anger.

"Well"—he looked at me with great, troubled eyes, as though, in fact, he hoped his eyes would tell me things he could never otherwise say—"they *think* so. And *if* they think so—!"

"And what do *you* think?" I asked.

He sat on the sofa and put his can of beer on the floor. "I don't know," he said, and I couldn't be sure if he were answering my question or pursuing his thoughts. His face didn't tell me. "It's not so much to *play.* It's to *stand* it, to be able to make it at all. On any level." He frowned and smiled: "In order to keep from shaking to pieces."

"But these friends of yours," I said, "they seem to shake themselves to pieces pretty goddamn fast."

"Maybe." He played with the notebook. And something told me that I should curb my tongue, that Sonny was doing his best to talk, that I should listen. "But of course you only know the ones that've gone to pieces. Some don't—or at least they haven't *yet* and that's just about all *any* of us can say." He paused. "And then there are some who just live, really, in hell, and they know it and they see what's happening and they go right on. I don't know." He sighed, dropped the notebook, folded his arms. "Some guys, you can tell from the way they play, they on something *all* the time. And you can see that, well, it makes something real for them. But of course," he picked up his beer from the floor and sipped it and put the can down again, "they *want* to, too, you've got to see that. Even some of them that say they don't—*some,* not all."

"And what about you?" I asked—I couldn't help it. "What about you? Do *you* want to?"

He stood up and walked to the window and remained silent for a long time. Then he sighed. "Me," he said. Then: "While I was downstairs before, on my way here, listening to that woman sing, it struck me all of a sudden how much suffering she must have had to go through—to sing like that. It's *repulsive* to think you have to suffer that much."

I said: "But there's no way not to suffer—is there, Sonny?"

"I believe not," he said and smiled, "but that's never stopped anyone from trying." He looked at me. "Has it?" I realized, with this mocking look, that there stood between us, forever, beyond the power of time or forgiveness, the fact that I had held silence—so long!—when he had needed human speech to help him. He turned back to the window. "No, there's no way not to suffer. But you try all kinds of ways to keep from drowning in it, to keep on top of it, and to make it seem—well, like *you.* Like you did something, all right, and now you're suffering for it. You know?" I said nothing. "Well you know," he said, impatiently, "why *do* people suffer? Maybe it's better to do something to give it a reason, *any* reason."

"But we just agreed," I said, "that there's no way not to suffer. Isn't it better, then, just to—take it?"

"But nobody just takes it," Sonny cried, "that's what I'm telling you! *Everybody* tries not to. You're just hung up on the *way* some people try—it's not *your* way!"

The hair on my face began to itch, my face felt wet. "That's not true," I said, "that's not true. I don't give a damn what other people do, I don't even care how they suffer. I just care how *you* suffer." And he looked at me. "Please believe me," I said, "I don't want to see you—die—trying not to suffer."

"I won't," he said, flatly, "die trying not to suffer. At least, not any faster than anybody else."

"But there's no need," I said, trying to laugh, "is there? in killing yourself."

I wanted to say more, but I couldn't. I wanted to talk about will power and how life could be—well, beautiful. I wanted to say that it was all within; but was it? or, rather, wasn't that exactly the trouble? And I wanted to promise that I would never fail him again. But it would all have sounded—empty words and lies.

So I made the promise to myself and prayed that I would keep it.

"It's terrible sometimes, inside," he said, "that's what's the trouble. You walk these streets, black and funky and cold, and there's not really a living ass to talk to, and there's nothing shaking, and there's no way of getting it out—that storm inside. You can't talk it and you can't make love with it, and when you finally try to get with it and play it, you realize *nobody's* listening. So *you've* got to listen. You got to find a way to listen."

And then he walked away from the window and sat on the sofa again, as though all the wind had suddenly been knocked out of him. "Sometimes you'll do *anything* to play, even cut your mother's throat." He laughed and looked at me. "Or your brother's." Then he sobered. "Or your own." Then: "Don't worry. I'm all right now and I think I'll *be* all right. But I can't forget—where I've been. I don't mean just the physical place I've been, I mean where I've *been.* And *what* I've been."

"What have you been, Sonny?" I asked.

He smiled—but sat sideways on the sofa, his elbow resting on the back, his fingers playing with his mouth and chin, not looking at me. "I've been something I didn't recognize, didn't know I could be. Didn't know anybody could be." He stopped, looking inward, looking helplessly young, looking old. "I'm not talking about it now because I feel *guilty* or anything like that—maybe it would be better if I did, I don't know. Anyway, I can't really talk about it. Not to you, not to anybody," and now he turned and faced me. "Sometimes, you know, and it was actually when I was most *out* of the world, I felt that I was in it, that I was *with* it, really, and I could play or I didn't really have to *play,* it just came out of me, it was there. And I don't know how I played, thinking about it now, but I know I did awful things, those times, sometimes, to people. Or it wasn't that I *did* anything to them—it was that they weren't real." He picked up the beer can; it was empty; he rolled it between his palms: "And other times—well, I needed a fix, I needed to find a place to lean, I needed to clear a space to *listen*—and I couldn't find it, and I—went crazy, I did terrible things to *me,* I was terrible *for* me." He began pressing the beer can between his hands, I watched the metal begin to give. It glittered, as he played with it, like a knife, and I was afraid he would cut himself, but I said nothing. "Oh well. I can never tell you. I was all by myself at the bottom of something, stinking and sweating and crying and shaking, and I smelled it, you know? *my* stink, and I thought I'd die if I couldn't get away from it and yet, all the same, I knew that everything I was doing was just locking me in with it. And I didn't know," he paused, still flattening the beer can, "I didn't know, I still *don't* know, something kept telling me that maybe it was good to smell your own stink, but I didn't think that *that* was what I'd been trying to do—and—who can stand it?" and he abruptly dropped the ruined beer can, looking at me with a small, still smile, and then rose, walking to the window as though it were the lodestone rock. I watched his face, he watched the avenue. "I couldn't tell you when Mama died—but the reason I wanted to leave Harlem so bad was to get away from drugs. And then, when I ran away, that's what I was running from—really. When I came back, nothing had changed, *I* hadn't changed, I was just—older." And he stopped, drumming with his fingers on the windowpane. The sun had vanished, soon darkness would fall. I watched his face. "It can come again," he said, almost as though speaking to himself. Then he turned to me. "It can come again," he repeated. "I just want you to know that."

"All right," I said, at last. "So it can come again, All right."

He smiled, but the smile was sorrowful. "I had to try to tell you," he said.

"Yes," I said. "I understand that."

"You're my brother," he said, looking straight at me, and not smiling at all.

"Yes," I repeated, "yes. I understand that."

He turned back to the window, looking out. "All that hatred down there," he said, "all that hatred and misery and love. It's a wonder it doesn't blow the avenue apart."

We went to the only nightclub on a short, dark street, downtown. We squeezed through the narrow, chattering, jampacked bar to the entrance of the big room, where the bandstand was. And we stood there for a moment, for the lights were very dim in this room and we couldn't see. Then, "Hello, boy," said a voice and an enormous black man, much older than Sonny or myself, erupted out of all that atmospheric lighting and put an arm around Sonny's shoulder. "I been sitting right here," he said, "waiting for you."

He had a big voice, too, and heads in the darkness turned toward us. Sonny grinned and pulled a little away, and said, "Creole, this is my brother. I told you about him."

Creole shook my hand. "I'm glad to meet you, son," he said, and it was clear that he was glad to meet me *there,* for Sonny's sake. And he smiled, "You got a real musician in *your* family," and he took his arm from Sonny's shoulder and slapped him, lightly, affectionately, with the back of his hand.

"Well. Now I've heard it all," said a voice behind us. This was another musician, and a friend of Sonny's, a coal-black, cheerful-looking man, built close to the ground. He immediately began confiding to me, at the top of his lungs, the most terrible things about Sonny, his teeth gleaming like a lighthouse and his laugh coming up out of him like the beginning of an earthquake. And it turned out that everyone at the bar knew Sonny, or almost everyone; some were musicians, working there, or nearby, or not working, some were simply hangers-on, and some were there to hear Sonny play. I was introduced to all of them and they were all very polite to me. Yet, it was clear that, for them, I was only Sonny's brother. Here, I was in Sonny's world. Or, rather: his kingdom. Here, it was not even a question that his veins bore royal blood.

They were going to play soon and Creole installed me, by myself, at a table in a dark corner. Then I watched them, Creole, and the little black man, and Sonny, and the others, while they horsed around, standing just below the bandstand. The light from the bandstand spilled just a little short of them and, watching them laughing and gesturing and moving about, I had the feeling that they, nevertheless, were being most careful not to step into that circle of light too suddenly: that if they moved into the light too suddenly, without thinking, they would perish in flame. Then, while I watched, one of them, the small, black man, moved into the light and crossed the bandstand and started fooling around with his drums. Then—being funny and being, also, extremely ceremonious—Creole took Sonny by the arm and led him to the piano. A woman's voice called Sonny's name and a few hands started clapping. And Sonny, also being funny and being ceremonious, and so touched, I think, that he could have cried, but neither hiding it nor showing it, riding it like a man, grinned, and put both hands to his heart and bowed from the waist.

Creole then went to the bass fiddle and a lean, very bright-skinned brown man jumped up on the bandstand and picked up his horn. So there they were, and the atmosphere on the bandstand and in the room began to change and tighten. Someone stepped up to the microphone and announced them. Then there were all kinds of murmurs. Some people at the bar shushed others. The waitress ran around, frantically getting in the last orders, guys and chicks got closer to each other, and the lights on the bandstand, on the quartet, turned to a kind of indigo. Then they all looked different there. Creole looked about him for the last time, as though he were making certain that all his chickens were in the coop, and then he—jumped and struck the fiddle. And there they were.

All I know about music is that not many people ever really hear it. And even then, on the rare occasions when something opens within, and the music enters, what we mainly hear, or hear corroborated, are personal, private, vanishing evocations. But the man who creates the music is hearing something else, is dealing with the roar rising from the void and imposing order on it as it hits the air. What is evoked in him, then, is of another order, more terrible because it has no words, and triumphant, too, for that same reason. And his triumph, when he triumphs, is ours. I just watched Sonny's face. His face was troubled, he was working hard, but he wasn't with it. And I had the feeling that, in a way, everyone on the bandstand was waiting for him, both waiting for him and pushing him along. But as I began to watch Creole, I realized that it was Creole who held them all back. He had them on a short rein. Up there, keeping the beat with his whole body, wailing on the fiddle, with his eyes half closed, he was listening to everything, but he was listening to Sonny. He was having a dialogue with Sonny. He wanted Sonny to leave the shoreline and strike out for the deep water. He was Sonny's witness that deep water and drowning were not the same thing—he had been there, and he knew. And he wanted Sonny to know. He was waiting for Sonny to do the things on the keys which would let Creole know that Sonny was in the water.

And, while Creole listened, Sonny moved, deep within, exactly like someone in torment. I had never before thought of how awful the relationship must be between the musician and his instrument. He has to fill it, this instrument, with the breath of life, his own. He has to make it do what he wants it to do. And a piano is just a piano. It's made out of so much wood and wires and little hammers and big ones, and ivory. While there's only so much you can do with it, the only way to find this out is to try; to try and make it do everything.

And Sonny hadn't been near a piano for over a year. And he wasn't on much better terms with his life, not the life that stretched before him now. He and the piano stammered, started one way, got scared, stopped; started another way, panicked, marked time, started again; then seemed to have found a direction, panicked again, got stuck. And the face I saw on Sonny I'd never seen before. Everything had been burned out of it, and, at the same time, things usually hidden were being burned in, by the fire and fury of the battle which was occurring in him up there.

Yet, watching Creole's face as they neared the end of the first set, I had the feeling that something had happened, something I hadn't heard. Then they finished, there was scattered applause and then, without an instant's warning, Creole started into something else, it was almost sardonic, it was *Am I Blue*. And, as though he commanded, Sonny began to play. Something began to happen. And Creole let out the reins. The dry, low, black man said something awful on the drums, Creole answered, and the drums talked back. Then the horn insisted, sweet and high, slightly detached perhaps, and Creole listened, commenting now and then, dry, and driving, beautiful and calm and old. Then they all came together again, and Sonny was part of the family again. I could tell this from his face. He seemed to have found, right there beneath his fingers, a damn brand-new piano. It seemed that he couldn't get over it. Then, for awhile, just being happy with Sonny, they seemed to be agreeing with him that brand-new pianos certainly were a gas.

Then Creole stepped forward to remind them that what they were playing was the blues. He hit something in all of them, he hit something in me, myself, and the music tightened and deepened, apprehension began to beat the air. Creole began to tell us what the blues were all about. They were not about anything very new. He and his boys up there were keeping it new, at the risk of ruin, destruction, madness, and death, in order to find new ways to make us listen. For, while the tale of how we suffer, and how we are delighted, and how we may triumph is never new, it always must be heard. There isn't any other tale to tell, it's the only light we've got in all this darkness.

And this tale, according to that face, that body, those strong hands on those strings, has another aspect in every country, and a new depth in every generation. Listen, Creole seemed to be saying, listen. Now these are Sonny's blues. He made the little black man on the drums know it, and the bright, brown man on the horn. Creole wasn't trying any longer to get Sonny in the water. He was wishing him Godspeed. Then he stepped back, very slowly, filling the air with the immense suggestion that Sonny speak for himself.

Then they all gathered around Sonny and Sonny played. Every now and again one of them seemed to say, amen. Sonny's fingers filled the air with life, his life. But that life contained so many others. And Sonny went all the way back, he really began with the spare, flat statement of the opening phrase of the song. Then he began to make it his. It was very beautiful because it wasn't hurried and it was no longer a lament. I seemed to hear with what burning he had made it his, with what burning we had yet to make it ours, how we could cease lamenting. Freedom lurked around us and I understood, at last, that he could help us to be free if we would listen, that he would never be free until we did. Yet, there was no battle in his face now. I heard what he had gone through, and would continue to go through until he came to rest in earth. He had made it his: that long line, of which we knew only Mama and Daddy. And he was giving it back, as everything must be given back, so that, passing through death, it can live forever. I saw my mother's face again, and felt, for the first time, how the stones of the road she had walked on must have bruised her feet. I saw the moonlit road where my father's brother died. And it brought something else back to me, and carried me past it, I saw my little girl again and felt Isabel's tears again, and I felt my own tears begin to rise. And I was yet aware that this was only a moment, that the world waited outside, as hungry as a tiger, and that trouble stretched above us, longer than the sky.

Then it was over. Creole and Sonny let out their breath, both soaking wet, and grinning. There was a lot of applause and some of it was real. In the dark, the girl came by and I asked her to take drinks to the bandstand. There was a long pause, while they talked up there in the indigo light and after awhile I saw the girl put a Scotch and milk on top of the piano for Sonny. He didn't seem to notice it, but just before they started playing again, he sipped from it and looked toward me, and nodded. Then he put it back on top of the piano. For me, then, as they began to play again, it glowed and shook above my brother's head like the very cup of trembling.

The Gilded Six-Bits

Zora Neale Hurston

It was a Negro yard around a Negro house in a Negro settlement that looked to the payroll of the G and G Fertilizer works for its support.

But there was something happy about the place. The front yard was parted in the middle by a sidewalk from gate to door-step, a sidewalk edged on either side by quart bottles driven neck down to the ground on a slant. A mess of homey flowers planted without a plan but blooming cheerily from their helter-skelter places. The fence and house were white-washed. The porch and steps scrubbed white.

The front door stood open to the sunshine so that the floor of the front room could finish drying after its weekly scouring. It was Saturday. Everything clean from the front gate to the privy house. Yard raked so that the strokes of the rake would make a pattern. Fresh newspaper cut in fancy edge on the kitchen shelves.

Missie May was bathing herself in the galvanized washtub in the bedroom. Her dark-brown skin glistened under the soapsuds that skittered down from her wash rag. Her stiff young breasts thrust forward aggressively like broad-based cones with the tips lacquered in black.

She heard men's voices in the distance and glanced at the dollar clock on the dresser.

"Humph! Ah'm way behind time t'day! Joe gointer be heah 'fore Ah git mah clothes on if Ah don't make haste."

She grabbed the clean meal sack at hand and dried herself hurriedly and began to dress. But before she could tie her slippers, there came the ring of singing metal on wood. Nine times.

Missie May grinned with delight. She had not seen the big tall man come stealing in the gate and creep up the walk grinning happily at the joyful mischief he was about to commit. But she knew that it was her husband throwing silver dollars in the door for her to pick up and pile beside her plate at dinner. It was this way every Saturday afternoon. The nine dollars hurled into the open door, he scurried to a hiding place behind the cape jasmine bush and waited.

Missie May promptly appeared at the door in mock alarm.

"Who dat chunkin' money in mah do'way?" she demanded. No answer from the yard. She leaped off the porch and began to search the shrubbery. She peeped under the porch and hung over the gate to look up and down the road. While she did this, the man behind the jasmine darted to the chinaberry tree. She spied him and gave chase.

"Nobody ain't gointer be chunkin' money at me and Ah not do 'em nothin'," she shouted in mock anger. He ran around the house with Missie May at his heels. She overtook him at the kitchen door. He ran inside but could not close it after him before she crowded in and locked with him in a rough and tumble. For several minutes the two were a furious mass of male and female energy. Shouting, laughing, twisting, turning, tussling, tickling each other in the ribs; Missie May clutching onto Joe and Joe trying, but not too hard, to get away.

"Missie May, take yo' hand out mah pocket!" Joe shouted out between laughs.

"Ah ain't, Joe, not lessen you gwine gimme whateve' it is good you got in yo' pocket. Turn it go, Joe, do Ah'll tear yo' clothes."

"Go on tear 'em. You de one dat pushes de needles round heah. Move yo' hand, Missie May."

"Lemme git dat paper sack out yo' pocket. Ah bet it's candy kisses."

"Tain't. Move yo' hand. Woman ain't got no business in a man's clothes nohow. Go away."

Missie May gouged way down and gave an upward jerk and triumphed.

"Unhhunh! Ah got it. It 'tis so candy kisses. Ah knowed you had somethin' for me in yo' clothes. Now Ah got to see whut's in every pocket you got."

Joe smiled indulgently and let his wife go through all of his pockets and take out the things that he had hidden there for her to find. She bore off the chewing gum, the cake of sweet soap, the pocket handkerchief as if she had wrested them from him, as if they had not been bought for the sake of the friendly battle.

"Whew! dat play-fight done got me all warmed up," Joe exclaimed. "Got me some water in de kittle?"

"Yo' water is on de fire and yo' clean things is cross de bed. Hurry up and wash yo'self and git changed so we kin eat. Ah'm hongry." As Missie said this, she bore the steaming kettle into the bedroom.

"You ain't hongry, sugar." Joe contradicted her. "Youse jes' a little empty. Ah'm de one whut's hongry. Ah could eat up camp meetin', back off 'ssociation, and drink Jurdan dry. Have it on de table when Ah git out de tub."

"Don't you mess wid mah business, man. You git in yo' clothes. Ah'm a real wife, not no dress and breath. Ah might not look lak one, but if you burn me, you won't git a thing but wife ashes."

Joe splashed in the bedroom and Missie May fanned around in the kitchen. A fresh red and white checked cloth on the table. Big pitcher of buttermilk beaded with pale drops of butter from the churn. Hot fried mullet, crackling bread, ham hock atop a mound of string beans and new potatoes, and perched on the window-sill a pone of spicy potato pudding.

Very little talk during the meal but that little consisted of banter that pretended to deny affection but in reality flaunted it. Like when

Missie May reached for a second helping of the tater pone. Joe snatched it out of her reach.

After Missie May had made two or three unsuccessful grabs at the pan, she begged, "Aw, Joe gimme some mo' dat tater pone."

"Nope. Sweetenin' is for us men-folks. Y'all pritty li'l frail eels don't need nothin' lak dis. You too sweet already."

"Please, Joe."

"Naw, naw. Ah don't want you to git no sweeter than whut you is already. We goin' down de road a li'l piece t'night so you go put on yo' Sunday-go-to-meetin' things."

Missie May looked at her husband to see if he was playing some prank. "Sho nuff, Joe?"

"Yeah. We goin' to de ice cream parlor."

"Where de ice cream parlor at, Joe?"

"A new man done come heah from Chicago and he done got a place and took and opened it up for a ice cream parlor, and bein' as it's real swell, Ah wants you to be one de first ladies to walk in dere and have some set down."

"Do Jesus, Ah ain't knowed nothin' 'bout it. Who de man done it?"

"Mister Otis D. Slemmons, of spots and places—Memphis, Chicago, Jacksonville, Philadelphia and so on."

"Dat heavy-set man wid his mouth full of gold teethes?"

"Yeah. Where did you see 'im at?

"Ah went down to de sto' tuh git a box of lye and Ah seen 'im standin' on de corner talkin' to some of de mens, and Ah come on back and went to scrubbin' de floor, and he passed and tipped his hat whilst Ah was scourin' de steps. Ah thought never Ah seen *him* befo'."

Joe smiled pleasantly. "Yeah, he's up to date. He got de finest clothes Ah ever seen on a colored man's back."

"Aw, he don't look no better in his clothes than you do in yourn. He got a puzzlegut on 'im and he so chuckle-headed, he got a pone behind his neck."

Joe looked down at his own abdomen and said wistfully. "Wisht Ah had a build on me lak he got. He ain't puzzle-gutted, honey. He jes' got a corperation. Dat make 'm look lak a rich white man. All rich mens is got some belly on 'em."

"Ah seen de pitchers of Henry Ford and he's a spare-built man and Rockefeller look lak he ain't got but one gut. But Ford and Rockefeller and dis Slemmons and all do rest kin be as many-gutted as dey please, Ah'm satisfied wid you jes' lak you is, baby. God took pattern after a pine tree and built you noble. Youse a pritty man, and if Ah knowed any way to make you mo' pritty still Ah'd take and do it."

Joe reached over gently and toyed with Missie May's ear. "You jes' say dat cause you love me, but Ah know Ah can't hold no light to Otis D. Slemmons. Ah ain't never been nowhere and Ah ain't got nothin' but you."

Missie May got on his lap and kissed him and he kissed back in kind. Then he went on. "All de womens is crazy 'bout 'im everywhere he go."

"How you know dat, Joe?"

"He tole us so hisself."

"Dat don't make it so. His mouf is cut cross-ways, ain't it? Well, he kin lie jes' lak anybody else."

"Good Lawd, Missie! You womens sho is hard to sense into things. He's got a five-dollar gold piece for a stick-pin and he got a ten-dollar gold piece on his watch chain and his mouf is jes' crammed full of gold teethes. Sho wisht it wuz mine. And whut make it so cool, he got money 'cumulated. And womens give it all to 'im."

"Ah don't see whut do womens see on 'im. Ah wouldn't give 'im a wink if de sheriff wuz after 'im."

"Well, he tole us how de white womens in Chicago give 'im all dat gold money. So he don't 'low nobody to touch it at all. Not even put dey finger on it. Dey tole 'im not to. You kin make 'miration at it, but don't tetch it."

"Whyn't he stay up dere where dey so crazy 'bout 'im?"

"Ah reckon dey done made 'im vast-rich and he wants to travel some. He say dey wouldn't leave 'im hit a lick of work. He got mo' lady people crazy 'bout him than he kin shake a stick at."

"Joe, Ah hates to see you so dumb. Dat stray nigger jes' tell y'all anything and y'all b'lieve it."

"Go 'head on now, honey and put on yo' clothes. He talkin' 'bout his pritty womens—Ah want 'im to see *mine.*"

Missie May went off to dress and Joe spent the time trying to make his stomach punch out like Slemmons' middle. He tried the rolling swagger of the stranger, but found that his tall bone-and-muscle stride fitted ill with it. He just had time to drop back into his seat before Missie May came in dressed to go.

On the way home that night Joe was exultant. "Didn't Ah say ole Otis was swell? Can't he talk Chicago talk? Wuzn't dat funny whut he said when great big fat ole Ida Armstrong come in? He asted me, 'Who is dat broad wid de forte shake?' Dat's a new word. Us always thought forty was a set of figgers but he showed us where it means a whole heap of things. Sometimes he don't say forty, he jes' say thirty-eight and two and dat mean de same thing. Know whut he tole me when Ah was payin for our ice cream? He say, 'Ah have to hand it to you, Joe. Dat wife of yours is jes' thirty-eight and two. Yessuh, she's forte!' Ain't he killin'?"

"He'll do in case of a rush. But he sho is got uh heap uh gold on 'im. Dat's de first time Ah ever seed gold money. It lookted good on him sho nuff, but it'd look a whole heap better on you."

"Who, me? Missie May youse crazy! Where would a po' man lak me git gold money from?"

Missie May was silent for a minute, then she said, "Us might find some goin' long de road some time. Us could."

"Who would be losin' gold money 'round heah? We ain't even seen none dese white folks wearin' no gold money on dey watch chain. You must be figgerin' Mister Packard or Mister Cadillac goin' pass through heah."

"You don't know whut been lost 'round heah. Maybe somebody way back in memorial times lost they gold money and went on off and it ain't never been found. And then if we wuz to find it, you could wear some 'thout havin' no gang of womens lak dat Slemmons say he got."

Joe laughed and hugged her. "Don't be so wishful 'bout me. Ah'm satisfied de way Ah is. So long as Ah be yo' husband, Ah don't keer 'bout nothin' else. Ah'd ruther all de other womens in de world to be dead than for you to have de toothache. Less we go to bed and git our night rest."

It was Saturday night once more before Joe could parade his wife in Slemmons' ice cream parlor again. He worked the night shift and Saturday was his only night off. Every other evening around six o'clock he left home, and dying dawn saw him hustling home around the lake where the challenging sun flung a flaming sword from east to west across the trembling water.

That was the best part of life—going home to Missie May. Their whitewashed house, the mock battle on Saturday, the dinner and ice cream parlor afterwards, church on Sunday nights when Missie outdressed any woman in town—all, everything was right.

One night around eleven the acid ran out at the G and G. The foreman knocked off the crew and let the steam die down. As Joe rounded the lake on his way home, a lean moon rode the lake in a silver boat. If anybody had asked Joe about the moon on the lake, he would have said he hadn't paid it any attention. But he saw it with his feelings. It made him yearn painfully for Missie. Creation obsessed him. He thought about children. They had been married for more than a year now. They had money put away. They ought to be making little feet for shoes. A little boy child would be about right.

He saw a dim light in the bedroom and decided to come in through the kitchen door. He could wash the fertilizer dust off himself before presenting himself to Missie May. It would be nice for her not to know that he was there until he slipped into his place in bed and hugged her back. She always liked that.

He eased the kitchen door open slowly and silently, but when he went to set his dinner bucket on the table he bumped it into a pile of dishes, and something crashed to the floor. He heard his wife gasp in fright and hurried to reassure her.

"Iss me, honey. Don't get skeered."

There was a quick. large movement in the bedroom. A rustle, a thud, and a stealthy silence. The light went out.

What? Robbers? Murderers? Some varmint attacking his helpless wife, perhaps. He struck a match, threw himself on guard and stepped over the door-sill into the bedroom.

The great belt on the wheel of Time slipped and eternity stood still. By the match light he could see the man's legs fighting with his breeches in his frantic desire to get them on. He had both chance and time to kill the intruder in his helpless condition—half in and half out of his pants—but he was too weak to take action. The shapeless enemies of humanity that live in the hours of Time had waylaid Joe. He was assaulted in his weakness. Like Samson awakening after his haircut. So he just opened his mouth and laughed.

The match went out and he struck another and lit the lamp. A howling wind raced across his heart, but underneath its fury he heard his wife sobbing and Slemmons pleading for his life. Offering to buy it with all that he had. "Please, suh, don't kill me. Sixty-two dollars at de sto'. Gold money."

Joe just stood. Slemmons looked at the window, but it was screened. Joe stood out like a rough-backed mountain between him and the door. Barring him from escape, from sunrise, from life.

He considered a surprise attack upon the big clown that stood there laughing like a chessy cat. But before his fist could travel an inch, Joe's own rushed out to crush him like a battering ram. Then Joe stood over him.

"Git into yo' damn rags, Slemmons, and dat quick."

Slemmons scrambled to his feet and into his vest and coat. As he grabbed his hat, Joe's fury overrode his intentions and he grabbed at Slemmons with his left hand and struck at him with his right. The right landed. The left grazed the front of his vest. Slemmons was knocked a somersault into the kitchen and fled through the open door. Joe found himself alone with Missie May, with the gold watch charm clutched in his left fist. A short bit of broken chain dangled between his fingers.

Missie May was sobbing. Wails of weeping without words. Joe stood, and after awhile she found out that he had something in his hand. And then he stood and felt without thinking and without seeing with his natural eyes. Missie May kept on crying and Joe kept on feeling so much and not knowing what to do with all his feelings, he put Slemmons' watch charm in his pants pocket and took a good laugh and went to bed.

"Missie May, whut you crying for?"

"Cause Ah love you so hard and Ah know you don't love *me* no mo'."

Joe sank his face into the pillow for a spell then he said huskily, "You don't know de feelings of dat yet, Missie May."

"Oh Joe, honey, he said he wuz gointer gimme dat gold money and he jes' kept on after me—"

Joe was very still and silent for a long time. Then he said, "Well, don't cry no mo, Missie May. Ah got yo' gold piece for you."

The hours went past on their rusty ankles. Joe still and quiet on one bed-rail and Missie May wrung dry of sobs on the other. Finally the sun's tide crept upon the shore of night and drowned all its hours. Missie May with her face stiff and streaked towards the window saw the dawn come into her yard. It was day. Nothing more. Joe wouldn't be coming home as usual. No need to fling open the front door and sweep off the porch, making it nice for Joe. Never no more breakfast to cook; no more washing and starching of Joe's jumper-jackets and pants. No more nothing. So why get up?

With this strange man in her bed, she felt embarrassed to get up and dress. She decided to wait till he had dressed and gone. Then she would get up, dress quickly and be gone forever beyond the reach of

Joe's looks and laughs. But he never moved. Red light turned to yellow, then white.

From beyond the no-man's land between them came a voice. A strange voice that yesterday had been Joe's.

"Missie May, ain't you gonna fix me no breakfus'?"

She sprang out of bed. "Yeah, Joe. Ah didn't reckon you wuz hongry."

No need to die today. Joe needed her for a few more minutes anyhow.

Soon there was a roaring fire in the cook stove. Water bucket full and two chickens killed. Joe loved fried chicken and rice. She didn't deserve a thing and good Joe was letting her cook him some breakfast. She rushed hot biscuits to the table as Joe took his seat.

He ate with his eyes on his plate. No laughter, no banter.

"Missie May, you ain't eatin' yo' breakfus'."

"Ah don't choose none, Ah thank yuh."

His coffee cup was empty. She sprang to refill it. When she turned from the stove and bent to set the cup beside Joe's plate, she saw the yellow coin on the table between them.

She slumped into her seat and wept into her arms.

Presently Joe said calmly, "Missie May, you cry too much. Don't look back lak Lot's wife and turn to salt."

The sun, the hero of every day, the impersonal old man that beams as brightly on death as on birth, came up every morning and raced across the blue dome and dipped into the sea of fire every evening. Water ran down hill and birds nested.

Missie knew why she didn't leave Joe. She couldn't. She loved him too much. But she couldn't understand why Joe didn't leave her. He was polite, even kind at times, but aloof.

There were no more Saturday romps. No ringing silver dollars to stack beside her plate. No pockets to rifle. In fact the yellow coin in his trousers was like a monster hiding in the cave of his pockets to destroy her.

She often wondered if he still had it, but nothing could have induced her to ask nor yet to explore his pockets to see for herself. Its shadow was in the house whether or no.

One night Joe came home around midnight and complained of pains in the back. He asked Missie to rub him down with liniment. It had been three months since Missie had touched his body and it all seemed strange. But she rubbed him. Grateful for the chance. Before morning, youth triumphed and Missie exulted. But the next day, as she joyfully made up their bed, beneath her pillow she found the piece of money with the bit of chain attached.

Alone to herself, she looked at the thing with loathing, but look she must. She took it into her hands with trembling and saw first thing that it was no gold piece. It was a gilded half dollar. Then she knew why Slemmons had forbidden anyone to touch his gold. He trusted village eyes at a distance not to recognize his stick-pin as a gilded quarter, and his watch charm as a four-bit piece.

She was glad at first that Joe had left it there. Perhaps he was through with her punishment. They were man and wife again. Then another thought came clawing at her. He had come home to buy from her as if she were any woman in the long house. Fifty cents for her love. As if to say that he could pay as well as Slemmons. She slid the coin into his Sunday pants pocket and dressed herself and left his house.

Halfway between her house and the quarters she met her husband's mother, and after a short talk she turned and went back home. Never would she admit defeat to that woman who prayed for it nightly. If she had not the substance of marriage, she had the outside show. Joe must leave *her*. She let him see she didn't want his old gold four-bits too.

She saw no more of the coin for sometime though she knew that Joe could not help finding it in his pocket. But his health kept poor, and he came home at least every ten days to be rubbed.

The sun swept around the horizon, trailing its robes of weeks and days. One morning as Joe came in from work, he found Missie May chopping wood. Without a word he took the ax and chopped a huge pile before he stopped.

"You ain't got not business choppin' wood, and you know it."

"How come? Ah been choppin' it for de last longest."

"Ah ain't blind. You makin' feet for shoes."

"Won't you be glad to have a li'l baby chile, Joe?"

"You know dat 'thout astin' me."

"Iss gointer be a boy chile and de very spit of you."

"You reckon, Missie May?"

"Who else could it look lak?"

Joe said nothing, but he thrust his hand deep into his pocket and fingered something there.

It was almost six months later Missie May took to bed and Joe went and got his mother to come wait on the house.

Missie May was delivered of a fine boy. Her travail was over when Joe came in from work one morning. His mother and the old women were drinking great bowls of coffee around the fire in the kitchen.

The minute Joe came into the room his mother called him aside.

"How did Missie May make out?" he asked quickly.

"Who, dat gal? She strong as a ox. She gointer have plenty mo'. We done fixed her wid de sugar and lard to sweeten her for de nex' one."

Joe stood silent awhile.

"You ain't ast 'bout de baby, Joe. You oughter be mighty proud cause he sho' is de spittin' image of yuh, son. Dat's yourn all right, if you never git another one, dat un is yourn. And you know Ah'm mighty proud too, son, cause Ah never thought well of you marryin' Missie May cause her ma used tuh fan her foot around right smart and Ah been mighty skeered dat Missie May wuz gointer git misput on her road."

Joe said nothing. He fooled around the house till late in the day then just before he went to work, he went and stood at the foot of the bed and asked his wife how she felt. He did this every day during the week.

On Saturday he went to Orlando to make his market. It had been a long time since he had done that.

Meat and lard, meal and flour, soap and starch. Cans of corn and tomatoes. All the staples. He fooled around town awhile and bought bananas and apples. Way after while he went around to the candy store.

"Hellow, Joe," the clerk greeted him. "Ain't seen you in a long time."

"Nope, Ah ain't been heah. Been round in spots and places."

"Want some of them molasses kisses you always buy?"

"Yessuh." He threw the gilded half-dollar on the counter. "Will dat spend?"

"Whut is it, Joe? Well, I'll be doggone! A gold-plated four-bit piece. Where'd you git it, Joe?"

"Offen a stray nigger dat come through Eatonville. He had it on his watch chain for a charm—goin round making out iss gold money. Ha ha! He had a quarter on his tie pin and it wuz all golded up too. Tryin' to fool people. Makin' out he so rich and everything. Ha! Ha! Tryin' to tole off folkses wives from home."

"How did you git it, Joe? Did he fool you, too?"

"Who, me? Naw suh! He ain't fooled me none. Know whut Ah done? He come round me wid his smart talk. Ah hauled off and knocked 'im down and took his old four-bits way from 'im. Gointer buy my wife some good ole lasses kisses wid it. Gimme fifty cents worth of dem candy kisses."

"Fifty cents buys a mightly lot of candy kisses, Joe. Why don't you split it up and take some chocolate bars, too. They eat good, too."

"Yessuh, dey do, but Ah wants all dat in kisses. Ah got a lil boy chile home now. Tain't a week old yet, but he kin suck a sugar tit and maybe eat one them kisses hisself."

Joe got his candy and left the store. The clerk turned to the next customer. "Wisht I could be like these darkies. Laughin' all the time. Nothin' worries 'em."

Back in Eatonville, Joe reached his own front door. There was the ring of singing metal on wood. Fifteen times. Missie May couldn't run to the door, but she crept there as quickly as she could.

"Joe Banks, Ah hear you chunkin' money in mah do'way. You wait till Ah got mah strength back and Ah'm gointer fix you for dat."

Barbados

Paule Marshall

Dawn, like the night which had preceded it, came from the sea. In a white mist tumbling like spume over the fishing boats leaving the island and the hunched, ghost shapes of the fishermen. In a white, wet wind breathing over the villages scattered amid the tall canes. The cabbage palms roused, their high headdresses solemnly saluting the wind, and along the white beach which ringed the island the casuarina trees began their moaning—a sound of women lamenting their dead within a cave.

The wind, smarting of the sea, threaded a wet skein through Mr. Watford's five hundred dwarf coconut trees and around his house at the edge of the grove. The house, Colonial American in design, seemed created by the mist—as if out of the dawn's formlessness had come, magically, the solid stone walls, the blind, broad windows and the portico of fat columns which embraced the main story. When the mist cleared, the house remained—pure, proud, a pristine white—disdaining the crude wooden houses in the village outside its high gate.

It was not the dawn settling around his house which awakened Mr. Watford, but the call of his Barbary doves from their hutch in the yard. And it was more the feel of that sound than the sound itself. His hands had retained, from the many times a day he held the doves, the feel of their throats swelling with that murmurous, mournful note. He lay abed now, his hand—as cracked and callused as a cane cutter's—filled with the sound, and against the white sheet which flowed out to the white walls he appeared profoundly alone, yet secure in loneliness, contained. His face was fleshless and severe, his black skin sucked deep into the hollow of his jaw, while under a high brow, which was like a bastion raised against the world, his eyes were indrawn and pure. It was as if during all his seventy years, Mr. Watford had permitted nothing to sight which could have affected him.

He stood up, and his body, muscular but stripped of flesh, appeared to he absolved from time, still young. Yet each clenched gesture of his arms, of his lean shank as he dressed in a faded shirt and work pants, each vigilant, snapping motion of his head betrayed tension. Ruthlessly he spurred his body to perform like a younger man's. Savagely he denied the accumulated fatigue of the years. Only sometimes when he paused in his grove of coconut trees during the day, his eyes tearing and the breath torn from his lungs, did it seem that if he could find a place hidden from the world and himself he would give way to exhaustion and weep from weariness.

Dressed, he strode through the house, his step tense, his rough hand touching the furniture from Grand Rapids which crowded each room. For some reason, Mr. Watford had never completed the house. Everywhere the walls were raw and unpainted, the furniture unarranged. In the drawing room with its coffered ceiling, he stood before his favorite piece, an old mantel clock which eked out the time. Reluctantly it whirred five and Mr. Watford nodded. His day had begun.

It was no different from all the days which made up the five years since his return to Barbados. Downstairs in the unfinished kitchen, he prepared his morning tea—tea with canned milk and fried bakes—and ate standing at the stove while lizards skittered over the unplastered walls. Then, belching and snuffling the way a child would, he put on a pith helmet, secured his pants legs with bicycle clasps and stepped into the yard. There he fed the doves, holding them so that their sound poured into his hands and laughing gently—but the laugh gave way to an irritable grunt as he saw the mongoose tracks under the hutch. He set the trap again.

The first heat had swept the island like a huge tidal wave when Mr. Watford, with that tense, headlong stride, entered the grove. He had planted the dwarf coconut trees because of their quick yield and because, with their stunted trunks, they always appeared young. Now as he worked, rearranging the complex of pipes which irrigated the land, stripping off the dead leaves, the trees were like cool, moving presences; the stiletto fronds wove a protective dome above him and slowly, as the day soared toward noon, his mind filled with the slivers of sunlight through the trees and the feel of earth in his hands, as it might have been filled with thoughts.

Except for a meal at noon, he remained in the grove until dusk surged up from the sea; then returning to the house, he bathed and dressed in a medical doctor's white uniform, turned on the lights in the parlor and opened the tall doors to the portico. Then the old women of the village on their way to church, the last hawkers caroling, "Fish, flying fish, a penny, my lady," the roistering saga-boys lugging their heavy steel drums to the crossroad where they would rehearse under the street lamp—all passing could glimpse Mr. Watford, stiff in his white uniform and with his head bent heavily over a Boston newspaper. The papers reached him weeks late but he read them anyway, giving a little savage chuckle at the thought that beyond his world that other world went its senseless way. As he read, the night sounds of the village welled into a joyous chorale against the sea's muffled cadence and the hollow, haunting music of the steel band. Soon the moths, lured in by the light, fought to die on the lamp, the beetles crashed drunkenly against the walls and the night—like a woman offering herself to him—became fragrant with the night-blooming cactus.

Even in America Mr. Watford had spent his evenings this way. Coming home from the hospital, where he worked in the boiler room, he would dress in his white uniform and read in the basement of the large rooming house he owned. He had lived closeted like this, detached, because America—despite the money and property he had slowly accumulated—had meant nothing to him. Each morning, walking to the hospital along the rutted Boston streets, through the smoky dawn light, he had known—although it had never been a thought—that his allegiance, his place, lay elsewhere. Neither had the few acquaintances he had made mattered. Nor the women he had occasionally kept as a younger man. After the first month their bodies would grow coarse to his hand and he would begin edging away. . . . So that he had felt no regret when, the year before his retirement, he resigned his job, liquidated his properties and, his fifty-year exile over, returned home.

The clock doled out eight and Mr. Watford folded the newspaper and brushed the burnt moths from the lamp base. His lips still shaped the last words he had read as he moved through the rooms, fastening the windows against the night air, which he had dreaded even as a boy. Something palpable but unseen was always, he believed, crouched in the night's dim recess, waiting to snare him. . . . Once in bed in his sealed room, Mr. Watford fell asleep quickly.

The next day was no different except that Mr. Goodman, the local shopkeeper, sent the boy for coconuts to sell at the racetrack and then came that evening to pay for them and to herald—although Mr. Watford did not know this—the coming of the girl.

That morning, taking his tea, Mr. Watford heard the careful tap of the mule's hoofs and looking out saw the wagon jolting through the dawn and the boy, still lax with sleep, swaying on the seat. He was perhaps eighteen and the muscles packed tightly beneath his lustrous black skin gave him a brooding strength. He came and stood outside the back door, his hands and lowered head performing the small, subtle rites of deference.

Mr. Watford's pleasure was full, for the gestures were those given only to a white man in his time. Yet the boy always nettled him. He sensed a natural arrogance like a pinpoint of light within his dark stare. The boy's stance exhumed a memory buried under the years. He remembered, staring at him, the time when he had worked as a yard boy for a white family, and had had to assume the same respectful pose while their flat, raw, Barbadian voices assailed him with orders. He remembered the muscles in his neck straining as he nodded deeply and a taste like alum on his tongue as he repeated the "Yes, please," as in a litany. But because of their whiteness and wealth, he had never dared hate them. Instead his rancor, like a boomerang, had rebounded, glancing past him to strike all the dark ones like himself, even his mother with her spindled arms and her stomach sagging with a child who was, invariably, dead at birth. He had been the only one of ten to live, the only one to escape. But he had never lost the sense of being pursued by the same dread presence which had claimed them. He had never lost the fear that if he lived too fully he would tire and death would quickly close the gap. His only defense had been a cautious life and work. He had been almost broken by work at the age of twenty when his parents died, leaving him enough money for the passage to America. Gladly had he fled the island. But nothing had mattered after his flight.

The boy's foot stirred the dust. He murmured, "Please, sir, Mr. Watford, Mr. Goodman at the shop send me to pick the coconut."

Mr. Watford's head snapped up. A caustic word flared, but died as he noticed a political button pinned to the boy's patched shirt with "Vote for the Barbados People's Party" printed boldly on it, and below that the motto of the party: "The Old Shall Pass." At this ludicrous touch (for what could this boy, with his splayed and shigoed feet and blunted mind, understand about politics?) he became suddenly nervous, angry. The button and its motto seemed, somehow, directed at him. He said roughly, "Well, come then. You can't pick any coconuts standing there looking foolish!"—and he led the way to the grove.

The coconuts, he knew, would sell well at the booths in the center of the track, where the poor were penned in like cattle. As the heat thickened and the betting grew desperate, they would clamor: "Man, how you selling the water coconuts?" and hacking off the tops they would pour rum into the water within the hollow centers, then tilt the coconuts to their heads so that the rum-sweetened water skimmed their tongues and trickled bright down their dark chins. Mr. Watford had stood among them at the track as a young man, as poor as they were, but proud. And he had always found something unutterably graceful and free in their gestures, something which had roused contradictory feelings in him: admiration, but just as strong, impatience at their easy ways, and shame. . . .

That night, as he sat in his white uniform reading, he heard Mr. Goodman's heavy step and went out and stood at the head of the stairs in a formal, proprietary pose. Mr. Goodman's face floated up into the light—the loose folds of flesh, the skin slick with sweat as if oiled, the eyes scribbled with veins and mottled, bold—as if each blemish there was a sin he proudly displayed or a scar which proved he had met life head-on. His body, unlike Mr. Watford's, was corpulent and, with the trousers caught up around his full crotch, openly concupiscent. He owned the one shop in the village which gave credit and a booth which sold coconuts at the race track, kept a wife and two outside women, drank a rum with each customer at his bar, regularly caned his fourteen children, who still followed him everywhere (even now they were waiting for him in the darkness beyond Mr. Watford's gate) and bet heavily at the races, and when he lost gave a loud hacking laugh which squeezed his body like a pain and left him gasping.

The laugh clutched him now as he flung his pendulous flesh into a chair and wheezed, "Watford, how? Man, I near lose house, shop, shirt and all at race today. I tell you, they got some horses from Trinidad in this meet that's making ours look like they running backwards. Be Jese, I wouldn't bet on a Bajan horse tomorrow if Christ heself was to give me the top. Those bitches might look good but they's nothing 'pon a track."

Mr. Watford, his back straight as the pillar he leaned against, his eyes unstained, his gaunt fare planed by contempt, gave Mr. Goodman his cold, measured smile, thinking that the man would be dead soon, bloated with rice and rum—and somehow this made his own life more certain.

Sputtering with his amiable laughter, Mr. Goodman paid for the coconuts, but instead of leaving then as he usually did, he lingered, his eyes probing for a glimpse inside the house. Mr. Watford waited, his head snapping warily; then, impatient, he started toward the door and Mr. Goodman said, "I tell you, your coconut trees bearing fast enough even for dwarfs. You's lucky, man."

Ordinarily Mr. Watford would have waved both the man and his remark aside, but repelled more than usual tonight by Mr. Goodman's gross form and immodest laugh, he said—glad of the cold edge his slight American accent gave the word—"What luck got to do with it? I does care the trees properly and they bear, that's all. Luck! People, especially this bunch around here, is always looking to luck when the only answer is a little brains and plenty of hard work. . . ." Suddenly remembering the boy that morning and the political button, he added in loud disgust, "Look that half-foolish boy you does send here to pick the coconuts. Instead of him learning a trade and going to England where he might find work he's walking about with a political button. He and all in politics now! But that's the way with these down here. They'll do some of everything but work. They don't want work!" He gestured violently, almost dancing in anger. "They too busy spreeing."

The chair creaked as Mr. Goodman sketched a pained and gentle denial. "No, man," he said, "you wrong. Things is different to before. I mean to say, the young people nowadays is different to how we was. They not just sitting back and taking things no more. They not so frighten for the white people as we was. No, man. Now take that said same boy, for an example. I don't say he don't like a spree, but he's serious, you see him there. He's a member of this new Barbados People's Party. He wants to see his own color running the government. He wants to be able to make a living right here in Barbados instead of going to any cold England. And he's right!" Mr. Goodman paused at a vehement pitch, then shrugged heavily. "What the young people must do, nuh? They got to look to something. . . ."

"Look to work!" And Mr. Watford thrust out a hand so that the horned knuckles caught the light.

"Yes, that's true—and it's up to we that got little something to give them work," Mr. Goodman said, and a sadness filtered among the dissipations in his eyes. "I mean to say we that got little something got to help out. In a manner of speaking, we's responsible . . ."

"Responsible!" The word circled Mr. Watford's head like a gnat and he wanted to reach up and haul it down, to squash it underfoot.

Mr. Goodman spread his hands; his breathing rumbled with a sigh. "Yes, in a manner of speaking. That's why, Watford man, you got to provide little work for some poor person down in here. Hire a servant at least! 'Cause I gon tell you something . . ." And he hitched forward his chair, his voice dropped to a wheeze. "People talking. Here you come back rich from big America and build a swell house and plant 'nough coconut trees and you still cleaning and cooking and thing like some woman. Man, it don't look good!" His face screwed in emphasis and he sat back. "Now, there's this girl, the daughter of a friend that just dead, and she need work bad enough. But I wouldn't like to see she working for these white people 'cause you know how those men will take advantage of she. And she'd make a good servant, man. Quiet and quick so, and nothing a-tall to feed and she can sleep anywhere about the place. And she don't have no boys always around her either. . . ." Still talking, Mr. Goodman eased from his chair and reached the stairs with surprising agility. "You need a servant," he whispered, leaning close to Mr. Watford as he passed. "It don't look good, man, people talking. I gon send she."

Mr. Watford was overcome by nausea. Not only from Mr. Goodman's smell—a stench of salt fish, rum and sweat—but from an outrage which was like a sediment in his stomach. For a long time he stood there almost kecking from disgust, until his clock struck eight, reminding him of the sanctuary within—and suddenly his cold laugh dismissed Mr. Goodman and his proposal. Hurrying in, he locked the doors and windows against the night air and, still laughing, he slept.

The next day, coming from the grove to prepare his noon meal, he saw her. She was standing in his driveway, her bare feet like strong dark roots amid the jagged stones, her face tilted toward the sun—and she might have been standing there always waiting for him. She seemed of the sun, of the earth. The folktale of creation might have been true with her: that along a riverbank a god had scooped up the earth—rich and black and warmed by the sun—and molded her poised head with its tufted braids and then with a whimsical touch crowned it with a sober brown felt hat which should have been worn by some stout English matron in a London suburb, had sculptured the passionless face and drawn a screen of gossamer across her eyes to hide the void behind. Beneath her bodice her small breasts were smooth at the crest. Below her waist, her hips branched wide, the place prepared for its load of life. But it was the bold and sensual strength of her legs which completely unstrung Mr. Watford. He wanted to grab a hoe and drive her off.

"What it 'tis you want?" he called sharply.

"Mr. Goodman send me."

"Send you for what?" His voice was shrill in the glare.

She moved. Holding a caved-in valise and a pair of white sandals, her head weaving slightly as though she bore a pail of water there or a tray of mangoes, she glided over the stones as if they were smooth ground. Her bland expression did not change, but her eyes, meeting his, held a vague trust. Pausing a few feet away, she curtsied deeply. "I's the new servant."

Only Mr. Watford's cold laugh saved him from anger. As always it raised him to a height where everything below appeared senseless and insignificant—especially his people, whom the girl embodied. From this height, he could even be charitable. And thinking suddenly of how she had waited in the brutal sun since morning without taking shelter under the nearby tamarind tree, he said, not unkindly, "Well, girl, go back and tell Mr. Goodman for me that I don't need no servant."

"I can't go back."

"How you mean can't?" His head gave its angry snap.

"I'll get lashes," she said simply, "My mother say I must work the day and then if you don't wish me, I can come back. But I's not to leave till night falling, if not I get lashes."

He was shaken by her dispassion. So much so that his head dropped from its disdaining angle and his hands twitched with helplessness. Despite anything he might say or do, her fear of the whipping would keep her there until nightfall, the valise and shoes in hand. He felt his day with its order and quiet rhythms threatened by her intrusion—and suddenly waving her off as if she were an evil visitation, he hurried into the kitchen to prepare his meal.

But he paused, confused, in front of the stove, knowing that he could not cook and leave her hungry at the door, nor could he cook and serve her as though he were the servant.

"Yes, please."

They said nothing more. She entered the room with a firm step and an air almost of familiarity, placed her valise and shoes in a corner and went directly to the larder. For a time Mr. Watford stood by, his muscles flexing with anger and his eyes bounding ahead of her every move, until feeling foolish and frighteningly useless, he went out to feed his doves.

The meal was quickly done and as he ate he heard the dry slap of her feet behind him—a pleasant sound—and then silence. When he glanced back she was squatting in the doorway, the sunlight aslant the absurd hat and her face bent to a bowl she held in one palm. She ate slowly, thoughtfully, as if fixing the taste of each spoonful in her mind.

It was then that he decided to let her work the day and at nightfall to pay her a dollar and dismiss her. His decision held when he returned later from the grove and found tea awaiting him, and then through the supper she prepared. Afterward, dressed in his white uniform, he patiently waited out the day's end on the portico, his face setting into a grim mold. Then just as dusk etched the first dark line between the sea and sky, he took out a dollar and went downstairs.

She was not in the kitchen, but the table was set for his morning tea. Muttering at her persistence, he charged down the corridor, which ran the length of the basement, flinging open the doors to the damp, empty rooms on either side, and sending the lizards and the shadows long entrenched there scuttling to safety.

He found her in the small slanted room under the stoop, asleep on an old cot he kept there, her suitcase turned down beside the bed, and the shoes, dress and the ridiculous hat piled on top. A loose nightshift muted the outline of her body and hid her legs, so that she appeared suddenly defenseless, innocent, with a child's trust in her curled hand and in her deep breathing. Standing in the doorway, with his own breathing snarled and his eyes averted, Mr. Watford felt like an intruder. She had claimed the room. Quivering with frustration, he slowly turned away, vowing that in the morning he would shove the dollar at her and lead her like a cow out of his house. . . .

Dawn brought rain and a hot wind which set the leaves rattling and swiping at the air like distraught arms. Dressing in the dawn darkness, Mr. Watford again armed himself with the dollar and, with his shoulders at an uncompromising set, plunged downstairs. He descended into the warm smell of bakes and this smell, along with the thought that she had been up before him, made his hand knot with exasperation on the banister. The knot tightened as he saw her, dust swirling at her feet as she swept the corridor, her face bent solemn to the task. Shutting her out with a lifted hand, he shouted, "Don't bother sweeping. Here's a dollar. G'long back."

The broom paused and although she did not raise her head, he sensed her groping through the shadowy maze of her mind toward his voice. Behind the dollar which he waved in her face, her eyes slowly cleared. And, surprisingly, they held no fear. Only anticipation and a tenuous trust. It was as if she expected him to say something kind.

"G'long back!" His angry cry was a plea.

Like a small, starved flame, her trust and expectancy died and she said, almost with reproof, "The rain falling."

To confirm this, the wind set the rain stinging across the windows and he could say nothing, even though the words sputtered at his lips. It was useless. There was nothing inside her to comprehend that she was not wanted. His shoulders sagged under the weight of her ignorance, and with a futile gesture he swung away, the dollar hanging from his hand like a small sword gone limp.

She became as fixed and familiar a part of the house as the stones—and as silent. He paid her five dollars a week, gave her Mondays off and in the evenings, after a time, even allowed her to sit in the alcove off the parlor, while he read with his back to her, taking no more notice of her than he did the moths on the lamp.

But once, after many silent evenings together, he detected a sound apart from the night murmurs of the sea and village and the metallic tuning of the steel band, a low, almost inhuman cry of loneliness which chilled him. Frightened, he turned to find her leaning hesitantly toward him, her eyes dark with urgency, and her face tight with bewilderment and a growing anger. He started, not understanding, and her arm lifted to stay him. Eagerly she bent closer. But as she uttered the low cry again, as her fingers described her wish to talk, he jerked around, afraid that she would be foolish enough to speak and that once she did they would be brought close. He would be forced then to acknowledge something about her which he refused to grant; above all, he would be called upon to share a little of himself. Quickly he returned to his newspaper, rustling it to settle the air, and after a time he felt her slowly, bitterly, return to her silence. . . .

Like sand poured in a careful measure from the hand, the weeks flowed down to August and on the first Monday, August Bank holiday, Mr. Watford awoke to the sound of the excursion buses leaving the village for the annual outing, their backfire pelleting the dawn calm and the ancient motors protesting the overcrowding. Lying there, listening, he saw with disturbing clarity his mother dressed for an excursion—the white headtie wound above her dark face and her head poised like a dancer's under the heavy outing basket of food. That set of her head had haunted his years, reappearing in the girl as she walked toward him the first day.

Aching with the memory, yet annoyed with himself for remembering, he went downstairs.

The girl had already left for the excursion, and although it was her day off, he felt vaguely betrayed by her eagerness to leave him. Somehow it suggested ingratitude. It was as if his doves were suddenly to refuse him their song or his trees their fruit, despite the care he gave them. Some vital past which shaped the simple mosaic of his life seemed suddenly missing. An alien silence curled like coal gas throughout the house. To escape it he remained in the grove all day and, upon his return to the house, dressed with more care than usual, putting on a fresh, starched uniform, and solemnly brushing his hair until it lay in a smooth bush above his brow. Leaning close to the mirror, but avoiding his eyes, he cleaned the white rheum at their corners, and afterward pried loose the dirt under his nails.

Unable to read his papers, he went out on the portico to escape the unnatural silence in the house, and stood with his hands clenched on the balustrade and his taut body straining forward. After a long wait he heard the buses return and voices in gay shreds upon the wind. Slowly his hands relaxed, as did his shoulders under the white uniform; for the first time that day his breathing was regular. She would soon come.

But she did not come and dusk bloomed into night, with a fragrant heat and a full moon which made the leaves glint as though touched with frost. The steel band at the crossroads began the lilting songs of sadness and seduction, and suddenly—like shades roused by the night and the music—images of the girl flitted before Mr. Watford's eyes. He saw her lost amid the carousings in the village, despoiled; he imagined someone like Mr. Goodman clasping her lewdly or tumbling her in the canebrake. His hand rose, trembling, to rid the air of her; he tried to summon his cold laugh. But, somehow, he could not dismiss her as he had always done with everyone else. Instead, he wanted to punish and protect her, to find and lead her back to the house.

As he leaned there, trying not to give way to the desire to go and find her, his fist striking the balustrade to deny his longing, he saw them. The girl first, with the moonlight like a silver patina on her skin, then the boy whom Mr. Goodman sent for the coconuts, whose easy strength and the political button—"The Old Order Shall Pass"—had always mocked and challenged Mr. Watford. They were joined in a tender battle: the boy in a sport shirt riotous with color was reaching for the girl as he leaped and spun, weightless, to the music, while she fended him off with a gesture which was lovely in its promise of surrender. Her protests were little scattered bursts: "But, man, why don't you stop, nuh . . .? But, you know, you getting on like a real-real idiot. . . ."

Each time she chided him he leaped higher and landed closer, until finally he eluded her arm and caught her by the waist. Boldly he pressed a leg between her tightly closed legs until they opened under his pressure. Their bodies cleaved into one whirling form and while he sang she laughed like a wanton, with her hat cocked over her ear. Dancing, the stones moiling underfoot, they claimed the night. More than the night. The steel band played for them alone. The trees were their frivolous companions, swaying as they swayed. The moon rode the sky because of them.

Mr. Watford, hidden by a dense shadow, felt the tendons which strung him together suddenly go limp; above all, an obscure belief which, like rare china, he had stored on a high shelf in his mind began to tilt. He sensed the familiar specter which hovered in the night reaching out to embrace him, just as the two in the yard were embracing. Utterly unstrung, incapable of either speech or action, he stumbled into the house, only to meet there an accusing silence from the clock, which had missed its eight o'clock winding, and his newspapers lying like ruined leaves over the floor.

He lay in bed in the white uniform, waiting for sleep to rescue him, his hands seeking the comforting sound of his doves. But sleep eluded him and instead of the doves, their throats tremulous with sound, his scarred hands filled with the shape of a woman he had once kept: her skin, which had been almost bruising in its softness; the buttocks and breasts spread under his hands to inspire both cruelty and tenderness. His hands closed to softly crush those forms, and the searing thrust of passion, which he had not felt for years, stabbed his dry groin. He imagined the two outside, their passion at a pitch by now, lying together behind the tamarind tree, or perhaps—and he sat up sharply—they had been bold enough to bring their lust into the house. Did he not smell their taint on the air? Restored suddenly, he rushed downstairs. As he reached the corridor, a thread of light beckoned him from her room and he dashed furiously toward it, rehearsing the angry words which would jar their bodies apart. He neared the door, glimpsed her through the small opening, and his step faltered; the words collapsed.

She was seated alone on the cot, tenderly holding the absurd felt hat in her lap, one leg tucked under her while the other trailed down. A white sandal, its strap broken, dangled from the foot and gently knocked the floor as she absently swung her leg. Her dress was twisted around her body—and pinned to the bodice, so that it gathered the cloth between her small breasts, was the political button the boy always wore. She was dreamily fingering it, her mouth shaped by a gentle, ironic smile and her eyes strangely acute and critical. What had transpired on the cot had not only, it seemed, twisted the dress around her, tumbled her hat and broken her sandal, but had also defined her and brought the blurred forms of life into focus for her. There was a woman's force in her aspect now, a tragic knowing and acceptance in her bent head, a hint about her of Cassandra watching the future wheel before her eyes.

Before those eyes which looked to another world, Mr. Watford's anger and strength failed him and he held to the wall for support. Unreasonably, he felt that he should assume some hushed and reverent pose, to bow as she had the day she had come. If he had known their names, he would have pleaded forgiveness for the sins he had committed against her and the others all his life, against himself. If he could have borne the thought, he would have confessed that it had been love, terrible in its demand, which he had always fled. And that love had been the reason for his return. If he had been honest, he would have whispered—his head bent and a hand shading his eyes—that unlike Mr. Goodman (whom he suddenly envied for his full life) and the boy with his political button (to whom he had lost the girl), he had not been willing to bear the weight of his own responsibility. . . . But all Mr. Watford could admit, clinging there to the wall, was, simply, that he wanted to live—and that the girl held life within her as surely as she held the hat in her hands. If he could prove himself better than the boy, he could win it, Only than, he dimly knew, would he shake off the pursuer which had given him no rest since birth. Hopefully, he staggered forward, his step cautious and contrite, his hands, quivering along the wall.

She did not see or hear him as he pushed the door wider. And for some time he stood there, his shoulders hunched in humility, his skin stripped away to reveal each flaw, his whole self offered in one outstretched hand. Still unaware of him, she swung her leg, and the dangling shoe struck a derisive note. Then, just as he had turned away that evening in the parlor when she had uttered her low call, she turned away now, refusing him.

Mr. Watford's body went slack and then stiffened ominously. He knew that be would have to wrest from her the strength needed to sustain him. Slamming the door, he cried, his voice cracked and strangled, "What you and him was doing in here? Tell me! I'll not have you bringing nastiness round here. Tell me!"

She did not start. Perhaps she had been aware of him all along and had expected his outburst. Or perhaps his demented eye and the desperation rising from him like a musk filled her with pity instead of fear. Whatever, her benign smile held and her eyes remained abstracted until his hand reached out to fling her back on the cot. Then, frowning, she stood up, wobbling a little on the broken shoe and holding the political button as if it was a new power which would steady and protect her. With a cruel flick of her arm she struck aside his hand and, in a voice as cruel, halted him. "But you best move and don't come holding on to me, you nasty, pissy old man. That's all you is, despite yuh big house and fancy furnitures and yuh newspapers from America. You ain't people, Mr. Watford, you ain't people!" And with a look and a lift of her head which made her condemnation final, she placed the hat atop her braids, and turning aside picked up the valise which had always lain, packed, beside the cot—as if even on the first day she had known that this night would come and had been prepared against it. . . .

Mr. Watford did not see her leave, for a pain squeezed his heart dry and the driven blood was a bright, blinding cataract over his eyes. But his inner eye was suddenly clear. For the first time it gazed mutely upon the waste and pretense which had spanned his years. Flung there against the door by the girl's small blow, his body slowly crumpled under the weariness he had long denied. He sensed that dark but unsubstantial figure which roamed the nights searching for him wind him in its chill embrace. He struggled against it, his hands clutching the air with the spastic eloquence of a drowning man. He moaned—and the anguished sound reached beyond the room to fill the house. It escaped to the yard and his doves swelled their throats, moaning with him.

A Worn Path

Eudora Welty

It was December—a bright frozen day in the early morning. Far out in the country there was an old Negro woman with her head tied in a red rag, coming along a path through the pinewoods. Her name was Phoenix Jackson. She was very old and small and she walked slowly in the dark pine shadows, moving a little from side to side in her steps, with the balanced heaviness and lightness of a pendulum in a grandfather clock. She carried a thin, small cane made from an umbrella, and with this she kept tapping the frozen earth in front of her. This made a grave and persistent noise in the still air, that seemed meditative like the chirping of a solitary little bird.

She wore a dark striped dress reaching down to her shoe tops, and an equally long apron of bleached sugar sacks, with a full pocket: all neat and tidy, but every time she took a step she might have fallen over her shoelaces, which dragged from her unlaced shoes. She looked straight ahead. Her eyes were blue with age. Her skin had a pattern all its own of numberless branching wrinkles and as though a whole little tree stood in the middle of her forehead, but a golden color ran underneath, and the two knobs of her cheeks were illumined by a yellow burning under the dark. Under the red rag her hair came down on her neck in the frailest of ringlets, still black, and with an odor like copper.

Now and then there was a quivering in the thicket. Old Phoenix said, "Out of my way, all you foxes, owls, beetles, jack rabbits, coons and wild animals! . . . Keep out from under these feet, little bob-whites. . . . Keep the big wild hogs out of my path. Don't let none of those come running my direction. I got a long way." Under her small black-freckled hand her cane, limber as a buggy whip, would switch at the brush as if to rouse up any hiding things.

On she went. The woods were deep and still. The sun made the pine needles almost too bright to look at, up where the wind rocked. The cones dropped as light as feathers. Down in the hollow was the mourning dove—it was not too late for him.

The path ran up a hill. "Seem like there is chains about my feet, time I get this far," she said, in the voice of argument old people keep to use with themselves. "Something always take a hold of me on this hill—pleads I should stay."

After she got to the top she turned and gave a full, severe look behind her where she had come. "Up through pines," she said at length. "Now down through oaks."

Her eyes opened their widest, and she started down gently. But before she got to the bottom of the hill a bush caught her dress.

Her fingers were busy and intent, but her skirts were full and long, so that before she could pull them free in one place they were caught in another. It was not possible to allow the dress to tear. "I in the thorny bush," she said. "Thorns, you doing your appointed work. Never want to let folks pass—no sir, Old eyes thought you was a pretty little *green* bush."

Finally, trembling all over, she stood free, and after a moment dared to stoop for her cane.

"Sun so high!" she cried, leaning back and looking, while the thick tears went over her eyes. "The time getting all gone here."

At the foot of this hill was a place where a log was laid across the creek.

"Now comes the trial," said Phoenix.

Putting her right foot out, she mounted the log and shut her eyes. Lifting her skirt, leveling her cane fiercely before her, like a festival figure in some parade, she began to march across. Then she opened her eyes and she was safe on the other side.

"I wasn't as old as I thought," she said.

But she sat down to rest. She spread her skirts on the bank around her and folded her hands over her knees. Up above her was a tree in a pearly cloud of mistletoe. She did not dare to close her eyes, and when a little boy brought her a little plate with a slice of marble-cake on it she spoke to him. "That would be acceptable," she said. But when she went to take it there was just her own hand in the air.

So she left that tree, and had to go through a barbed-wire fence. There she bad to creep and crawl, spreading her knees and stretching her fingers like a baby trying to climb the steps. But she talked loudly to herself: she could not let her dress be torn now, so late in the day, and she could not pay for having her arm or her leg sawed off if she got caught fast where she was.

At last she was safe through the fence and risen up out in the clearing. Big dead trees, like black men with one arm, were standing in the purple stalks of the withered cotton field. There sat a buzzard.

"Who you watching?"

In the furrow she made her way along.

"Glad this not the season for bulls," she said, looking sideways, "and the good Lord made his snakes to curl up and sleep in the winter. A pleasure I don't see no two-headed snake coming around that tree, where it come once. It took a while to get by him, back in the summer."

She passed through the old cotton and went into a field of dead corn. It whispered and shook and was taller than her head, "Through the maze now," she said, for there was no path.

Then there was something tall, black, and skinny there, moving before her.

At first she took it for a man. It could have been a man dancing in the field. But she stood still and listened, and it did not make a sound. It was as silent as a ghost.

"Ghost," she said sharply, "who be you the ghost of? For I have heard of nary death close by."

But there was no answer—only the ragged dancing in the wind.

She shut her eyes, reached out her hand, and touched a sleeve. She found a coat and inside that an emptiness, cold as ice.

"You scarecrow," she said. Her face lighted. "I ought to be shut up for good," she said with laughter. "My senses is gone. I too old. I the oldest people I ever know. Dance, old scarecrow," she said, "while I dancing with you."

She kicked her foot over the furrow, and with mouth drawn down, shook her head once or twice in a little strutting way. Some husks blew down and whirled in streamers about her skirts.

Then she went on, parting her way from side to side with the cane, through the whispering field. At last she came to the end, to a wagon track where the silver grass blew between the red ruts. The quail were walking around like pullets, seeming all dainty and unseen.

"Walk pretty," she said. "This the easy place. This the easy going."

She followed the track, swaying through the quiet bare fields, through the little strings of trees silver in their dead leaves, past cabins silver from weather, with the doors and windows boarded shut, all like old women under a spell sitting there. "I walking in their sleep," she said, nodding her head vigorously.

In a ravine she went where a spring was silently flowing through a hollow log. Old Phoenix bent and drank. "Sweet-gum makes the water sweet," she said, and drank more. "Nobody know who made this well, for it was here when I was born."

The track crossed a swampy part where the moss hung as white as lace from every limb. "Sleep on, alligators, and blow your bubbles." Then the track went into the road.

Deep, deep the road went down between the high green-colored banks. Overhead the live-oaks met, and it was as dark as a cave.

A black dog with a lolling tongue came up out of the weeds by the ditch. She was meditating, and not ready, and when he came at her she only hit him a little with her cane. Over she went in the ditch, like a little puff of milkweed.

Down there, her senses drifted away. A dream visited her, and she reached her hand up, but nothing reached down and gave her a pull. So she lay there and presently went to talking. "Old woman," she said to herself, "that black dog come up out of the weeds to stall you off, and now there he sitting on his fine tail, smiling at you."

A white man finally came along and found her—a hunter, a young man, with his dog on a chain.

"Well, Granny!" he laughed. "What are you doing there?"

"Lying on my back like a June-bug waiting to be turned over, mister," she said, reaching up her hand.

He lifted her up, gave her a swing in the air, and set her down. "Anything broken, Granny?"

"No, sir, them old dead weeds is springy enough," said Phoenix, when she had got her breath. "I thank you for your trouble."

"Where do you live, Granny?" he asked, while the two dogs were growling at each other.

"Away back yonder, sir, behind the ridge. You can't even see it from here."

"On your way home?"

"No sir, I going to town."

"Why, that's too far! That's as far as I walk when I come out myself, and I get something for my trouble." He patted the stuffed bag he carried, and there hung down a little closed claw. It was one of the bob-whites, with its beak hooked bitterly to show it was dead. "Now you go on home, Granny!"

"I bound to go to town, mister," said Phoenix. "The time come around."

He gave another laugh, filling the whole landscape. "I know you old colored people! Wouldn't miss going to town to see Santa Claus!"

But something held old Phoenix very still. The deep lines in her face went into a fierce and different radiation. Without warning, she had seen with her own eyes a flashing nickel fall out of the man's pocket onto the ground.

"How old are you, Granny?" he was saying.

"There is no telling, mister," she said, "no telling."

Then she gave a little cry and clapped her hands and said, "Git on away from here, dog! Look! Look at that dog!" She laughed as if in admiration. "He ain't scared of nobody. He a big black dog." She whispered, "Sic him!"

"Watch me get rid of that cur," said the man. "Sic him, Pete! Sic him!"

Phoenix heard the dogs fighting, and heard the man running and throwing sticks. She even heard a gunshot. But she was slowly bending forward by that time, further and further forward, the lids stretched down over her eyes, as if she were doing this in her sleep. Her chin was lowered almost to her knees. The yellow palm of her hand came out from the fold of her apron. Her fingers slid down and along the ground under the piece of money with the grace and care they would have in lifting an egg from under a setting hen. Then she slowly straightened up, she stood erect, and the nickel was in her apron pocket. A bird flew by. Her lips moved. "God watching me the whole time. I come to stealing."

The man came back, and his own dog panted about them. "Well, I scared him off that time," he said, and then he laughed and lifted his gun and pointed it at Phoenix,

She stood straight and faced him.

"Doesn't the gun scare you?" he said, still pointing it.

"No, sir, I seen plenty go off closer by, in my day, and for less than what I done," she said, holding utterly still.

He smiled, and shouldered the gun. "Well, Granny," he said, "you must be a hundred years old, and scared of nothing. I'd give you a dime if I had any money with me. But you take my advice and stay home, and nothing will happen to you.

"I bound to go on my way, mister," said Phoenix. She inclined her head in the red rag. Then they went in different directions, but she could hear the gun shooting again and again over the hill.

She walked on. The shadows hung from the oak trees to the road like curtains. Then she smelled wood-smoke, and smelled the river, and she saw a steeple and the cabins on their steep steps. Dozens of little black children whirled around her. There ahead was Natchez shining. Bells were ringing. She walked on.

In the paved city it was Christmas time. There were red and green electric lights strung and criss-crossed everywhere, and all turned on in the daytime. Old Phoenix would have been lost if she had not distrusted her eyesight and depended on her feet to know where to take her.

She paused quietly on the sidewalk where people were passing by. A lady came along in the crowd, carrying an armful of red-, green-, and silver-wrapped presents; she gave off perfume like the red roses in hot summer, and Phoenix stopped her.

"Please, missy, will you lace up my shoe?" She held up her foot.

"What do you want, Grandma?"

"See my shoe," said Phoenix. "Do all right for out in the country, but wouldn't look right to go in a big building."

"Stand still then, Grandma." said the lady. She put her packages down on the sidewalk beside her and laced and tied both shoes tightly.

"Can't lace 'em with a cane," said Phoenix. "Thank you, missy. I doesn't mind asking a nice lady to tie up my shoe, when I gets out on the street."

Moving slowly and from side to side, she went into the big building, and into a tower of steps, where she walked up and around and around until her feet knew to stop.

She entered a door, and there she saw nailed up on the wall the document that had been stamped with the gold seal and framed in the gold frame, which matched the dream that was hung up in her head.

"Here I be," she said. There was a fixed and ceremonial stiffness over her body.

"A charity case, I suppose," said an attendant who sat at the desk before her.

But Phoenix only looked above her head. There was sweat on her face, the wrinkles in her skin shone like a bright net.

"Speak up, Grandma," the woman said. "What's your name? We must have your history, you know. Have you been here before? What seems to be the trouble with you?"

Old Phoenix only gave a twitch to her face as if a fly were bothering her.

"Are you deaf?" cried the attendant.

But then the nurse came in.

"Oh, that's just old Aunt Phoenix," she said. "She doesn't come for herself—she has a little grandson. She makes these trips just as regular as clockwork. She lives away back off the Old Natchez Trace." She bent down. "Well, Aunt Phoenix, why don't you just take a seat? We won't keep you standing after your long trip." She pointed.

The old woman sat down, bolt upright in the chair.

"Now, how is the boy?" asked the nurse.

Old Phoenix did not speak.

"I said, how is the boy?"

But Phoenix only waited and stared straight ahead, her face very solemn and withdrawn into rigidity.

"Is his throat any better?" asked the nurse. "Aunt Phoenix, don't you hear me? Is your grandson's throat any better since the last time you came for the medicine?".

With her hands on her knees, the old woman waited, silent, erect and motionless, just as if she were in armor.

"You mustn't take up our time this way, Aunt Phoenix," the nurse said. "Tell us quickly about your grandson, and get it over. He isn't dead, is he?"

At last there came a flicker and then a flame of comprehension across her face, and she spoke.

"My grandson. It was my memory had left me. There I sat and forgot why I made my long trip."

"Forgot?" The nurse frowned. "After you came so far?"

Then Phoenix was like an old woman begging a dignified forgiveness for waking up frightened in the night. "I never did go to school, I was too old at the Surrender," she said in a soft voice. "I'm an old woman without an education. It was my memory fail me. My little grandson, he is just the same, and I forgot it in the coming."

"Throat never heals, does it?" said the nurse, speaking in a loud, sure voice to old Phoenix. By now she had a card with something written on it, a little list. "Yes. Swallowed lye. When was it?—January—two-three years ago—"

Phoenix spoke unasked now. "No, missy, he not dead, he just the same. Every little while his throat begin to close up again, and he not able to swallow. He not get his breath. He not able to help himself. So the time come around, and I go on another trip for the soothing medicine."

"All right. The doctor said as long as you came to get it, you could have it," said the nurse. "But it's an obstinate case."

"My little grandson, he sit up there in the house all wrapped up, waiting by himself." Phoenix went on. "We is the only two left in the world. He suffer and it don't seem to put him back at all. He got a sweet look. He going to last. He wear a little patch quilt and peep out holding his mouth open like a little bird. I remembers so plain now. I not going to forget him again, no, the whole enduring time. I could tell him from all the others in creation."

"All right." The nurse was trying to hush her now. She brought her a bottle of medicine. "Charity." she said, making a check mark in a book.

Old Phoenix held the bottle close to her eyes, and then carefully put it into her pocket.

"I thank you," she said.

"It's Christmas time, Grandma," said the attendant. "Could I give you a few pennies out of my purse?"

"Five pennies is a nickel," said Phoenix stiffly.

"Here's a nickel," said the attendant.

Phoenix rose carefully and held out her hand. She received the nickel and then fished the other nickel out of her pocket and laid it beside the new one. She stared at her palm closely, with her head on one side.

Then she gave a tap with her cane on the floor.

"This is what come to me to do," she said. "I going to the store and buy my child a little windmill they sells, made out of paper. He going to find it hard to believe there such a thing in the world. I'll march myself back where he waiting, holding it straight up in his hand."

She lifted her free hand, gave a little nod, turned around, and walked out of the doctor's office. Then her slow step began on the stairs, going down.

The Jilting of Granny Weatherall

Katherine Anne Porter

She flicked her wrist neatly out of Doctor Harry's pudgy careful fingers and pulled the sheet up to her chin. The brat ought to be in knee breeches. Doctoring around the country with spectacles on his nose! "Get along now, take your schoolbooks and go. There's nothing wrong with me."

Doctor Harry spread a warm paw like a cushion on her forehead where the forked green vein danced and made her eyelids twitch. "Now, now, be a good girl, and we'll have you up in no time."

"That's no way to speak to a woman nearly eighty years old just because she's down. I'd have you respect your elders, young man."

"Well, Missy, excuse me." Doctor Harry patted her cheek. "But I've got to warn you, haven't I? You're a marvel, but you must be careful or you're going to be good and sorry."

"Don't tell me what I'm going to be. I'm on my feet now, morally speaking. It's Cornelia. I had to go to bed to get rid of her."

Her bones felt loose, and floated around in her skin, and Doctor Harry floated like a balloon around the foot of the bed. He floated and pulled down his waistcoat and swung his glasses on a cord. "Well, stay where you are, it certainly can't hurt you."

"Get along and doctor your sick," said Granny Weatherall. "Leave a well woman alone. I'll call for you when I want you. . . . Where were you forty years ago when I pulled through milk-leg and double pneumonia? You weren't even born. Don't let Cornelia lead you on," she shouted, because Doctor Harry appeared to float up to the ceiling and out. "I pay my own bills, and I don't throw my money away on nonsense!"

She meant to wave good-by, but it was too much trouble. Her eyes closed of themselves, it was like a dark curtain drawn around the bed. The pillow rose and floated under her, pleasant as a hammock in a light wind. She listened to the leaves rustling outside the window. No, somebody was swishing newspapers: no, Cornelia and Doctor Harry were whispering together. She leaped broad awake, thinking they whispered in her ear.

"She was never like this, *never* like this!" "Well, what can we expect?" "Yes, eighty years old. . . ."

Well, and what if she was? She still had ears. It was like Cornelia to whisper around doors. She always kept things secret in such a public way. She was always being tactful and kind. Cornelia was dutiful; that was the trouble with her. Dutiful and good: "So good and dutiful," said Granny, "that I'd like to spank her." She saw herself spanking Cornelia and making a fine job of it.

"What'd you say, Mother?"

Granny felt her face tying up in hard knots.

"Can't a body think, I'd like to know?"

"I thought you might want something."

"I do. I want a lot of things. First off, go away and don't whisper."

She lay and drowsed, hoping in her sleep that the children would keep out and let her rest a minute. It had been a long day. Not that she was tired. It was always pleasant to snatch a minute now and then. There was always so much to be done, let me see: tomorrow.

Tomorrow was far away and there was nothing to trouble about. Things were finished somehow when the time came; thank God there was always a little margin over for peace: then a person could spread out the plan of life and tuck in the edges orderly. It was good to have everything clean and folded away, with the hair brushes and tonic bottles sitting straight on the white embroidered linen: the day started without fuss and the pantry shelves laid out with rows of jelly glasses and brown jugs and white stone-china jars with blue whirligigs and words painted on them: coffee, tea, sugar, ginger, cinnamon, allspice: and the bronze clock with the lion on top nicely dusted off. The dust that lion could collect in twenty-four hours! The box in the attic with all those letters tied up, well, she'd have to go through that tomorrow. All those letters—George's letters and John's letters and her letters to them both—lying around for the children to find afterwards made her uneasy. Yes, that would be tomorrow's business. No use to let them know how silly she had been once.

While she was rummaging around she found death in her mind and it felt clammy and unfamiliar. She had spent so much time preparing for death there was no need for bringing it up again. Let it take care of itself now. When she was sixty she had felt very old, finished, and went around making farewell trips to see her children and grandchildren, with a secret in her mind: This is the very last of your mother, children! Then she made her will and came down with a long fever. That was all just a notion like a lot of other things, but it was lucky too, for she had once for all got over the idea of dying for a long time. Now she couldn't be worried. She hoped she had better sense now. Her father had lived to be one hundred and two years old and had drunk a noggin of strong hot toddy on his last birthday. He told the reporters it was his daily habit, and he owed his long life to that. He had made quite a scandal and was very pleased about it. She believed she'd just plague Cornelia a little.

"Cornelia! Cornelia!" No footsteps, but a sudden hand on her cheek. "Bless you, where have you been?"

"Here, Mother."

"Well, Cornelia, I want a noggin of hot toddy."

"Are you cold, darling?"

"I'm chilly, Cornelia. Lying in bed stops the circulation. I must have told you that a thousand times."

Well, she could just hear Cornelia telling her husband that Mother was getting a little childish and they'd have to humor her. The thing that most annoyed her was that Cornelia thought she was deaf, dumb, and blind. Little hasty glances and tiny gestures tossed around her and over her head saying, "Don't cross her, let her have her way, she's eighty years old," and she sitting there as if she lived in a thin glass cage. Sometimes Granny almost made up her mind to pack up and move back to her own house where nobody could remind her every minute that she was old. Wait, wait, Cornelia, till your own children whisper behind your back!

In her day she had kept a better house and had got more work done. She wasn't too old yet for Lydia to be driving eighty miles for advice when one of the children jumped the track, and Jimmy still dropped in and talked things over: "Now, Mammy, you've a good business head, I want to know what you think of this? . . ." Old. Cornelia couldn't change the furniture around without asking. Little things, little things! They had been so sweet when they were little. Granny wished the old days were back again with the children young and everything to be done over. It had been a hard pull, but not too much for her. When she thought of all the food she had cooked, and all the clothes she had cut and sewed, and all the gardens she had made—well, the children showed it. There they were, made out of her, and they couldn't get away from that. Sometimes she wanted to see John again and point to them and say, Well, I didn't do so badly, did I? But that would have to wait. That was for tomorrow. She used to think of him as a man, but now all the children were older than their father, and he would be a child beside her if she saw him now. It seemed strange and there was something wrong in the idea. Why, he couldn't possibly recognize her. She had fenced in a hundred acres once, digging the post holes herself and clamping the wires with just a negro boy to help. That changed a woman. John would be looking for a young woman with the peaked Spanish comb in her hair and the painted fan. Digging post holes changed a woman. Riding country roads in the winter when women had their babies was another thing: sitting up nights with sick horses and sick negroes and sick children and hardly ever losing one. John, I hardly ever lost one of them! John would see that in a minute, that would be something he could understand, she wouldn't have to explain anything!

It made her feel like rolling up her sleeves and putting the whole place to rights again. No matter if Cornelia was determined to be everywhere at once, there were a great many things left undone on this place. She would start tomorrow and do them. It was good to be strong enough for everything, even if all you made melted and changed and slipped under your hands, so that by the time you finished you almost forgot what you were working for. What was it I set out to do? she asked herself intently, but she could not remember. A fog rose over the valley, she saw it marching across the creek swallowing the trees and moving up the hill like an army of ghosts. Soon it would be at the near edge of the orchard, and then it was time to go in and light the lamps. Come in, children, don't stay out in the night air.

Lighting the lamps had been beautiful. The children huddled up to her and breathed like little calves waiting at the bars in the twilight. Their eyes followed the match and watched the flame rise and settle in a blue curve, then they moved away from her. The lamp was lit, they didn't have to be scared and hang on to mother any more. Never, never, never more. God, for all my life I thank Thee. Without Thee, my God, I could never have, done it. Hail, Mary, full of grace.

I want you to pick all the fruit this year and see that nothing is wasted. There's always someone who can use it. Don't let good things rot for want of using. You waste life when you waste good food. Don't let things get lost. It's bitter to lose things. Now, don't let me get to thinking, not when I am tired and taking a little nap before supper. . . .

The pillow rose about her shoulders and pressed against her heart and the memory was being squeezed out of it: oh, push down the pillow, somebody: it would smother her if she tried to hold it. Such a fresh breeze blowing and such a green day with no threats in it. But he had not come, just the same. What does a woman do when she has put on the white veil and set out the white cake for a man and he doesn't come? She tried to remember. No, I swear he never harmed me but in that. He never harmed me but in that . . . and what if he did? There was the day, the day, but a whirl of dark smoke rose and covered it, crept up and over into the bright field where everything was planted so carefully in orderly rows. That was hell, she knew hell when she saw it. For sixty years she had prayed against remembering him and against losing her soul in the deep pit of hell, and now the two things were mingled in one and the thought of him was a smoky cloud from hell that moved and crept in her head when she had just got rid of Doctor Harry and was trying to rest a minute. Wounded vanity, Ellen, said a sharp voice in the top of her mind. Don't let your wounded vanity get the upper hand of you. Plenty of girls get jilted. You were jilted, weren't you? Then stand up to it. Her eyelids wavered and let in streamers of blue-gray light like tissue paper over her eyes. She must get up and pull the shades down or she'd never sleep. She was in bed again and the shades were not down. How could that happen? Better turn over, hide from the light, sleeping in the light gave you nightmares. "Mother, how do you feel now?" and a stinging wetness on her forehead. But I don't like having my face washed in cold water!

Hapsy? George? Lydia? Jimmy? No, Cornelia, and her features were swollen and full of little puddles. "They're coming, darling, they'll all be here soon." Go wash your face, child, you look funny.

Instead of obeying, Cornelia knelt down and put her head on the pillow. She seemed to be talking but there was no sound. "Well, are you tongue-tied? Whose birthday is it? Are you going to give a party?"

Cornelia's mouth moved urgently in strange shapes. "Don't do that, you bother me, daughter."

"Oh, no, Mother. Oh, no. . . ."

Nonsense. It was strange about children. They disputed your every word. "No what, Cornelia?"

"Here's Doctor Harry."

"I won't see that boy again. He just left five minutes ago."

"That was this morning, Mother. It's night now. Here's the nurse."

"This is Doctor Harry, Mrs. Weatherall. I never saw you look so young and happy!"

"Ah, I'll never be young again—but I'd be happy if they'd let me lie in peace and get rested."

She thought she spoke up loudly, but no one answered. A warm weight on her forehead, a warm bracelet on her wrist, and a breeze went on whispering, trying to tell her something. A shuffle of leaves in the everlasting hand of God, He blew on them and they danced and rattled. "Mother, don't mind, we're going to give you a little hypodermic." "Look here, daughter, how do ants get in this bed? I saw sugar ants yesterday." Did you send for Hapsy too?

It was Hapsy she really wanted. She had to go a long way back through a great many rooms to find Hapsy standing with a baby on her arm. She seemed to herself to be Hapsy also, and the baby on Hapsy's arm was Hapsy and himself and herself, all at once, and there was no surprise in the meeting. Then Hapsy melted from within and turned flimsy as gray gauze and the baby was a gauzy shadow, and Hapsy came up close and said, "I thought you'd never come, and looked at her very searchingly and said, "You haven't changed a bit!" They leaned forward to kiss, when Cornelia began whispering from a long way off, "Oh, is there anything you want to tell me? Is there anything I can do for you?"

Yes, she had changed her mind after sixty years and she would like to see George. I want you to find George. Find him and be sure to tell him I forgot him. I want him to know I had my husband just the same and my children and my house like any other woman. A good house too and a good husband that I loved and fine children out of him. Better than I hoped for even. Tell him I was given back everything he took away and more. Oh, no, oh, God, no, there was something else besides the house and the man and the children. Oh, surely they were not all? What was it? Something not given back. . . . Her breath crowded down under her ribs and grew into a monstrous frightening shape with cutting edges; it bored up into her head, and the agony was unbelievable: Yes, John, get the Doctor now, no more talk, my time has come.

When this one was born it should be the last. The last. It should have been born first, for it was the one she had truly wanted. Everything came in good time. Nothing left out, left over. She was strong, in three days she would be as well as ever. Better. A woman needed milk in her to have her full health.

"Mother, do you hear me?"

"I've been telling you—"

"Mother, Father Connolly's here."

"I went to Holy Communion only last week. Tell him I'm not so sinful as all that."

"Father just wants to speak to you."

He could speak as much as he pleased. It was like him to drop in and inquire about her soul as if it were a teething baby, and then stay on for a cup of tea and a round of cards and gossip. He always had a funny story of some sort, usually about an Irishman who made his little mistakes and confessed them, and the point lay in some absurd thing he would blurt out in the confessional showing his struggles between native piety and original sin. Granny felt easy about her soul. Cornelia, where are your manners? Give Father Connolly a chair. She had her secret comfortable understanding with a few favorite saints who cleared a straight road to God for her. All as surely signed and sealed as the papers for the new Forty Acres. Forever . . . heirs and assigns forever. Since the day the wedding cake was not cut, but thrown out and wasted. The whole bottom dropped out of the world, and there she was blind and sweating with nothing under her feet and the walls falling away. His hand had caught her under the breast, she had not fallen, there was the freshly polished floor with the green rug on it, just as before. He had cursed like a sailor's parrot and said, "I'll kill him for you." Don't lay a hand on him, for my sake leave something to God. "Now, Ellen, you must believe what I tell you . . . ."

So there was nothing, nothing to worry about any more, except sometimes in the night one of the children screamed in a nightmare, and they both hustled out shaking and hunting for the matches and calling, "There, wait a minute, here we are!" John, get the doctor now, Hapsy's time has come. But there was Hapsy standing by the bed in a white cap. "Cornelia, tell Hapsy to take off her cap. I can't see her plain."

Her eyes opened very wide and the room stood out like a picture she had seen somewhere. Dark colors with the shadows rising towards the ceiling in long angles. The tall black dresser gleamed with nothing on it but John's picture, enlarged from a little one, with John's eyes very black when they should have been blue. You never saw him, so how do you know how he looked? But the man insisted the copy was perfect, it was very rich and handsome. For a picture, yes, but it's not my husband. The table by the bed had a linen cover and a candle and a crucifix. The light was blue from Cornelia's silk lampshades. No sort of light at all, just frippery. You had to live forty years with kerosene lamps to appreciate honest electricity. She felt very strong and she saw Doctor Harry with a rosy nimbus around him.

"You look like a saint, Doctor Harry, and I vow that's as near as you'll ever come to it."

"She's saying something."

"I heard you, Cornelia. What's all this carrying-on?"

"Father Connolly's saying—"

Cornelia's voice staggered and bumped like a cart in a bad road. It rounded corners and turned back again and arrived nowhere. Granny stepped up in the cart very lightly and reached for the reins, but a man sat beside her and she knew him by his hands, driving the cart. She did not look in his face, for she knew without seeing, but looked instead down the road where the trees leaned over and bowed to each other and a thousand birds were singing a Mass. She felt like singing too, but she put her hand in the bosom of her dress and pulled out a rosary, and Father Connolly murmured Latin in a very solemn voice and tickled her feet. My God, will you stop that nonsense? I'm a married woman. What if he did run away and leave me to face the priest by myself? I found another a whole world better. I wouldn't have exchanged my husband for anybody except St. Michael himself, and you may tell him that for me with a thank you in the bargain.

Light flashed on her closed eyelids, and a deep roaring shook her. Cornelia, is that lightning? I hear thunder. There's going to be a storm. Close all the windows. Call the children in. . . . "Mother, here we are, all of us." "Is that you, Hapsy?" "Oh, no, I'm Lydia. We drove as fast as we could." Their faces drifted above her, drifted away. The rosary fell out of her hands and Lydia put it back. Jimmy tried to help, their hands fumbled together, and Granny closed two fingers around Jimmy's thumb. Beads wouldn't do, it must be something alive. She was so amazed her thoughts ran round and round. So, my dear Lord, this is my death and I wasn't even thinking about it. My children have come to see me die. But I can't, it's not time. Oh, I always hated surprises. I wanted to give Cornelia the amethyst set—Cornelia, you're to have the amethyst set, but Hapsy's to wear it when she wants, and, Doctor Harry, do shut up. Nobody sent for you. Oh, my dear Lord, do wait a minute. I meant to do something about the Forty Acres, Jimmy doesn't need it and Lydia will later on, with that worthless husband of hers. I meant to finish the altar cloth and send six bottles of wine to Sister Borgia for her dyspepsia. I want to send six bottles of wine to Sister Borgia, Father Connolly, now don't let me forget.

Cornelia's voice made short turns and tilted over and crashed. "Oh, Mother, oh, Mother, oh, Mother. . . ."

"I'm not going, Cornelia. I'm taken by surprise. I can't go."

You'll see Hapsy again. What about her? "I thought you'd never come." Granny made a long journey outward, looking for Hapsy. What if I don't find her? What then? Her heart sank down and down, there was no bottom to death, she couldn't come to the end of it. The blue light from Cornelia's lampshade drew into a tiny point in the center of her brain, it flickered and winked like an eye, quietly it fluttered and dwindled. Granny lay curled down within herself, amazed and watchful, staring at the point of light that was herself; her body was now only a deeper mass of shadow in an endless darkness and this darkness would curl around the light and swallow it up. God, give a sign!

For the second time there was no sign. Again no bridegroom and the priest in the house. She could not remember any other sorrow because this grief wiped them all away. Oh, no, there's nothing more cruel than this—I'll never forgive it. She stretched herself with a deep breath and blew out the light.

The Red Convertible

Louise Erdrich

Lyman Lamartine

I was the first one to drive a convertible on my reservation. And of course it was red, a red Olds. I owned that car along with my brother Henry Junior. We owned it together until his boots filled with water on a windy night and he bought out my share. Now Henry owns the whole car, and his younger brother Lyman (that's myself), Lyman walks everywhere he goes.

How did I earn enough money to buy my share in the first place? My one talent was I could always make money. I had a touch for it, unusual in a Chippewa. From the first I was different that way, and everyone recognized it. I was the only kid they let in the American Legion Hall to shine shoes, for example, and one Christmas I sold spiritual bouquets for the mission door to door. The nuns let me keep a percentage. Once I started, it seemed the more money I made the easier the money came. Everyone encouraged it. When I was fifteen I got a job washing dishes at the Joliet Café, and that was where my first big break happened.

It wasn't long before I was promoted to bussing tables, and then the short-order cook quit and I was hired to take her place. No sooner than you know it I was managing the Joliet. The rest is history. I went on managing. I soon became part owner, and of course there was no stopping me then. It wasn't long before the whole thing was mine.

After I'd owned the Joliet for one year, it blew over in the worst tornado ever seen around here. The whole operation was smashed to bits. A total loss. The fryalator was up in a tree, the grill torn in half like it was paper. I was only sixteen. I had it all in my mother's name, and I lost it quick, but before I lost it I had every one of my relatives, and their relatives, to dinner, and I also bought that red Olds I mentioned, along with Henry.

The first time we saw it! I'll tell you when we first saw it. We had gotten a ride up to Winnipeg, and both of us had money. Don't ask me why, because we never mentioned a car or anything, we just had all our money. Mine was cash, a big bankroll from the Joliet's insurance. Henry had two checks—a week's extra pay for being laid off, and his regular check from the Jewel Bearing Plant.

We were walking down Portage anyway, seeing the sights, when we saw it. There it was, parked, large as life. Really as *if* it was alive. I thought of the word *repose,* because the car wasn't simply stopped, parked, or whatever. That car reposed, calm and gleaming, a FOR SALE sign in its left front window. Then, before we had thought it over at all, the car belonged to us and our pockets were empty. We had just enough money for gas back home.

We went places in that car, me and Henry. We took off driving all one whole summer. We started off toward the Little Knife River and Mandaree in Fort Berthold and then we found ourselves down in Wakpala somehow, and then suddenly we were over in Montana on the Rocky Boys, and yet the summer was not even half over. Some people hang on to details when they travel, but we didn't let them bother us and just lived our everyday lives here to there.

I do remember this one place with willows. I remember I laid under those trees and it was comfortable. So comfortable. The branches bent down all around me like a tent or a stable. And quiet, it was quiet, even though there was a powwow close enough so I could see it going on. The air was not too still, not too windy either. When the dust rises up and hangs in the air around the dancers like that, I feel good. Henry was asleep with his arms thrown wide. Later on, he woke up and we started driving again. We were somewhere in Montana, or maybe on the Blood Reserve—it could have been anywhere. Anyway it was where we met the girl.

All her hair was in buns around her ears, that's the first thing I noticed about her. She was posed alongside the road with her arm out, so we stopped. That girl was short, so short her lumber shirt looked comical on her, like a nightgown. She had jeans on and fancy moccasins and she carried a little suitcase.

"Hop on in," says Henry. So she climbs in between us.

"We'll take you home," I says. "Where do you live?"

"Chicken," she says.

"Where the hell's that?" I ask her.

"Alaska."

"Okay," says Henry, and we drive.

We got up there and never wanted to leave. The sun doesn't truly set there in summer, and the night is more a soft dusk. You might doze off, sometimes, but before you know it you're up again, like an animal in nature. You never feel like you have to sleep hard or put away the world. And things would grow up there. One day just dirt or moss, the next day flowers and long grass. The girl's name was Susy. Her family really took to us. They fed us and put us up. We had our own tent to live in by their house, and the kids would be in and out of there all day and night. They couldn't get over me and Henry being brothers, we looked so different. We told them we knew we had the same mother, anyway.

One night Susy came in to visit us. We sat around in the tent talking of this and that. The season was changing. It was getting darker by that time, and the cold was even getting just a little mean. I told her it was time for us to go. She stood up on a chair.

"You never seen my hair," Susy said.

That was true. She was standing on a chair, but still, when she unclipped her buns the hair reached all the way to the ground. Our eyes opened. You couldn't tell how much hair she had when it was rolled up so neatly. Then my brother Henry did something funny. He went up to the chair and said, "Jump on my shoulders." So she did that, and her hair reached down past his waist, and he started twirling, this way and that, so her hair was flung out from side to side.

"I always wondered what it was like to have long pretty hair," Henry says. Well we laughed. It was a funny sight, the way he did it. The next morning we got up and took leave of those people.

On to greener pastures, as they say. It was down through Spokane and across Idaho then Montana and very soon we were racing the weather right along under the Canadian border through Columbus, Des Lacs, and then we were in Bottineau County and soon home. We'd made most of the trip, that summer, without putting up the car hood at all. We got home just in time, it turned out, for the army to remember Henry had signed up to join it.

I don't wonder that the army was so glad to get my brother that they turned him into a Marine. He was built like a brick outhouse anyway. We liked to tease him that they really wanted him for his Indian nose. He had a nose big and sharp as a hatchet, like the nose on Red Tomahawk, the Indian who killed Sitting Bull, whose profile is on signs all along the North Dakota highways. Henry went off to training camp, came home once during Christmas, then the next thing you know we got an overseas letter from him. It was 1970, and he said he was stationed up in the northern hill country. Whereabouts I did not know. He wasn't such a hot letter writer, and only got off two before the enemy caught him. I could never keep it straight, which direction those good Vietnam soldiers were from.

I wrote him back several times, even though I didn't know if those letters would get through. I kept him informed all about the car. Most of the time I had it up on blocks in the yard or half taken apart, because that long trip did a hard job on it under the hood.

I always had good luck with numbers, and never worried about the draft myself. I never even had to think about what my number was. But Henry was never lucky in the same way as me. It was at least three years before Henry came home. By then I guess the whole war was solved in the government's mind, but for him it would keep on going. In those years I'd put his car into almost perfect shape. I always thought of it as his car while he was gone, even though when he left he said, "Now it's yours," and threw me his key.

"Thanks for the extra key," I'd said. "I'll put it up in your drawer just in case I need it." He laughed.

When he came home, though, Henry was very different, and I'll say this: the change was no good. You could hardly expect him to change for the better, I know. But he was quiet, so quiet, and never comfortable sitting still anywhere but always up and moving around. I thought back to times we'd sat still for whole afternoons, never moving a muscle, just shifting our weight along the ground, talking to whoever sat with us, watching things. He'd always had a joke, then, too, and now you couldn't get him to laugh, or when he did it was more the sound of a man choking, a sound that stopped up the throats of other people around him. They got to leaving him alone most of the time, and I didn't blame them. It was a fact: Henry was jumpy and mean.

I'd bought a color TV set for my mom and the rest of us while Henry was away. Money still came very easy. I was sorry I'd ever bought it though, because of Henry. I was also sorry I'd bought color, because with black-and-white the pictures seem older and farther away. But what are you going to do? He sat in front of it, watching it, and that was the only time he was completely still. But it was the kind of stillness that you see in a rabbit when it freezes and before it will bolt. He was not easy. He sat in his chair gripping the armrests with all his might, as if the chair itself was moving at a high speed and if he let go at all he would rocket forward and maybe crash right through the set.

Once I was in the room watching TV with Henry and I heard his teeth click at something. I looked over, and he'd bitten through his lip. Blood was going down his chin. I tell you right then I wanted to smash that tube to pieces. I went over to it but Henry must have known what I was up to. He rushed from his chair and shoved me out of the way, against the wall. I told myself he didn't know what he was doing.

My mom came in, turned the set off real quiet, and told us she had made something for supper. So we went and sat down. There was still blood going down Henry's chin, but he didn't notice it and no one said anything, even though every time he took a bite of his bread his blood fell onto it until he was eating his own blood mixed in with the food.

While Henry was not around we talked about what was going to happen to him. There were no Indian doctors on the reservation, and my mom was afraid of trusting Old Man Pillager because he courted her long ago and was jealous of her husbands. He might take revenge through her son. We were afraid that if we brought Henry to a regular hospital they would keep him.

"They don't fix them in those places," Mom said; "they just give them drugs."

"We wouldn't get him there in the first place," I agreed, "so let's just forget about it."

Then I thought about the car.

Henry had not even looked at the car since he'd gotten home, though like I said, it was in tip-top condition and ready to drive. I thought the car might bring the old Henry back somehow. So I bided my time and waited for my chance to interest him in the vehicle.

One night Henry was off somewhere. I took myself a hammer. I went out to that car and I did a number on its underside. Whacked it up. Bent the tail pipe double. Ripped the muffler loose. By the time I was done with the car it looked worse than any typical Indian car that has been driven all its life on reservation roads, which they always say are like government promises—full of holes. It just about hurt me, I'll tell you that! I threw dirt in the carburetor and I ripped all the electric tape off the seats. I made it look just as beat up as I could. Then I sat back and waited for Henry to find it.

Still, it took him over a month. That was all right, because it was just getting warm enough, not melting, but warm enough to work outside.

"Lyman," he says, walking in one day, "that red car looks like shit."

"Well it's old," I says. "You got to expect that."

"No way!" says Henry. "That car's a classic! But you went and ran the piss right out of it, Lyman, and you know it don't deserve that. I kept that car in A-one shape. You don't remember. You're too young. But when I left, that car was running like a watch. Now I don't even know if I can get it to start again, let alone get it anywhere near its old condition."

"Well you try," I said, like I was getting mad, "but I say it's a piece of junk."

Then I walked out before he could realize I knew he'd strung together more than six words at once.

After that I thought he'd freeze himself to death working on that car. He was out there all day, and at night he rigged up a little lamp, ran a cord out the window, and had himself some light to see by while he worked. He was better than he had been before, but that's still not saying much. It was easier for him to do the things the rest of us did. He ate more slowly and didn't jump up and down during the meal to get this or that or look out the window. I put my hand in the back of the TV set, I admit, and fiddled around with it good, so that it was almost impossible now to get a clear picture. He didn't look at it very often anyway. He was always out with that car or going off to get parts for it. By the time it was really melting outside, he had it fixed.

I had been feeling down in the dumps about Henry around this time. We had always been together before. Henry and Lyman. But he was such a loner now that I didn't know how to take it. So I jumped at the chance one day when Henry seemed friendly. It's not that he smiled or anything. He just said, "Let's take that old shitbox for a spin." Just the way he said it made me think he could be coming around.

We went out to the car. It was spring. The sun was shining very bright. My only sister, Bonita, who was just eleven years old, came out and made us stand together for a picture. Henry leaned his elbow on the red car's windshield, and he took his other arm and put it over my shoulder, very carefully, as though it was heavy for him to lift and he didn't want to bring the weight down all at once.

"Smile." Bonita said, and he did.

That picture. I never look at it anymore. A few months ago, I don't know why, I got his picture out and tacked it on the wall. I felt good about Henry at the time, close to him. I felt good having his picture on the wall, until one night when I was looking at television. I was a little drunk and stoned. I looked up at the wall and Henry was staring at me. I don't know what it was, but his smile had changed, or maybe it was gone. All I know is I couldn't stay in the same room with that picture. I was shaking. I got up, closed the door, and went into the kitchen. A little later my friend Ray came over and we both went back into that room. We put the picture in a brown bag, folded the bag over and over tightly, then put it way back in a closet.

I still see that picture now, as if it tugs at me, whenever I pass that closet door. The picture is very clear in my mind. It was so sunny that day Henry had to squint against the glare. Or maybe the camera Bonita held flashed like a mirror, blinding him, before she snapped the picture. My face is right out in the sun, big and round. But he might have drawn back, because the shadows on his face are deep as holes. There are two shadows curved like little hooks around the ends of his smile, as if to frame it and try to keep it there—that one, first smile that looked like it might have hurt his face. He has his field jacket on and the worn-in clothes he'd come back in and kept wearing ever since. After Bonita took the picture, she went into the house and we got into the car. There was a full cooler in the trunk. We started off, east, toward Pembina and the Red River because Henry said he wanted to see the high water.

The trip over there was beautiful. When everything starts changing, drying up, clearing off, you feel like your whole life is starting. Henry felt it, too. The top was down and the car hummed like a top. He'd really put it back in shape, even the tape on the seats was very carefully put down and glued back in layers. It's not that he smiled again or even joked, but his face looked to me as if it was clear, more peaceful. It looked as though he wasn't thinking of anything in particular except the bare fields and windbreaks and houses we were passing.

The river was high and full of winter trash when we got there. The sun was still out, but it was colder by the river. There were still little clumps of dirty snow here and there on the banks. The water hadn't gone over the banks yet, but it would, you could tell. It was just at its limit, hard swollen, glossy like an old gray scar. We made ourselves a fire, and we sat down and watched the current go. As I watched it I felt something squeezing inside me and tightening and trying to let go all at the same time. I knew I was not just feeling it myself; I knew I was feeling what Henry was going through at that moment. Except that I couldn't stand it, the closing and opening. I jumped to my feet. I took Henry by the shoulders and I started shaking him. "Wake up," I says, "wake up, wake up, wake up!" I didn't know what had come over me. I sat down beside him again.

His face was totally white and hard. Then it broke, like stones break all of a sudden when water boils up inside them.

"I know it," he says. "I know it. I can't help it. It's no use."

We start talking. He said he knew what I'd done with the car. It was obvious it had been whacked out of shape and not just neglected. He said he wanted to give the car to me for good now, it was no use. He said he'd fixed it just to give it back and I should take it.

"No way," I says. "I don't want it."

"That's okay," he says, "you take it."

"I don't want it, though," I says back to him, and then to emphasize, just to emphasize, you understand, I touch his shoulder. He slaps my hand off.

"Take that car," he says.

"No," I say. "Make me," I say, and then he grabs my jacket and rips the arm loose. That jacket is a class act, suede with tags and zippers. I push Henry backwards, off the log. He jumps up and bowls me over. We go down in a clinch and come up swinging hard, for all we're worth, with our fists. He socks my jaw so hard I feel like it swings loose. Then I'm at his ribcage and land a good one under his chin so his head snaps back. He's dazzled. He looks at me and I look at him and then his eyes are full of tears and blood and at first I think he's crying. But no, he's laughing. "Ha! Ha!" he says. "Ha! Ha! Take good care of it."

"Okay," I says. "Okay, no problem. Ha! Ha!"

I can't help it, and I start laughing, too. My face feels fat and strange, and after a while I get a beer from the cooler in the trunk, and when I hand it to Henry he takes his shirt and wipes my germs off. "Hoof-and-mouth disease," he says. For some reason this cracks me up, and so we're really laughing for a while, and then we drink all the rest of the beers one by one and throw them in the river and see how far, how fast, the current takes them before they fill up and sink.

"You want to go on back?" I ask after a while. "Maybe we could snag a couple nice Kashpaw girls."

He says nothing. But I can tell his mood is turning again.

"They're all crazy, the girls up here, every damn one of them."

"You're crazy too," I say, to jolly him up. "Crazy Lamartine boys!"

He looks as though he will take this wrong at first. His face twists, then clears, and he jumps up on his feet. "That's right!" he says. "Crazier 'n hell. Crazy Indians!"

I think it's the old Henry again. He throws off his jacket and starts swinging his legs out from the knees like a fancy dancer. He's down doing something between a grouse dance and a bunny hop, no kind of dance I ever saw before, but neither has anyone else on all this green growing earth. He's wild. He wants to pitch whoopee! He's up and at me and all over. All this time I'm laughing so hard, so hard my belly is getting tied up in a knot.

"Got to cool me off!" he shouts all of a sudden. Then he runs over to the river and jumps in.

There's boards and other things in the current. It's so high. No sound comes from the river after the splash he makes, so I run right over. I look around. It's getting dark. I see he's halfway across the water already, and I know he didn't swim there but the current took him. It's far. I hear his voice, though, very clearly across it.

"My boots are filling," he says.

He says this in a normal voice, like he just noticed and he doesn't know what to think of it. Then he's gone. A branch comes by. Another branch. And I go in.

By the time I get out of the river, off the snag I pulled myself onto, the sun is down. I walk back to the car, turn on the high beams, and drive it up the bank. I put it in first gear and then I take my foot off the clutch. I get out, close the door, and watch it plow softly into the water. The headlights reach in as they go down, searching, still lighted even after the water swirls over the back end. I wait. The wires short out. It is all finally dark. And then there is only the water, the sound of it going and running and going and running and running.

The Story of an Hour

Kate Chopin

Knowing that Mrs. Mallard was afflicted with a heart trouble, great care was taken to break to her as gently as possible the news of her husband's death.

It was her sister Josephine who told her, in broken sentences, veiled hints that revealed in half concealing. Her husband's friend Richards was there, too, near her. It was he who had been in the newspaper office when intelligence of the railroad disaster was received, with Brently Mallard's name leading the list of "killed." He had only taken the time to assure himself of its truth by a second telegram, and had hastened to forestall any less careful, less tender friend in bearing the sad message.

She did not hear the story as many women have heard the same, with a paralyzed inability to accept its significance. She wept at once, with sudden, wild abandonment, in her sister's arms. When the storm of grief had spent itself she went away to her room alone. She would have no one follow her.

There stood, facing the open window, a comfortable, roomy armchair. Into this she sank, pressed down by a physical exhaustion that haunted her body and seemed to reach into her soul.

She could see in the open square before her house the tops of trees that were all aquiver with the new spring life. The delicious breath of rain was in the air. In the street below a peddler was crying his wares. The notes of a distant song which some one was singing reached her faintly, and countless sparrows were twittering in the eaves.

There were patches of blue sky showing here and there through the clouds that had met and piled above the other in the west facing her window.

She sat with her head thrown back upon the cushion of the chair quite motionless, except when a sob came up into her throat and shook her, as a child who has cried itself to sleep continues to sob in its dreams.

She was young, with a fair, calm face, whose lines bespoke repression and even a certain strength. But now there was a dull stare in her eyes, whose gaze was fixed away off yonder on one of those patches of blue sky. It was not a glance of reflection, but rather indicated a suspension of intelligent thought.

There was something coming to her and she was waiting for it, fearfully. What was it? She did not know; it was too subtle and elusive to name. But she felt it, creeping out of the sky, reaching toward her through the sounds, the scents, the color that filled the air.

Now her bosom rose and fell tumultuously. She was beginning to recognize this thing that was approaching to possess her, and she was striving to beat it back with her will—as powerless as her two white slender hands would have been.

When she abandoned herself a little whispered word escaped her slightly parted lips. She said it over and over under her breath: "Free, free, free!" The vacant stare and the look of terror that had followed it went from her eyes. They stayed keen and bright. Her pulses beat fast, and the coursing blood warmed and relaxed every inch of her body.

She did not stop to ask if it were not a monstrous joy that held her. A clear and exalted perception enabled her to dismiss the suggestion as trivial.

She knew that she would weep again when she saw the kind, tender hands folded in death; the face that had never looked save with love upon her, fixed and gray and dead. But she saw beyond that bitter moment a long procession of years to come that would belong to her absolutely. And she opened and spread her arms out to them in welcome.

There would be no one to live for during those coming years; she would live for herself. There would be no powerful will bending her in that blind persistence with which men and women believe they have a right to impose a private will upon a fellow creature. A kind intention or a cruel intention made the act seem no less a crime as she looked upon it in that brief moment of illumination.

And yet she had loved him—sometimes. Often she had not. What did it matter! What could love, the unsolved mystery, count for in face of this possession of self-assertion which she suddenly recognized as the strongest impulse of her being.

"Free! Body and soul free!" she kept whispering.

Josephine was kneeling before the closed door with her lips to the keyhole, imploring for admission. "Louise, open the door! I beg; open the door—you will make yourself ill. What are you doing, Louise? For heaven's sake open the door."

"Go away. I am not making myself ill." No; she was drinking in a very elixir of life through that open window.

Her fancy was running riot along those days ahead of her. Spring days, and summer days, and all sorts of days that would be her own. She breathed a quick prayer that life might be long. It was only yesterday she had thought with a shudder that life might be long.

She arose at length and opened the door to her sister's importunities. There was a feverish triumph in her eyes, and she carried herself unwittingly like a goddess of Victory. She clasped her sister's waist, and together they descended the stairs. Richards stood waiting for them at the bottom.

Some one was opening the front door with a latchkey. It was Brently Mallard who entered, a little travel-stained, composedly carrying his grip-sack and umbrella. He had been far from the scene of accident, and did not even know there had been one. He stood amazed at Josephine's piercing cry; at Richards' quick motion to screen him from the view of his wife.

But Richards was too late.

When the doctors came they said she had died of heart disease—of joy that kills.

Something for the Time Being

Nadine Gordimer

He thought of it as discussing things with her; but the truth was that she did not help him out at all. She said nothing, while she ran her hand up the ridge of bone behind the rim of her child-sized yellow-brown ear, and raked her fingers tenderly into her hairline along the back of her neck as if feeling out some symptom in herself. Yet her listening was very demanding; when he stopped at the end of a supposition or a suggestion, her silence made the stop inconclusive. He had to take up again what he had said, carry it—where?

"Ve vant to give you a tsance, but you von't let us," he mimicked; and made a loud glottal click, half-angry, resentfully amused. He knew it wasn't because Kalzin Brothers were Jews that he had lost his job at last, but just because he had lost it, Mr. Solly's accent suddenly presented to him the irresistibly vulnerable. He had come out of prison nine days before, after spending three months as an awaiting-trial prisoner in a political case that had just been quashed—he was one of those who would not accept bail. He had been in prison three or four times since 1952; his wife Ella and the Kalzin Brothers were used to it. Until now, his employers had always given him his job back when he came out. They were importers of china and glass and he was head packer in a team of black men who ran the dispatch department. "Well, what the hell, I'll get something else," he said. "Hey?"

She stopped the self-absorbed examination of the surface of her skin for a slow moment and shrugged, looking at him.

He smiled.

Her gaze loosened hold like hands falling away from grasp. The ends of her nails pressed at small imperfections in the skin of her neck. He drank his tea and tore off pieces of bread to dip in it; then he noticed the tin of sardines she had opened and sopped up the pale matrix of oil in which ragged flecks of silver were suspended. She offered him more tea, without speaking.

They lived in one room of a decent, three-roomed house belonging to someone else; it was better for her that way, since he was often likely to have to be away for long stretches. She worked in a factory that made knitted socks; there was no one at home to look after their one child, a girl, and the child lived with a grandmother in a dusty, peaceful village a day's train-journey from the city.

He said, dismissing it as of no importance, "I wonder what chance they meant? You can imagine. I don't suppose they were going to give me an office with my name on it." He spoke as if she would appreciate the joke. She had known when she married him that he was a political man; she had been proud of him because he didn't merely want something for himself, like the other young men she knew, but everything, and for *the people.* It had excited her, under his influence, to change her awareness of herself as a young black girl to awareness of herself belonging to the people. She new that everything wasn't like something—a handout, a wangled privilege, a trinket you could hold. She would never get something from him.

Her hand went on searching over her skin as if it must come soon, come anxiously, to the flaw, the sickness, the evidence of what was wrong with her; for on this Saturday afternoon all these things that she knew had deserted her. She had lost her wits. All that she could understand was the one room, the child growing up far away in the mud house, and the fact that you couldn't keep a job if you kept being away from work for weeks at a time.

"I think I'd better look up Flora Donaldson," he said. Flora Donaldson was a white woman who had set up an office to help political prisoners. "Sooner the better: Perhaps she'll dig up something for me by Monday. It's the beginning of the month."

He got on all right with those people. Ella had met Flora Donaldson once; she was a pretty white woman who looked just like any white woman who would automatically send a black face around to the back door, but she didn't seem to know that she was white and you were black.

He pulled the curtain that hung across one corner of the room and took out his suit. It was a thin suit, of the kind associated with holiday-makers in American clothing advertisements, and when he was dressed in it, with a sharp-brimmed grey hat tilted slightly back on his small head, he looked a wiry, boyish figure, rather like one of these boy-men who sing and shake before a microphone, and whose clothes admirers try to touch as a talisman.

He kissed her goodbye, obliging her to put down, the lowering of a defense, the piece of sewing she held. She had cleared away the dishes from the table and set up the sewing-machine, and he saw that the shapes of cut material that lay on the table were the parts of a small girl's dress.

She spoke suddenly, "And when the next lot gets tired of you?"

"When that lot gets tired of me, I'll get another job again, that's all.

She nodded, very slowly, and her hand crept back to her neck.

"Who was that?" Madge Chadders asked.

Her husband had been out into the hall to answer the telephone. "Flora Donaldson. I wish you'd explain to these people exactly what sort of factory I've got. It's so embarrassing. She's trying to find a job for some chap, he's a skilled packer. There's no skilled packing done in my workshop, no skilled jobs at all done by black men. What on earth can I offer the fellow? She says he's desperate and anything will do.

Madge had the broken pieces of a bowl on a newspaper spread on the Persian carpet. "Mind the glue, darling! There, just next to your foot. Well, anything is better than nothing, I suppose it's someone who was in the Soganiland sedition case. Three months awaiting trial taken out of their lives, and now they're chucked back to fend for themselves."

William Chadders had not had New York black friends or mixed with coloured people on any but master-servant terms until he married Madge, but his views on the immorality and absurdity of the colour bar were sound; sounder, she often felt, than her own, for they were backed by the impersonal authority of a familiarity with the views of great thinkers, saints and philosophers, with history, political economy, sociology and anthropology. She knew only what she felt. And she always did something, at once, to express what she felt. She never measured the smallness of her personal protest against the establishment she opposed; she marched with Flora and eight hundred black women in a demonstration against African women being forced to carry passes; outside the university where she had once been a student, she stood between sandwich-boards bearing messages of mourning because a Bill had been passed closing the university, for the future, to all but white students; she had living in the house for three months a young African who wanted to write and hadn't the peace or space to get on with it in a location. She did not stop to consider the varying degree of usefulness of the things she did, and if others pointed this out to her and suggested that she might make up her mind to throw her weight on the side either of politics or philanthropy, she was not resentful but answered candidly that there was so little it was possible to do that she simply took any and every chance to get off her chest her disgust at the colour bar: when she had married William Chadders, her friends had thought that her protestant activities would stop; they underestimated not only Madge, but also William, who, although he was a wealthy businessman, subscribed to the view of absolute personal freedom as strictly as any bohemian. Besides he was not fool enough to want to change in any way the person who had enchanted him just as she was.

She reacted upon him, rather than he upon her; she, of course, would not hesitate to go ahead and change anybody. (But why not? she would have said, astonished. If it's to the good?) The attitude she sought to change would occur to her as something of independent existence, she would not see it as a cell in the organism of personality whose whole structure would have to regroup itself round the change. She had the boldness of being unaware of these consequences.

William did not carry a banner in the streets, of course; he worked there, among his first principles and historical precedents and economic necessities, but now they were translated from theory to practice of an anonymous, large-scale and behind-the-scenes sort—he was the brains and part of the money in a scheme to get Africans some economic power besides consumer power, through the setting up of an all-African trust company and investment corporation. A number of Madge's political friends, both white and black (like her activities, her friends were mixed, some political, some dogooders), thought this was putting the middle-class cart before the proletarian horse, but most of the African leaders welcomed the attempt as an essential backing to popular movements on other levels—something to count on outside the unpredictability of mobs. Sometimes it amused Madge to think that William, making a point at a meeting in a board room, fifteen floors above life in the streets, might achieve in five minutes something of more value than she did in all her days of turning her hand to anything—from sorting old clothes to roneoing[1] a manifesto or driving people during a bus boycott. Yet this did not knock the meaning out of her own life, for her; she knew that she had to see, touch and talk to people in order to care about them, that's all there was to it.

Before she and her husband dressed to go out that evening she finished sticking together the broken Chinese bowl and showed it to him with satisfaction. To her, it was whole again. But it was one of a set, that had belonged together and whose unity had illustrated certain philosophical concepts. William had bought them long ago, in London; for him, the whole set was damaged for ever.

He said nothing to her, but he was thinking of the bowls when she said to him as they drove off, "Will you see that chap, on Monday, yourself?"

He changed gear deliberately, attempting to follow her out of his preoccupation. But she said, "The man Flora's sending. What was his name?"

He opened his hand on the steering wheel, indicating that the name escaped him.

"See him yourself!"

"I'll have to leave it to the works manager to find something for him to do," he said.

"Yes, I know. But see him yourself, too?"

Her anxious voice made him feel very fond of her. He turned and smiled at her suspiciously. "Why?"

She was embarrassed at his indulgent manner: She said, frank and wheedling, "Just to show him. You know. That you know about him and it's not much of a job."

"All right," he said. "I'll see him myself."

He met her in town straight from the office on Monday and they went to the opening of an exhibition of paintings and on to dinner and to see a play, with friends. He had not been home at all, until they returned after midnight. It was a summer night and they sat for a few minutes on their terrace, where it was still mild with the warmth of the day's sun coming from the walls of darkness, and drank lime juice and water to quench the thirst that wine and the stuffy theater had given them. Madge made gasps and groans of pleasure at the release from the pressures of company and noise. Then she lay quiet for a while, her voice lifting now and then in fragments of unrelated comment on the evening—the occasional chirp of a bird that has already put its head under its wing for the night.

By the time they went in, they were free of the evening. Her black dress, her ear-rings and her bracelets felt like fancy-dress; she shed the character and sat on the bedroom carpet, and, passing her, he said, "Oh—that chap of Flora's came today, but I don't think he'll last. I explained to him that I didn't have the sort of job he was looking for."

"Well, that's all right, then," she said enquiringly. "What more could you do?

"Yes," he said, deprecating, "But I could see he didn't like the idea much. It's a cleaner's job; nothing for him. He's an intelligent chap, I didn't like having to offer it to him."

She was moving about her dressing table, piling out upon it the contents of her handbag. "Then I'm sure he'll understand. It'll give him something for the time being, anyway, darling. You can't help it if you don't need the sort of work he does."

"Huh, he won't last. I could see that. He accepted it, but only with his head. He'll get fed up. Probably won't turn up tomorrow, I had to speak to him about his Congress[2] button, too. The works manager came to me."

"What about his Congress button?" she said.

He was unbuttoning his shirt and his eyes were on the unread evening paper that lay folded on the bed. "He was wearing one," he said inattentively.

"I know, but what did you have to speak to him about it for?"

"He was wearing it in the workshop all day."

"Well, what about it?" She was sitting at her dressing-table, legs spread, as if she had sat heavily and suddenly. She was not looking at him, but at her own face.

He gave the paper a push and drew his pajamas from under the pillow. Vulnerable and naked, he said authoritatively, "You can't wear a button like that among the men in the workshop."

"Good heavens," she said, almost in relief, laughing, backing away from the edge of tension, chivvying him out of a piece of stuffiness. "And why can't you?"

"You can't have someone clearly representing a political organization like Congress."

"But he's not there *representing* anything, he's there as a workman?" Her mouth was still twitching with something between amusement and nerves.

"Exactly."

"Then why can't he wear a button that signifies his allegiance to an organization in his private life outside the workshop? There's no rule about not wearing tie-pins or club buttons or anything, in the workshop, is there?"

"No, there isn't, but that's not quite the same thing."

"My dear William," she said, "it is exactly the same. It's nothing to do with the works manager whether the man wears a Rotary button,[3] or an Elvis Presley button, or an African National Congress button. It's damn all his business."

"No, Madge, I'm sorry," William said, patient, "but it's not the same. I can give the man a job because I feel sympathetic towards the struggle he's in, but I can't put him in the workshop as a Congress man. I mean that wouldn't be fair to Fowler. That I can't do to Fowler." He was smiling as he went towards the bathroom, but his profile, as he turned into the doorway, was incisive.

She sat on at her dressing-table, pulling a comb through her hair, dragging it down through knots. Then she rested her face on her palms, caught sight of herself and became aware, against her fingers, of the curving shelf of bone, like the lip of a strong shell, under each eye. Everyone has his own intimations of mortality. For her, the feel of the bone beneath the face, in any living creature, brought her the message of the skull. Once hollowed out of this, outside the world, too. For what it's worth. It's worth a lot, the world, she affirmed, as she always did, life rising at once in her as a fish opens its jaws to a fly. It's worth a lot; and she sighed and got up with the sigh.

She went into the bathroom and sat down on the edge of the bath. He was lying there in the water, his chin relaxed on his chest, and he smiled at her. She said, "You mean you don't want Fowler to know."

"Oh," he said, seeing where they were again. "What is it I don't want Fowler to know?"

"You don't want your partner to know that you slip black men with political ideas into your workshop. Cheeky kaffir agitators. Specially a man who's been in jail for getting people to defy the government! What was his name; you never said?"

"Daniel something. I don't know. Mongoma or Ngoma. Something like that."

A line like a cut appeared between her eyebrows. "Why can't you remember his name?" Then she went on at once, "You don't want Fowler to know what you think, do you? That's it? You want to pretend you're like him, you don't mind the native in his place. You want to pretend that to please Fowler. You don't want Fowler to think you're cracked or Communist or whatever it is that good-natured, kind, jolly rich people like old Fowler think about people like us."

"I couldn't have less interest in what Fowler thinks outside our board room. And inside it, he never thinks about anything but how to sell more earth-moving gear.

"I don't mind the native in his place. You want him to think you go along with all that." She spoke aloud, but she seemed to be telling herself rather than him,

"Fowler and I run a factory. Our one common interest is the efficient running of that factory. Our *only* one. The factory depends on a stable, satisfied black labour-force, and that we've got. Right, you and I know that the whole black wage standard is too low, right, we know that they haven't a legal union to speak for them, right, we know that the conditions they live under make it impossible for them really to be stable. All that. But the fact is, so far as accepted standards go in this crazy country, they're a stable, satisfied labour-force with better working conditions than most. So long as I'm a partner in a business that lives by them, I can't officially admit an element that represents dissatisfaction with their lot."

"A green badge with a map of Africa on it," she said,

"If you make up your mind not to understand, you don't, and there it is," he said indulgently.

"You give him a job but you make him hide his Congress button."

He began to soap himself. She wanted everything to stop while she inquired into things, she could not go on while a remark was unexplained or a problem unsettled, but he represented a principle she subscribed to but found so hard to follow, that life must go on, trivially, commonplace, the trailing hem of the only power worth clinging to. She smoothed the film of her nightgown over the shape of her knees, over and over, and presently she said, in exactly the flat tone of statement that she had used before, the flat tone that was the height of belligerence in her, "He can say and do what he likes, he can call for strikes and boycotts and anything he likes, outside the factory, but he mustn't wear his Congress button at work."

He was standing up, washing his body that was full of scars; she knew them all, from the place on his left breast where a piece of shrapnel had gone in, all the way back to the place under his arm where he had torn himself on barbed wire as a child. "Yes, of course, anything he likes."

"Anything except his self-respect," she grumbled to herself. "Pretend, pretend. Pretend he doesn't belong to a political organization. Pretend he doesn't want to be a man. Pretend he hasn't been to prison for what he believes." Suddenly she spoke to her husband: "You'll let him have anything except the one thing worth giving."

They stood in uncomfortable proximity to each other, in the smallness of the bathroom. They were at once aware of each other as people who live in intimacy are only when hostility returns each to the confines of himself. He felt himself naked before her, where he had stepped out onto the towelling mat, and he took a towel and slowly covered himself, pushing the free end in round his waist. She felt herself an intrusion and, in silence, went out.

Her hands were tingling as if she were coming round from a faint. She walked up and down the bedroom floor like someone waiting to be summoned, called to account. I'll forget about it, she kept thinking, very fast. I'll forget about it again. Take a sip of water. Read another chapter. Don't call a halt. Let things flow, cover up, go on.

But when he came into the room with his wet hair combed and his stranger's face, and he said, "You're angry," it came from her lips, a black bird in the room, before she could understand what she had released—"I'm not angry. I'm beginning to get to know you."

Ella Mngoma knew he was going to a meeting that evening and didn't expect him home early. She put the paraffin lamp on the table so that she could see to finish the child's dress. It was done, buttons and all, by the time he came in at half past ten.

"Well, now we'll see what happens, I've got them to accept, *in principle,* that in future we won't take bail. You should have seen Ben Tsolo's face when I said that we lent the government our money interest-free when we paid bail. That really hit him. That was language he understood." He laughed, and did not seem to want to sit down, the heat of the meeting still upon him. "*In principle.* Yes, it's easy to accept in principle. We'll see."

She pumped the primus[4] and set a pot of stew to warm up for him. "Ah, that's nice." He saw the dress. "Finished already?" And she nodded vociferously in pleasure; but at once she noticed his forefinger run lightly along the braid round the neck, and the traces of failure that were always at the bottom of her cup tasted on her tongue again. Probably he was not even aware of it, or perhaps his instinct for what was true—the plumb line, the coin with the right ring—led him absently to it, but the fact was that she had botched the neck.

She had an almost Oriental delicacy about not badgering him and she waited until he had washed and sat down to eat before she asked, "How did the job go?"

"Oh that," he said. "It went." He was eating quickly, moving his tongue strongly round his mouth to marshal the bits of meat that escaped his teeth. She was sitting with him, feeling, in spite of herself, the rest of satisfaction in her evening's work. "Didn't you get it?"

"It *got me.* But I got loose again, all right."

She watched his face to see what he meant. "They don't want you to come back tomorrow?"

He shook his head, no, no, no, to stem the irritation of her suppositions. He finished his mouthful and said, "Everything very nice. Boss takes me into his office, apologizes for the pay, he knows it's not the sort of job I should have and so forth. So I go off and clean up in the assembly shop. Then at lunchtime he calls me into the office again: they don't want me to wear my A.N.C. badge at work. Flora Donaldson's sympathetic white man who's going to do me the great favour of paying me three pounds a week." He laughed. "Well, there you are."

She kept on looking at him. Her eyes widened and her mouth tightened; she was trying to prime herself to speak, or was trying not to cry. The idea of tears exasperated him and he held her with a firm, almost belligerently inquiring gaze. Her hand went up round the back of her neck under her collar, anxiously exploratory. "Don't do that!" he said. "You're like a monkey catching lice."

She took her hand down swiftly and broke into trembling, like a sweat. She began to breathe hysterically. "You couldn't put it in your pocket, for the day," she said wildly, grimacing at the bitterness of malice towards him.

He jumped up from the table. "Christ! I knew you'd say it! I've been waiting for you to say it. You've been wanting to say it for five years. Well, now it's out. Out with it. Spit it out!" She began to scream softly as if he were hitting her. The impulse to cruelty left him and he sat down before his dirty plate, where the battered spoon lay among bits of gristle and potato-eyes. Presently he spoke. "You come out and you think there's everybody waiting for you. The truth is, there isn't anybody. You think straight in prison because you've got nothing to lose. Nobody thinks straight outside. They don't want to hear you. What are you all going to do with me, Ella? Send me back to prison as quickly as possible? Perhaps I'll get a banishment order next time. That'd do. That's what you've got for me. I must keep myself busy with that kind of thing."

He went over to her and said, in a kindly voice, kneading her shoulder with spread fingers, "Don't cry. Don't cry. You're just like any other woman."

Notes

1. A kind of duplicating.
2. African National Congress, South African anti-apartheid organization.
3. A civic organization.
4. Portable stove which requires oil as fuel.

Samuel

Grace Paley

Some boys are very tough. They're afraid of nothing. They are the ones who climb a wall and take a bow at the top. Not only are they brave on the roof, but they make a lot of noise in the darkest part of the cellar where even the super hates to go. They also jiggle and hop on the platform between the locked doors of the subway cars.

Four boys are jiggling on the swaying platform. Their names are Alfred, Calvin, Samuel, and Tom. The men and the women in the cars on either side watch them. They don't like them to jiggle or jump but don't want to interfere. Of course some of the men in the cars were once brave boys like these. One of them had ridden the tail of a speeding truck from New York to Rockaway Beach without getting off, without his sore fingers losing hold. Nothing happened to him then or later. He had made a compact with other boys who preferred to watch: Starting at Eighth Avenue and Fifteenth Street, he would get to some specified place, maybe Twenty-third and the river, by hopping the tops of the moving trucks. This was hard to do when one truck turned a corner in the wrong direction and the nearest truck was a couple of feet too high. He made three or four starts before succeeding. He had gotten his idea from a film at school called *The Romance of Logging.* He had finished high school, married a good friend, was in a responsible job and going to night school.

These two men and others looked at the four boys jumping and jiggling on the platform and thought, It must be fun to ride that way, especially now the weather is nice and we're out of the tunnel and way high over the Bronx. Then they thought, These kids do seem to be acting sort of stupid. They *are* little. Then they thought of some of the brave things they had done when they were boys and jiggling didn't seem so risky.

The ladies in the car became very angry when they looked at the four boys. Most of them brought their brows together and hoped the boys could see their extreme disapproval. One of the ladies wanted to get up and say, Be careful you dumb kids, get off that platform or I'll call a cop. But three of the boys were Negroes and the fourth was something else she couldn't tell for sure. She was afraid they'd be fresh and laugh at her and embarrass her. She wasn't afraid they'd hit her, but she was afraid of embarrassment. Another lady thought, Their mothers never know where they are. It wasn't true in this particular case. Their mothers all knew that they had gone to see the missile exhibit on Fourteenth Street.

Out on the platform, whenever the train accelerated, the boys would raise their hands and point them up to the sky to act like rockets going off, then they rat-tat-tatted the shatterproof glass pane like machine guns, although no machine guns had been exhibited.

For some reason known only to the motorman, the train began a sudden slowdown. The lady who was afraid of embarrassment saw the boys jerk forward and backward and grab the swinging guard chains. She had her own boy at home. She stood up with determination and went to the door. She slid it open and said, "You boys will be hurt. You'll be killed. I'm going to call the conductor if you don't just go into the next car and sit down and be quiet."

Two of the boys said, "Yes'm," and acted as though they were about to go. Two of them blinked their eyes a couple of times and pressed their lips together. The train resumed its speed. The door slid shut, parting the lady and the boys. She leaned against the side door because she had to get off at the next stop.

The boys opened their eyes wide at each other and laughed. The lady blushed. The boys looked at her and laughed harder. They began to pound each other's back. Samuel laughed the hardest and pounded Alfred's back until Alfred coughed and the tears came. Alfred held tight to the chain hook. Samuel pounded him even harder when he saw the tears. He said, "Why you bawling? You a baby, huh?" and laughed. One of the men whose boyhood had been more watchful than brave became angry. He stood up straight and looked at the boys for a couple of seconds. Then he walked in a citizenly way to the end of the car, where he pulled the emergency cord. Almost at once, with a terrible hiss, the pressure of air abandoned the brakes and the wheels were caught and held.

People standing in the most secure places fell forward, then backward. Samuel had let go of his hold on the chain so he could pound Tom as well as Alfred. All the passengers in the cars whipped back and forth, but he pitched only forward and fell head first to be crushed and killed between the cars.

The train had stopped hard, halfway into the station, and the conductor called at once for the trainmen who knew about this kind of death and how to take the body from the wheels and brakes. There was silence except for passengers from other cars who asked, What happened! What happened! The ladies waited around wondering if he might be an only child. The men recalled other afternoons with very bad endings. The little boys stayed close to each other, leaning and touching shoulders and arms and legs.

When the policeman knocked at the door and told her about it, Samuel's mother began to scream. She screamed all day and moaned all night, though the doctors tried to quiet her with pills.

Oh, oh, she hopelessly cried. She did not know how she could ever find another boy like that one. However, she was a young woman and she became pregnant. Then for a few months she was hopeful. The child born to her was a boy. They brought him to be seen and nursed. She smiled. But immediately she saw that this baby wasn't Samuel. She and her husband together have had other children, but never again will a boy exactly like Samuel be known.

The Yellow Wallpaper

Charlotte Perkins Gilman

It is very seldom that mere ordinary people like John and myself secure ancestral halls for the summer.

A colonial mansion, a hereditary estate, I would say a haunted house, and reach the height of romantic felicity—but that would be asking too much of fate!

Still I will proudly declare that there is something queer about it.

Else, why should it be let so cheaply? And why have stood so long untenanted?

John laughs at me, of course, but one expects that in marriage.

John is practical in the extreme. He has no patience with faith, an intense horror of superstition, and he scoffs openly at any talk of things not to be felt and seen and put down in figures.

John is a physician, and *perhaps*—(I would not say it to a living soul, of course, but this is dead paper and a great relief to my mind—) *perhaps* that is one reason I do not get well faster.

You see he does not believe I am sick!

And what can one do?

If a physician of high standing, and one's own husband, assures friends and relatives that there is really nothing the matter with one but temporary nervous depression—a slight hysterical tendency—what is one to do?

My brother is also a physician, and also of high standing, and he says the same thing.

So I take phosphates or phosphites—whichever it is, and tonics, and journeys, and air, and exercise, and am absolutely forbidden to "work" until I am well again.

Personally, I disagree with their ideas.

Personally, I believe that congenial work, with excitement and change, would do me good.

But what is one to do?

I did write for a while in spite of them; but it *does* exhaust me a good deal—having to be so sly about it, or else meet with heavy opposition.

I sometimes fancy that in my condition if I had less opposition and more society and stimulus—but John says the very worst thing I can do is to think about my condition, and I confess it always makes me feel bad.

So I will let it alone and talk about the house.

The most beautiful place! It is quite alone, standing well back from the road, quite three miles from the village. it makes me think of English places that you read about, for there are hedges and walls and gates that lock, and lots of separate little houses for the gardeners and people.

There is a *delicious* garden! I never saw such a garden—large and shady, full of box-bordered paths, and lined with long grape-covered arbors with seats under them.

There were greenhouses, too, but they are all broken now.

There was some legal trouble, I believe, something about the heirs and coheirs; anyhow, the place has been empty for years.

That spoils my ghostliness, I am afraid, but I don't care—there is something strange about the house—I can feel it.

I even said so to John one moonlight evening, but he said what I felt was a *draught*, and shut the window.

I get unreasonably angry with John sometimes. I'm sure I never used to be so sensitive. I think it is due to this nervous condition.

But John says if I feel so, I shall neglect proper self-control; so I take pains to control myself—before him, at least, and that makes me very tired.

I don't like our room a bit. I wanted one downstairs that opened on the piazza and had roses all over the window, and such pretty old-fashioned chintz hangings! but John would not hear of it.

He said there was only one window and not room for two beds, and no near room for him if he took another.

He is very careful and loving, and hardly lets me stir without special direction.

I have a schedule prescription for each hour in the day; he takes all care from me, and so I feel basely ungrateful not to value it more.

He said we came here solely on my account, that I was to have perfect rest and all the air I could get. "Your exercise depends on your strength, my dear," said he, "and your food somewhat on your appetite; but air you can absorb all the time." So we took the nursery at the top of the house.

It is a big, airy room, the whole floor nearly, with windows that look all ways, and air and sunshine galore. It was nursery first and then playroom and gymnasium, I should judge; for the windows are barred for little children, and there are rings and things in the walls.

The paint and paper look as if a boys' school had used it. It is stripped off—the paper—in great patches all around the head of my bed, about as far as I can reach. and in a great place on the other side of the room low down. I never saw a worse paper in my life.

One of those sprawling flamboyant patterns committing every artistic sin.

It is dull enough to confuse the eye in following, pronounced enough to constantly irritate and provoke study, and when you follow the lame uncertain curves for a little distance they suddenly commit suicide—

plunge off at outrageous angles, destroy themselves in unheard of contradictions.

The color is repellent, almost revolting; a smoldering unclean yellow, strangely faded by the slow-turning sunlight.

It is a dull yet lurid orange in some places, a sickly sulphur tint in others.

No wonder the children hated it! I should hate it myself if I had to live in this room long.

There comes John, and I must put this away,—he hates to have me write a word.

I

We have been here two weeks, and I haven't felt like writing before, since that first day.

I am sitting by the window now, up in this atrocious nursery, and there is nothing to hinder my writing as much as I please, save lack of strength.

John is away all day, and even some nights when his cases are serious.

I am glad my case is not serious!

But these nervous troubles are dreadfully depressing.

John does not know how much I really suffer. He knows there is no *reason* to suffer, and that satisfies him.

Of course it is only nervousness. It does weigh on me so not to do my duty in any way!

I meant to be such a help to John, such a real rest and comfort, and here I am a comparative burden already!

Nobody would believe what an effort it is to do what little I am able,—to dress and entertain, and order things.

It is fortunate Mary is so good with the baby. Such a dear baby!

And yet I *cannot* be with him, it makes me so nervous.

I suppose John never was nervous in his life. He laughs at me so about this wallpaper!

At first he meant to repaper the room, but afterwards he said that I was letting it get the better of me, and that nothing was worse for a nervous patient than to give way to such fancies.

He said that after the wallpaper was changed it would be the heavy bedstead, and then the barred windows, and then that gate at the head of the stairs, and so on.

"You know the place is doing you good," he said, "and really, dear, I don't care to renovate the house just for a three months' rental."

"Then do let us go downstairs," I said, "there are such pretty rooms there."

Then he took me in his arms and called me a blessed little goose, and said he would go down cellar, if I wished, and have it whitewashed into the bargain.

But he is right enough about the beds and windows and things.

It is as airy and comfortable room as any one need wish, and, of course, I would not be so silly as to make him uncomfortable just for a whim.

I'm really getting quite fond of the big room, all but that horrid paper.

Out of one window I can see the garden, those mysterious deep-shaded arbors, the riotous old-fashioned flowers, and bushes and gnarly trees.

Out of another I get a lovely view of the bay and a little private wharf belonging to the estate. There is a beautiful shaded lane that runs down there from the house. I always fancy I see people walking in these numerous paths and arbors, but John has cautioned me not to give way to fancy in the least. He says that with my imaginative power and habit of story-making, a nervous weakness like mine is sure to lead to all manner of excited fancies, and that I ought to use my will and good sense to check the tendency. So I try.

I think sometimes that if I were only well enough to write a little it would relieve the press of ideas and rest me.

But I find I get pretty tired when I try.

It is so discouraging not to have any advice and companionship about my work. When I get really well, John says we will ask Cousin Henry and Julia down for a long visit; but he says he would as soon put fireworks in my pillow-case as to let me have those stimulating people about now.

I wish I could get well faster.

But I must not think about that. This paper looks to me as if it *knew* what a vicious influence it had!

There is a recurrent spot where the pattern lolls like a broken neck and two bulbous eyes stare at you upside down.

I get positively angry with the impertinence of it and the everlastingness. Up and down and sideways they crawl, and those absurd, unblinking eyes are everywhere. There is one place where two breadths didn't match, and the eyes go all up and down the line, one a little higher than the other.

I never saw so much expression in an inanimate thing before, and we all know how much expression they have! I used to lie awake as a child and get more entertainment and terror out of blank walls and plain furniture than most children could find in a toy-store.

I remember what a kindly wink the knobs of our big, old bureau used to have, and there was one chair that always seemed like a strong friend.

I used to feel that if any of the other things looked too fierce I could always hop into that chair and be safe.

The furniture in this room is no worse than inharmonious, however, for we had to bring it all from downstairs. I suppose when this was used as a playroom they had to take the nursery things out, and no wonder! I never saw such ravages as the children have made here.

The wallpaper, as I said before, is torn off in spots, and it sticketh closer than a brother—they must have had perseverance as well as hatred.

Then the floor is scratched and gouged and splintered, the plaster itself is dug out here and there, and this great heavy bed which is all we found in the room, looks as if it had been through the wars.

But I don't mind it a bit—only the paper.

There comes John's sister. Such a dear girl as she is, and so careful of me! I must not let her find me writing.

She is a perfect and enthusiastic housekeeper, and hopes for no better profession. I verily believe she thinks it is the writing which made me sick!

But I can write when she is out, and see her a long way off from these windows.

There is one that commands the road, a lovely shaded winding road, and one that just looks off over the country. A lovely country, too, full of great elms and velvet meadows.

This wallpaper has a kind of subpattern in a different shade, a particularly irritating one, for you can only see it in certain lights, and not clearly then.

But in the places where it isn't faded and where the sun is just so—I can see a strange, provoking, formless sort of figure, that seems to skulk about behind that silly and conspicuous front design.

There's sister on the stairs!

II

Well, the Fourth of July is over! The people are all gone and I am tired out. John thought it might do me good to see a little company, so we just had mother and Nellie and the children down for a week.

Of course I didn't do a thing. Jennie sees to everything now.

But it tired me all the same.

John says if I don't pick up faster he shall send me to Weir Mitchell in the fall.

But I don't want to go there at all. I had a friend who was in his hands once, and she says he is just like John and my brother, only more so!

Besides, it is such an undertaking to go so far.

I don't feel as it was worth while to turn my hand over for anything, and I'm getting dreadfully fretful and querulous.

I cry at nothing, and cry most of the time.

Of course I don't when John is here, or anybody else, but when I am alone.

And I am alone a good deal just now. John is kept in town very often by serious cases, and Jennie is good and lets me alone when I want her to.

So I walk a little in the garden or down the lovely lane, sit on the porch under the roses, and lie down up here a good deal.

I'm getting really fond of the room in spite of the wallpaper. Perhaps *because* of the wallpaper.

It dwells in my mind so!

I lie here on this great immovable bed—it is nailed down, I believe—and follow that pattern about by the hour. It is as good as gymnastics, I assure you. I start, we'll say, at the bottom, down in the corner over there where it has not been touched, and I determine for the thousandth time that I *will* follow that pointless pattern to some sort of a conclusion.

I know a little of the principle of design, and I know this thing was not arranged on any laws of radiation, or alternation, or repetition, or symmetry, or anything else that I ever heard of.

It is repeated, of course, by the breadths, but not otherwise.

Looked at in one way each breadth stands alone, the bloated curves and flourishes—a kind of "debased Romanesque" with *delirium tremens* —go waddling up and down in isolated columns of fatuity.

But, on the other hand, they connect diagonally, and the sprawling outlines run off in great slanting waves of optic horror, like a lot of wallowing seaweeds in full chase.

The whole thing goes horizontally, too, at least it seems so, and I exhaust myself in trying to distinguish the order of its going in that direction.

They have used a horizontal breadth for a frieze, and that adds wonderfully to the confusion.

There is one end of the room where it is almost intact, and there, when the crosslights fade and the low sun shines directly upon it, I can almost fancy radiation after all,—the interminable grotesques seem to form around a common center and rush off in headlong plunges of equal distraction.

It makes me tired to follow it. I will take a nap I guess.

III

I don't know why I should write this.

I don't want to.

I don't feel able.

And I know John would think it absurd. But I *must* say what I feel and think in some way—it is such a relief!

But the effort is getting to be greater than the relief.

Half the time now I am awfully lazy, and lie down ever so much.

John says I mustn't lose my strength, and has me take cod liver oil and lots of tonics and things, to say nothing of ale and wine and rare meat.

Dear John! He loves me very dearly, and hates to have me sick. I tried to have a real earnest reasonable talk with him the other day, and tell him how I wish he would let me go and make a visit to Cousin Henry and Julia.

But he said I wasn't able to go, nor able to stand it after I got there; and I did not make out a very good case for myself, for I was crying before I had finished.

It is getting to be a great effort for me to think straight. Just this nervous weakness I suppose.

And dear John gathered me up in his arms, and just carried me upstairs and laid me on the bed, and sat by me and read to me till it tired my head.

He said I was his darling and his comfort and all he had, and that I must take care of myself for his sake, and keep well.

He says no one but myself can help me out of it, that I must use my will and self-control and not let any silly fancies run away with me.

There's one comfort, the baby is well and happy, and does not have to occupy this nursery with the horrid wallpaper.

If we had not used it, that blessed child would have! What a fortunate escape! Why, I wouldn't have a child of mine, an impressionable little thing, live in such a room for worlds.

I never thought of it before, but it is lucky that John kept me here after all, I can stand it so much easier than a baby, you see.

Of course I never mention it to them any more—I am too wise,—but I keep watch of it all the same.

There are things in that paper that nobody knows but me, or ever will.

Behind that outside pattern the dim shapes get clearer every day.

It is always the same shape, only very numerous.

And it is like a woman stooping down and creeping about behind that pattern. I don't like it a bit. I wonder—I begin to think—I wish John would take me away from here!

IV

It is so hard to talk with John about my case, because he is so wise, and because he loves me so.

But I tried it last night.

It was moonlight. The moon shines in all around just as the sun does.

I hate to see it sometimes, it creeps so slowly, and always comes in by one window or another.

John was asleep and I hated to waken him, so I kept still and watched the moonlight on the undulating wallpaper till I felt creepy.

The faint figure behind seemed to shake the pattern, just as if she wanted to get out.

I got up softly and went to feel and see if the paper *did* move, and when I came back John was awake.

"What is it, little girl?" he said, "Don't go walking about like that—you'll get cold."

I thought it was a good time to talk, so I told him that I really was not gaining here, and that I wished he would take me away.

"Why, darling!" said he, "our lease will be up in three weeks, and I can't see how to leave before.

"The repairs are not done at home, and I cannot possibly leave town just now. Of course if you were in any danger, I could and would, but you really are better, dear, whether you can see it or not. I am a doctor, dear, and I know. You are gaining flesh and color, your appetite is better, I feel really much easier about you."

"I don't weigh a bit more," said I, "nor as much; and my appetite may be better in the evening when you are here, but is worse in the morning when you are away!"

"Bless her little heart!" said he with a big hug, "she shall be as sick as she pleases! But now let's improve the shining hours by going to sleep, and talk about it in the morning!"

"And you won't go away?" I asked gloomily.

"Why, how can I, dear? It is only three weeks more and then we will take a nice little trip of a few days while Jennie is getting the house ready. Really dear you are better!"

"Better in body perhaps—" I began, and stopped short, for he sat up straight and looked at me with such a stern, reproachful look that I could not say another word.

"My darling," said he, "I beg of you, for my sake and for our child's sake, as well as for your own, that you will never for one instant let that idea enter your mind! There is nothing so dangerous, so fascinating, to a temperament like yours. It is a false and foolish fancy. Can you not trust me as a physician when I tell you so?"

So of course I said no more on that score, and we went to sleep before long. he thought I was asleep first, but I wasn't and lay there for hours trying to decide whether that front pattern and the back pattern really did move together or separately.

V

On a pattern like this, by daylight, there is a lack of sequence, a defiance of law, that is a constant irritant to a normal mind.

The color is hideous enough, and unreliable enough, and infuriating enough, but the pattern is torturing.

You think you have mastered it, but just as you get well underway in following, it turns a back-somersault and there you are. It slaps you in the face, knocks you down, and tramples upon you. It is like a bad dream.

The outside pattern is a florid arabesque, reminding one of a fungus. If you can imagine a toadstool in joints, an interminable string of toadstools, budding and sprouting in endless convolutions—why, that is something like it.

That is, sometimes!

There is one marked peculiarity about this paper, a thing nobody seems to notice but myself, and that is that it changes as the light changes.

When the sun shoots in through the east window—I always watch for that first long, straight ray—it changes so quickly that I never can quite believe it.

That is why I watch it always.

By moonlight—the moon shines in all night when there is a moon—I wouldn't know it was the same paper.

At night in any kind of light, in twilight, candlelight, lamplight, and worst of all by moonlight, it becomes bars! The outside pattern I mean, and the women behind it is as plain as can be.

I didn't realize for a long time what the thing was that showed behind, that dim subpattern, but now I am quite sure it is a woman.

By daylight she is subdued, quiet. I fancy it is the pattern that keeps her so still. It is so puzzling. It keeps me quiet by the hour.

I lie down ever so much now. John says it is good for me, and to sleep all I can.

Indeed he started the habit by making me lie down for an hour after each meal.

It is a very bad habit, I am convinced, for you see I don't sleep.

And that cultivates deceit, for I don't tell them I'm awake—O no!

The fact is I am getting a little afraid of John.

He seems very queer sometimes, and even Jennie has an inexplicable look.

It strikes me occasionally, just as a scientific hypothesis,—that perhaps it is the paper!

I have watched John when he did not know I was looking, and come into the room suddenly on the most innocent excuses, and I've caught him several times *looking at the paper!* And Jennie too. I caught Jennie with her hand on it once.

She didn't know I was in the room, and when I asked her in a quiet, a very quiet voice, with the most restrained manner possible, what she was doing with the paper—she turned around as if she had been caught stealing, and looked quite angry—asked me why I should frighten her so!

Then she said that the paper stained everything it touched, that she had found yellow smooches on all my clothes and John's, and she wished we would be more careful!

Did not that sound innocent? But I know she was studying that pattern, and I am determined that nobody shall find it out but myself!

VI

Life is very much more exciting now than it used to be. You see I have something more to expect, to look forward to, to watch. I really do eat better, and am more quiet than I was.

John is so pleased to see me improve! He laughed a little the other day, and said I seemed to be flourishing in spite of my wallpaper.

I turned it off with a laugh. I had no intention of telling him it was *because* of the wallpaper—he would make fun of me. He might even want to take me away.

I don't want to leave now until I have found it out. There is a week more, and I think that will be enough.

VII

I'm feeling ever so much better! I don't sleep much at night, for it is so interesting to watch developments; but I sleep a good deal in the daytime.

In the daytime it is tiresome and perplexing.

There are always new shoots on the fungus, and new shades of yellow all over it. I cannot keep count of them, though I have tried conscientiously.

It is the strangest yellow, that wall-paper! it makes me think of all the yellow things I ever saw—not beautiful ones like buttercups, but old foul, bad yellow things.

But there is something else about that paper—the smell! I noticed it the moment we came into the room, but with so much air and sun it was not bad. Now we have had a week of fog and rain, and whether the windows are open or not, the smell is here.

It creeps all over the house.

I find it hovering in the dining-room, skulking in the parlor, hiding in the hall, lying in wait for me on the stairs.

It gets into my hair.

Even when I go to ride, if I turn my head suddenly and surprise it—there is that smell!

Such a peculiar odor, too! I have spent hours in trying to analyze it, to find what it smelled like.

It is not bad—at first, and very gentle, but quite the subtlest, most enduring odor I ever met.

In this damp weather it is awful, I wake up in the night and find it hanging over me.

It used to disturb me at first. I thought seriously of burning the house—to reach the smell.

But now I am used to it. The only thing I can think of that is like is the *color* of the paper! A yellow smell.

There is a very funny mark on this wall, low down, near the mopboard. A streak that runs round the room. It goes behind every piece of furniture, except the bed, a long, straight, even *smooch*, as if it had been rubbed over and over.

I wonder how it was done and who did it, and what they did it for. Round and round and round—round and round and round—it makes me *dizzy*!

VIII

I really have discovered something at last.

Through watching so much at night, when it changes so, I have finally found out.

The front pattern *does* move—and no wonder! The woman behind shakes it!

Sometimes I think there are a great many women behind, and sometimes only one, and she crawls around fast, and her crawling shakes it all over.

Then in the very bright spots she keeps still, and in the very shady spots she just takes hold of the bars and shakes them hard.

And she is all the time trying to climb through. But nobody could climb through that pattern—it strangles so; I think that is why it has so many heads.

They get through and then the pattern strangles them off and turns them upside down, and makes their eyes white!

If those heads were covered or taken off it would not be half so bad.

IX

I think that woman gets out in the daytime!

And I'll tell you why—privately—I've seen her!

I can see her out of every one of my windows!

It is the same woman, I know, for she is always creeping, and most women do not creep by daylight.

I see her in that long shaded lane, creeping up and down. I see her in those dark grape arbors, creeping all around the garden.

I see her on that long road under the trees, creeping along, and when a carriage comes she hides under the blackberry vines.

I don't blame her a bit. It must be very humiliating to be caught creeping by daylight!

I always lock the door when I creep by daylight. I can't do it at night, for I know John would suspect something at once.

And John is so queer now, that I don't want to irritate him. I wish he would take another room! Besides, I don't want anybody to get that woman out at night buy myself.

I often wonder if I could see her out of all the windows at once.

But, turn as fast as I can, I can only see out of one at one time.

And though I always see her, she *may* be able to creep faster than I can turn!

I have watched her sometimes away off in the open country, creeping as fast as a cloud shadow in a high wind.

X

If only that top pattern could be gotten off from the under one! I mean to try it, little by little.

I have found out another funny thing, but I shan't tell it this time! It does not do to trust people too much.

There are only two more days to get this paper off, and I believe John is beginning to notice. I don't like the look in his eyes.

And I heard him ask Jennie a lot of professional questions about me. She had a very good report to give.

She said I slept a good deal in the daytime.

John knows I don't sleep very well at night, for all I'm so quiet!

He asked me all sorts of questions, too, and pretended to be very loving and kind.

As if I couldn't see through him!

Still, I don't wonder he acts so, sleeping under this paper for three months.

It only interests me, but I feel sure John and Jennie are secretly affected by it.

XI

Hurrah! This is the last day, but it is enough. John to stay in town over night, and won't be out until this evening.

Jennie wanted to sleep with me—the sly thing! but I told her I should undoubtedly rest better for a night all alone.

That was clever, for really I wasn't alone a bit! As soon as it was moonlight and that poor thing began to crawl and shake the pattern, I got up and ran to help her.

I pulled and she shook, I shook and she pulled, and before morning we had peeled yards of that paper.

A strip about as high as my head and half around the room.

And then when the sun came and that awful pattern began to laugh at me, I declared I would finish it today!

We go away tomorrow, and they are moving all my furniture down again to leave things as they were before.

Jennie looked at the wall in amazement, but I told her merrily that I did it out of pure spite at the vicious thing.

She laughed and said she wouldn't mind doing it herself, but I must not get tired.

How she betrayed herself that time!

But I am here, and no person touches this paper but me,—not *alive!*

She tried to get me out of the room—it was too patent! But I said it was so quiet and empty and clean now that I believed I would lie down again and sleep all I could; and not to wake me even for dinner—I would call when I woke.

So now she is gone, and the servants are gone, and the things are gone, and there is nothing left but that great bedstead nailed down, with the canvas mattress we found on it.

We shall sleep downstairs tonight, and take the boat home tomorrow.

I quite enjoy the room, now it is bare again.

How those children did tear about here!

This bedstead is fairly gnawed!

But I must get to work.

I have locked the door and thrown the key down into the front path.

I don't want to go out, and I don't want to have anybody come in, till John comes.

I want to astonish him.

I've got a rope up here that even Jennie did not find. If that woman does get out, and tries to get away, I can tie her!

But I forgot I could not reach far without anything to stand on!

This bed will *not* move!

I tried to lift and push it until I was lame, and then I got so angry I bit off a little piece at one corner—but it hurt my teeth.

Then I peeled off all the paper I could reach standing on the floor. It sticks horribly and the pattern just enjoys it! All those strangled heads and bulbous eyes and waddling fungus growths just shriek with derision!

I am getting angry enough to do something desperate. To jump out of the window would be admirable exercise, but the bars are too strong even to try.

Besides I wouldn't do it. Of course not. I know well enough that a step like that is improper and might be misconstrued.

I don't like to *look* out of the windows even—there are so many of those creeping women, and they creep so fast.

I wonder if they all come out of that wallpaper as I did?

But I am securely fastened now by my well-hidden rope—you don't get *me* out in the road there!

I suppose I shall have to get back behind the pattern when it comes night, and that is hard!

It is so pleasant to be out in this great room and creep around as I please!

I don't want to go outside. I won't, even if Jennie asks me to.

For outside you have to creep on the ground, and everything is green instead of yellow.

But here I can creep smoothly on the floor, and my shoulders just fits in that long smooch around the wall, so I cannot lose my way.

Why there's John at the door!

It is no use, young man, you can't open it!

How he does call and pound!

Now he's crying for an axe.

It would be a shame to break down that beautiful door!

"John dear!" said I in the gentlest voice, "the key is down by the front steps, under a plantain leaf!"

That silenced him for a few moments.

Then he said—very quietly indeed, "Open the door, my darling!"

"I can't," said I. "The key is down by the front door under a plantain leaf!"

And then I said it again, several times, very gently and slowly, and said it so often that he had to go and see, and he got it of course, and came in. He stopped short by the door.

"What is the matter?" he cried. "For God's sake, what are you doing!"

I kept on creeping just the same, but I looked at him over my shoulder.

"I've got out at last," said I, "in spite of you and Jennie! And I've pulled off most of the paper, so you can't put me back!"

Now why should that man have fainted? But he did, and right across my path by the wall, so that I had to creep over him every time!

Death of a Son

Njabulo S. Ndebele

Njabulo S. Ndebele was born and grew up in Charterston Location, a township in South Africa. Educated at Cambridge University where he received an M.A., and at the University of Denver where he earned a Ph.D., he currently teaches African, African-American, and English literature at the University College of Roma, Lesotho. His volume of short stories, Fools *(1983) is set in a town like Charterston Location and portrays his childhood world. In a 1984 essay, Ndebele urges black writers to create stories that abjure polemic and stereotyping and that represent "lived experience in all its complexities."*

This story exhibits Ndebele's concern both with the "lived experience" of men and women and with the injustice of apartheid. The story also portrays the difference between a man's and a woman's responses to the tragedy of the death of a son.

At last we got the body. Wednesday. Just enough time for a Saturday funeral. We were exhausted. Empty. The funeral still ahead of us. We had to find the strength to grieve. There had been no time for grief, really. Only much bewilderment and confusion. Now grief. For isn't grief the awareness of loss?

That is why when we finally got the body, Buntu said: "Do you realize our son is dead?" I realized. Our awareness of the death of our first and only child had been displaced completely by the effort to get his body. Even the horrible events that caused the death: we did not think of them, as such. Instead, the numbing drift of things took over our minds: the pleas, letters to be written, telephone calls to be made, telegrams to be dispatched, lawyers to consult, "influential" people to "get in touch with," undertakers to be contacted, so much walking and driving. That is what suddenly mattered: the irksome details that blur the goal (no matter how terrible it is), each detail becoming a door which, once unlocked, revealed yet another door. Without being aware of it, we were distracted by the smell of the skunk and not by what the skunk had done.

We realized something too, Buntu and I, that during the two-week effort to get our son's body, we had drifted apart. For the first time in our marriage, our presence to each other had become a matter of habit. He was there. He'll be there. And I'll be there. But when Buntu said: "Do you realize our son is dead?" he uttered a thought that suddenly brought us together again. It was as if the return of the body of our son was also our coming together. For it was only at that moment that we really began to grieve; as if our lungs had suddenly begun to take air when just before, we were beginning to suffocate. Something with meaning began to emerge.

We realized. We realized that something else had been happening to us, adding to the terrible events. Yes, we had drifted apart. Yet, our estrangement, just at that moment when we should have been together, seemed disturbingly comforting to me. I was comforted in a manner I did not quite understand.

The problem was that I had known all along that we would have to buy the body anyway. I had known all along. Things would end that way. And when things turned out that way, Buntu could not look me in the eye. For he had said: "Over my dead body! Over my dead body!" as soon as we knew we would be required to pay the police or the government for the release of the body of our child.

"Over my dead body! Over my dead body!" Buntu kept on saying.

Finally, we bought the body. We have the receipt. The police insisted we take it. That way, they would be "protected." It's the law, they said.

I suppose we could have got the body earlier. At first I was confused, for one is supposed to take comfort in the heroism of one's man. Yet, inwardly, I could draw no comfort from his outburst. It seemed hasty. What sense was there to it when all I wanted was the body of my child? What would happen if, as events unfolded, it became clear that Buntu would not give up his life? What would happen? What would happen to him? To me?

For the greater part of two weeks, all of Buntu's efforts, together with friends, relatives, lawyers and the newspapers, were to secure the release of the child's body without the humiliation of having to pay for it. A "fundamental principle."

Why was it difficult for me to see the wisdom of the principle? The worst thing, I suppose, was worrying about what the police may have been doing to the body of my child. How they may have been busy prying it open "to determine the cause of death"?

Would I want to look at the body when we finally got it? To see further mutilations in addition to the "cause of death"? What kind of mother would not want to look at the body of her child? people will ask. Some will say: "It's grief." She is too grief-stricken.

"But still . . . ," they will say. And the elderly among them may say: "Young people are strange."

But how can they know? It was not that I would not want to see the body of my child, but that I was too afraid to confront the horrors of my own imagination. I was haunted by the thought of how useless it had been to have created something. What had been the point of it all? This body filling up with a child. The child steadily growing into something that could be seen and felt. Moving, as it always did, at that time of day when I was all alone at home waiting for it. What had been the point of it all?

How can they know that the mutilation to determine "the cause of death" ripped my own body? Can they think of a womb feeling hunted? Disgorged?

And the milk that I still carried. What about it? What had been the point of it all?

Even Buntu did not seem to sense that that principle, the "fundamental principle," was something too intangible for me at that moment, something that I desperately wanted should assume the form of my child's body. He still seemed far from ever knowing.

I remember one Saturday morning early in our courtship, as Buntu and I walked hand-in-hand through town, window-shopping. We cannot even be said to have been window-shopping, for we were aware of very little that was not ourselves. Everything in those windows was merely an excuse for words to pass between us.

We came across three girls sitting on the pavement, sharing a packet of fish and chips after they had just bought it from a nearby Portuguese cafe. Buntu said: "I want fish and chips too." I said: "So seeing is desire." I said: "My man is greedy!" We laughed. I still remember how he tightened his grip on my hand. The strength of it!

Just then, two white boys coming in the opposite direction suddenly rushed at the girls, and, without warning, one of them kicked the packet of fish and chips out of the hands of the girl who was holding it. The second boy kicked away the rest of what remained in the packet. The girl stood up, shaking her hand as if to throw off the pain in it. Then she pressed it under her armpit as if to squeeze the pain out of it. Meanwhile, the two boys went on their way laughing. The fish and chips lay scattered on the pavement and on the street like stranded boats on a river that had gone dry.

"Just let them do that to you!" said Buntu, tightening once more his grip on my hand as we passed on like sheep that had seen many of their own in the flock picked out for slaughter. We would note the event and wait for our turn. I remember I looked at Buntu, and saw his face was somewhat glum. There seemed no connection between that face and the words of reassurance just uttered. For a while, we went on quietly. It was then that I noticed his grip had grown somewhat limp. Somewhat reluctant. Having lost its self-assurance, it seemed to have been holding on because it had to, not because of a confident sense of possession.

It was not to be long before his words were tested. How could fate work this way, giving to words meanings and intentions they did not carry when they were uttered? I saw that day, how the language of love could so easily be trampled underfoot or scattered like fish and chips on the pavement, and left stranded and abandoned like boats in a river that suddenly went dry. Never again was love to be confirmed with words. The world around us was too hostile for vows of love. At any moment, the vows could be subjected to the stress of proof. And love died. For words of love need not be tested.

On that day, Buntu and I began our silence. We talked and laughed, of course, but we stopped short of words that would demand proof of action. Buntu knew. He knew the vulnerability of words. And so he sought to obliterate words with acts that seemed to promise redemption.

On that day, as we continued with our walk in town, that Saturday morning, coming up towards us from the opposite direction, was a burly Boer[1] walking with his wife and two children. They approached Buntu and me with an ominously determined advance. Buntu attempted to pull me out of the way, but I never had a chance. The Boer shoved me out of the way, as if clearing a path for his family. I remember, I almost crashed into a nearby fashion display window. I remember, I glanced at the family walking away, the mother and the father each dragging a child. It was for one of those children that I had been cleared away. Remember, also, that as my tears came out, blurring the Boer family and everything else, I saw and felt deeply what was inside of me: a desire to be avenged.

But nothing happened. All I heard was Buntu say: "The dog!" At that very moment, I felt my own hurt vanish like a wisp of smoke. And as my hurt vanished, it was replaced, instead, by a tormenting desire to sacrifice myself for Buntu. Was it something about the powerlessness of the curse and the desperation with which it had been made? The filling of stunned silence with an utterance? Surely it ate into him, revealing how incapable he was of meeting the call of his words.

And so it was, that that afternoon, back in the township, left to ourselves at Buntu's home, I gave in to him for the first time. Or should I say I offered myself to him? Perhaps from some vague sense of wanting to heal something in him? Anyway, we were never to talk about that event. Never. We buried it alive deep inside of me that afternoon. Would it ever be exhumed? All I vaguely felt and knew was that I had the keys to the vault. That was three years ago, a year before we married.

The cause of death? One evening I returned home from work, particularly tired after I had been covering more shootings by the police in the East Rand. Then I had hurried back to the office in Johannesburg to piece together on my typewriter the violent scenes of the day, and then to file my report to meet the deadline. It was late when I returned home, and when I got there. I found a crowd of people in the yard. They were those who could not get inside. I panicked. What had happened? I did not ask those who were outside, being desperate to get into the house. They gave way easily when they recognized me.

Then I heard my mother's voice. Her cry rose well above the noise. It turned into a scream when she saw me. "'What is it, mother?" I asked, embracing her out of a vaguely despairing sense of terror. But she pushed me away with an hysterical violence that astounded me.

"What misery have I brought you, my child?" she cried. At that point, many women in the room began to cry too. Soon, there was much wailing in the room, and then all over the house. The sound of it! The anguish! Understanding, yet eager for knowledge, I became desperate. I had to hold onto something. The desire to embrace my mother no longer had anything to do with comforting her, for whatever she had done, whatever its magnitude, had become inconsequential. I needed to embrace her for all the anguish that tied everyone in the house into a knot.

I wanted to be part of that knot, yet I wanted to know what had brought it about.

Eventually, we found each other, my mother and I, and clasped each other tightly. When I finally released her, I looked around at the neighbors and suddenly had a vision of how that anguish had to be turned into a simmering kind of indignation. The kind of indignation that had to be kept at bay only because there was a higher purpose at that moment: the sharing of concern.

Slowly and with a calmness that surprised me, I began to gather the details of what had happened. Instinctively, I seemed to have been gathering notes for a news report.

It happened during the day, when the soldiers and the police that had been patrolling the township in their Casspirs[2] began to shoot in the streets at random. Need I describe what I did not see? How did the child come to die just at that moment when the police and the soldiers began to shoot at random, at any moving thing? That was how one of our windows was shattered by a bullet. And that was when my mother, who looked after her grandchild when we were away at work, panicked. She picked up the child and ran to the neighbors. It was only when she entered the neighbor's house that she noticed the wetness of the blanket that covered the child she held to her chest as she ran for the sanctuary of neighbors. She had looked at her unaccountably bloody hand, then she noted the still bundle in her arms, and began at that moment to blame herself for the death of her grandchild . . .

Later, the police, on yet another round of shooting, found people gathered at our house. They stormed in, saw what had happened. At first, they dragged my mother out, threatening to take her away unless she agreed not to say what had happened. But then they returned and, instead, took the body of the child away. By what freak of logic did they hope that by this act their carnage would never be discovered?

That evening, I looked at Buntu closely. He appeared suddenly to have grown older. We stood alone in an embrace in our bedroom. I noticed, when I kissed his face, how his once lean face had grown suddenly puffy.

At that moment, I felt the familiar impulse come upon me once more, the impulse I always felt when I sensed that Buntu was in some kind of danger, the impulse to yield something of myself to him. He wore the look of someone struggling to gain control of something. Yet, it was clear he was far from controlling anything. I knew that look. Had seen it many times. It came at those times when I sensed that he faced a wave that was infinitely stronger than he, that it would certainly sweep him away, but that he had to seem to be struggling. I pressed myself tightly to him as if to vanish into him; as if only the two of us could stand up to the wave.

"Don't worry," he said. "Don't worry. I'll do everything in my power to right this wrong. Everything. Even if it means suing the police!" We went silent.

I knew that silence. But I knew something else at that moment: that I had to find a way of disengaging myself from the embrace.

Suing the police? I listened to Buntu outlining his plans. "Legal counsel. That's what we need." he said. "I know some people in Pretoria,"[3] he said. As he spoke. I felt the warmth of intimacy between us cooling. When he finished, it was cold. I disengaged from his embrace Why had Buntu spoken?

Later he was to speak again, when all his plans had failed to work: "Over my dead body! Over my dead body!"

He sealed my lips. I would wait for him to feel and yield one day to all the realities of misfortune.

Ours was a home, it could be said. It seemed a perfect life for a young couple: I, a reporter; Buntu, a personnel officer at an American factory manufacturing fanning implements. He had traveled to the United States and returned with a mind fired with dreams. We dreamed together. Much time we spent, Buntu and I, trying to make a perfect home. The occasions are numerous on which we paged through *Femina, Fair Lady, Cosmopolitan, Home Garden, Car*, as if somehow we were going to surround our lives with the glossiness in the magazines. Indeed, much of our time was spent window-shopping through the magazines. This time, it was different from the window-shopping we did that Saturday when we courted. This time our minds were consumed by the things we saw and dreamed of owning: the furniture, the fridge, TV, videocassette recorders, washing machines, even a vacuum cleaner and every other imaginable thing that would ensure a comfortable modern life.

Especially when I was pregnant. What is it that Buntu did not buy, then? And when the boy was born, Buntu changed the car. A family, he would say, must travel comfortably.

The boy became the center of Buntu's life. Even before he was born. Buntu had already started making inquiries at white private schools. That was where he would send his son, the bearer of his name.

Dreams! It is amazing how the horrible findings of my newspaper reports often vanished before the glossy magazines of our dreams, how I easily forgot that the glossy images were concocted out of the keys of typewriters, made by writers whose business was to sell dreams at the very moment that death pervaded the land. So powerful are words and pictures that even their makers often believe in them.

Buntu's ordeal was long. So it seemed. He would get up early every morning to follow up the previous day's leads regarding the body of our son. I wanted to go with him, but each time I prepared to go he would shake his head.

"It's my task," he would say. But every evening he returned, empty-handed, while with each day that passed and we did not know where the body of my child was, I grew restive and hostile in a manner that gave me much pain. Yet Buntu always felt compelled to give a report on each day's events. I never asked for it. I suppose it was his way of dealing with my silence.

One day he would say: "The lawyers have issued a court order that the body be produced. The writ of *habeas corpus*."[4]

On another day he would say: "We have petitioned the Minister of Justice."

On yet another he would say: I was supposed to meet the Chief Security Officer. Waited the whole day. At the end of the day they said I would see him tomorrow if he was not going to be too busy. They are stalling."

Then he would say: "The newspapers, especially yours, are raising the hue and cry. The government is bound to be embarrassed. It's a matter of time."

And so it went on. Every morning he got up and left. Sometimes alone, sometimes with friends. He always left to bear the failure alone.

How much did I care about lawyers, petitions and Chief Security Officers? A lot. The problem was that whenever Buntu spoke about his efforts, I heard only his words. I felt in him the disguised hesitancy of someone who wanted reassurance without asking for it. I saw someone who got up every morning and left not to look for results, but to search for something he could only have found with me.

And each time he returned, I gave my speech to my eyes. And he answered without my having parted my lips. As a result, I sensed, for the first time in my life, a terrible power in me that could make him do anything. And he would never ever be able to deal with that power as long as he did not silence my eyes and call for my voice.

And so, he had to prove himself. And while he left each morning, I learned to be brutally silent. Could he prove himself without me? Could he? Then I got to know, those days, what I'd always wanted from him. I got to know why I have always drawn him into me whenever I sensed his vulnerability.

I wanted him to be free to fear. Wasn't there greater strength that way? Had he ever lived with his own feelings? And the stress of life in this land: didn't it call out for men to be heroes? And should they live up to it even though the details of the war to be fought may often be blurred? They should.

Yet it is precisely for that reason that I often found Buntu's thoughts lacking in strength. They lacked the experience of strife that could only come from a humbling acceptance of fear and then, only then, the need to fight it.

Me? In a way, I have always been free to fear. The prerogative of being a girl. It was always expected of me to scream when a spider crawled across the ceiling. It was known I would jump onto a chair whenever a mouse blundered into the room.

Then, once more, the Casspirs came. A few days before we got the body back, I was at home with my mother when we heard the great roar of truck engines. There was much running and shouting in the streets. I saw them, as I've always seen them on my assignments: the Casspirs. On five occasions they ran down our street at great speed, hurling teargas canisters at random. On the fourth occasion, they got our house. The canister shattered another window and filled the house with the terrible pungent choking smoke that I had got to know so well. We ran out of the house gasping for fresh air.

So, this was how my child was killed? Could they have been the same soldiers? Now hardened to their tasks,? Or were they new ones being hardened to their tasks? Did they drive away laughing? Clearing paths for their families? What paths?

And was this our home? It couldn't be. It had to be a little bird's nest waiting to be plundered by a predator bird. There seemed no sense to the wedding pictures on the walls, the graduation pictures, birthday pictures, pictures of relatives and paintings of lush landscapes. There seemed to be no sense anymore to what seemed recognizably human in our house. It took only a random swoop to obliterate personal worth, to blot out any value there may have been to the past. In desperation, we began to live only for the moment. I do feel hunted.

It was on the night of the tear gas that Buntu came home, saw what had happened, and broke down in tears. They had long been in the coming . . .

My own tears welled out too. How much did we have to cry to refloat stranded boats? I was sure they would float again.

A few nights later, on the night of the funeral, exhausted, I lay on my bed, listening to the last of the mourners leaving. Slowly I became conscious of returning to the world. Something came back after it seemed not to have been there for ages. It came as a surprise, as a reminder that we will always live around what will happen. The sun will rise and set, and the ants will do their endless work, until one day the clouds turn gray and rain falls. and even in the township, the ants will fly out into the sky. Come what may.

My moon came, in a heavy surge of blood. And, after such a long time, I remembered the thing Buntu and I had buried in me. I felt it as if it had just entered. I felt it again as it floated away on the surge. I would be ready for another month. Ready as always, each and every month, for new beginnings.

And Buntu? I'll be with him, now. Always. Without our knowing, all the trying events had prepared for us new beginnings. Shall we not prevail?

Notes

1. A South African of Dutch lineage.
2. Armored trucks.
3. The administrative capital of the Republic of South Africa.
4. Literally translated—to produce the body. Legal jurisdiction to prevent false imprisonment.

The Wild Man of the Green Swamp

Maxine Hong Kingston

Before Reading

Connecting: Can you remember a time that in encountering a culture different from your own, you misinterpreted someone else's behavior? Or a time that someone unfamiliar with your culture misinterpreted your behavior?

Anticipating: How does Kingston invite us to "see" the Wild Man? Do we see him differently as the narrative unfolds?

For eight months in 1975, residents on the edge of Green Swamp, Florida, had been reporting to the police that they had seen a Wild Man. When they stepped toward him, he made strange noises as in a foreign language and ran back into the saw grass. At first, authorities said the Wild Man was a mass hallucination. Man-eating animals lived in the swamp, and a human being could hardly find a place to rest without sinking. Perhaps it was some kind of a bear the children had seen.

In October, a game officer saw a man crouched over a small fire, but as he approached, the figure ran away. It couldn't have been a bear because the Wild Man dragged a burlap bag after him. Also, the fire was obviously man-made.

The fish-and-game wardens and the sheriff's deputies entered the swamp with dogs but did not search for long; no one could live in the swamp. The mosquitoes alone would drive him out.

The Wild Man made forays out of the swamp. Farmers encountered him taking fruit and corn from the turkeys. He broke into a house trailer, but the occupant came back, and the Wild Man escaped out a window. The occupant said that a bad smell came off the Wild Man. Usually, the only evidence of him were his abandoned campsites. At one he left the remains of a four-foot-long alligator, of which he had eaten the feet and tail.

In May a posse made an air and land search; the plane signaled down to the hunters on the ground, who circled the Wild Man. A fish-and-game warden "brought him down with a tackle," according to the news.

The Wild Man fought, but they took him to jail. He looked Chinese, so they found a Chinese in town to come translate.

The Wild Man talked a lot to the translator. He told him his name. He said he was thirty-nine years old, the father of seven children, who were in Taiwan. To support them, he had shipped out on a Liberian freighter. He had gotten very homesick and asked everyone if he could leave the ship and go home. But the officers would not let him off. They sent messages to China to find out about him. When the ship landed, they took him to the airport and tried to put him on an airplane to some foreign place. Then, he said, the white demons took him to Tampa Hospital, which is for insane people, but he escaped, just walked out and went into the swamp.

The interpreter asked how he lived in the swamp. He said he ate snakes, turtles, armadillos, and alligators. The captors could tell how he lived when they opened up his bag, which was not burlap but a pair of pants with the legs knotted. Inside, he had carried a pot, a piece of sharpened tin, and a small club, which he had made by sticking a railroad spike into a section of aluminum tubing.

The sheriff found the Liberian freighter that the Wild Man had been on. The ship's officers said that they had not tried to stop him from going home. His shipmates had decided that there was something wrong with his mind. They had bought him a plane ticket and arranged his passport to send him back to China. They had driven him to the airport, but there he began screaming and weeping and would not get on the plane. So they had found him a doctor, who sent him to Tampa Hospital.

Now the doctors at the jail gave him medicine for the mosquito bites, which covered his entire body, and medicine for his stomachache. He was getting better, but after he'd been in jail for three days, the U.S. Border Patrol told him they were sending him back. He became hysterical. That night, he fastened his belt to the bars, wrapped it around his neck, and hung himself.

In the newspaper picture he did not look very wild, being led by the posse out of the swamp. He did not look dirty, either. He wore a checkered shirt unbuttoned at the neck, where his white undershirt showed; his shirt was tucked into his pants; his hair was short. He was surrounded by men in cowboy hats. His fingers stretching open, his wrists pulling apart to the extent of the handcuffs, he lifted his head, his eyes screwed shut, and cried out.

There was a Wild Man in our slough too, only he was a black man. He wore a shirt and no pants, and some mornings when we walked to school, we saw him asleep under the bridge. The police came and took him away. The newspaper said he was crazy; it said the police had been on the lookout for him for a long time, but we had seen him every day.

Kubota

Garrett Hongo

On December 8, 1941, the day after the Japanese attack on Pearl Harbor in Hawaii, my grandfather barricaded himself with his family—my grandmother, my teenage mother, her two sisters and two brothers—inside of his home in La'ie, a sugar plantation village on Oahu's North Shore. This was my maternal grandfather, a man most villagers called by his last name, Kubota. It could mean either "Wayside Field" or else "Broken Dreams," depending on which ideograms he used. Kubota ran La'ie's general store, and the previous night, after a long day of bad news on the radio, some locals had come by, pounded on the front door, and made threats. One was said to have brandished a machete. They were angry and shocked, as the whole nation was in the aftermath of the surprise attack. Kubota was one of the few Japanese Americans in the village and president of the local Japanese language school. He had become a target for their rage and suspicion. A wise man, he locked all his doors and windows and did not open his store the next day, but stayed closed and waited for news from some official.

He was a *kibei,* a Japanese American born in Hawaii (a U.S. territory then, so he was thus a citizen) but who was subsequently sent back by his father for formal education in Hiroshima, Japan, their home province. *Kibei* is written two ideograms in Japanese: one is the word for "return" and the other is the word for "rice." Poetically, it means one who returns from America, known as the Land of Rice in Japanese (by contrast, Chinese immigrants called their new home Mountain of Gold).

Kubota was graduated from a Japanese high school and then came back to Hawaii as a teenager. He spoke English—and a Hawaiian creole version of it at that—with a Japanese accent. But he was well liked and good at numbers, scrupulous and hard working like so any immigrants and children of immigrants. Castle & Cook, a grower's company that ran the sugarcane business along the North Shore, hired him on first as a stock boy and then appointed him to run one of its company stores. He did well, had the trust of management and labor—not an easy accomplishment in any day—married, had children, and had begun to exert himself in community affairs and excel in his own recreations. He put together a Japanese community organization that backed a Japanese language school for children and sponsored teachers from Japan. Kubota boarded many of them, in succession, in his own home. This made dinners a silent affair for his talkative, Hawaiian-bred children, as their stern *sensei,* or teacher, was nearly always at the table and their own abilities in the Japanese language were as delinquent as their attendance. While Kubota and the *sensei* rattled on about things Japanese, speaking Japanese, his children hurried through their suppers and tried to run off early to listen to the radio shows.

After dinner, while the *sensei* graded exams seated in a wicker chair in the spare room and his wife and children gathered around the radio in the front parlor, Kubota sat on the screened porch outside, reading the local Japanese newspapers. He finished reading about the same time as he finished the tea he drank for his digestion—a habit he'd learned in Japan—and then he'd get out his fishing gear and spread it out on the plank floors. The wraps on his rods needed to be redone, gears in his reels needed oil, and, once through with those tasks, he'd painstakingly wind on hundreds of yards of new line. Fishing was his hobby and his passion. He spent weekends camping along the North Shore beaches with his children, setting up umbrella tents, packing a rice pot and hibachi along for meals. And he caught fish. *Ulu'a* mostly, the huge surf-feeding fish known on the mainland as the jack crevalle, but he'd go after almost anything in its season. In Kawela, a plantation-owned bay nearby, he fished for mullet Hawaiian-style with a throw net, stalking the bottom-hugging, gray-backed schools as they gathered at the stream mouths and in the freshwater springs. In an outrigger out beyond the reef, he'd try for *aku*—the skipjack tuna prized for steaks and, sliced raw and mixed with fresh seaweed and cut onions, for *sashimi* salad. In Kahaluu and Ka'awa and on an offshore rock locals called Goat Island, he loved to go torching, stringing lanterns on bamboo poles stuck in the sand to attract *kumu'u,* the red goatfish, as they schooled at night just inside the reef. But in La'ie on Laniloa Point near Kahuku, the northernmost tip of Oahu, he cast twelve- and fourteen-foot surf rods for the huge, varicolored, and fast-running ulu'a as they ran for schools of squid and baitfish just beyond the biggest breakers and past the low sand flats wadeable from the shore to nearly a half mile out. At sunset, against the western light, he looked as if he walked on water as he came back, fish and rods slung over his shoulders, stepping along the rock and coral path just inches under the surface of a running tide.

When it was torching season, in December or January, he'd drive out the afternoon before and stay with old friends, the Tanakas or Yoshikawas, shopkeepers like him who ran stores near the fishing grounds. They'd have been preparing for weeks, selecting and cutting their bamboo poles, cleaning the hurricane lanterns, tearing up burlap sacks for the cloths they'd soak with kerosene and tie onto sticks they'd poke into the soft sand of the shallows. Once lit, touched off with a Zippo lighter, these would be the torches they'd use as beacons to attract the schooling fish. In another time, they might have made up a dozen paper lanterns of the kind mostly used for decorating the summer folk dances outdoors on the grounds of the Buddhist church during O-Bon, the Festival for the Dead. But now, wealthy and modern and efficient killers of fish, Tanaka and Kubota used rag torches and Colemans and cast rods with tips made of Tonkin bamboo and butts of American-spun fiberglass. After just one good night, they might bring back a prize bounty of a dozen burlap bags filled with scores of bloody, rigid fish delicious to eat and even better to give away as gifts to friends, family, and special customers.

It was a Monday night, the day after Pearl Harbor, and there was a rattling knock on the front door. Two FBI agents presented themselves, showed identification, and took my grandfather in for questioning in Honolulu. He didn't return home for days. No one knew what had happened or what was wrong. But there was a roundup going on of all those in the Japanese-American community suspected of sympathizing with the enemy and worse. My grandfather was suspected of espionage, of communicating with offshore Japanese submarines launched from the attack fleet days before war began. Torpedo planes and escort fighters, decorated with the insignia of the Rising Sun, had taken an approach route from northwest of Oahu directly across Kahuku Point and on toward Pearl. They had strafed an auxiliary air station near the fishing grounds my grandfather loved and destroyed a small gun battery there, killing three men. Kubota was known to have sponsored and harbored Japanese nationals in his own home. He had a radio. He had wholesale access to firearms. Circumstances and an undertone of racial resentment had combined with wartime hysteria in the aftermath of the tragic naval battle to cast suspicion on the loyalties of my grandfather and all other Japanese-Americans. The FBI reached out and pulled hundreds of them in for questioning in dragnets cast throughout the West Coast and Hawaii.

My grandfather was lucky; he'd somehow been let go after only a few days. Others were not as fortunate. Hundreds, from small communities in Washington, California, Oregon, and Hawaii, were rounded up and, after what appeared to be routine questioning, shipped off under Justice Department orders to holding centers in Leuppe on the Navaho reservation in Arizona, in Fort Missoula in Montana, and on Sand Island in Honolulu Harbor. There were other special camps on Maui in Ha'iku and on Hawaii—the Big Island—in my own home village of Volcano.

Many of these men—it was exclusively the Japanese-American men suspected of ties to Japan who were initially rounded up—did not see their families again for more than four years. Under a suspension of due process that was only after the fact ruled as warranted by military necessity, they were, if only temporarily, "disappeared" in Justice Department prison camps scattered in particularly desolate areas of the United States designated as militarily "safe." These were grim forerunners of the assembly centers and concentration camps for the 120,000 Japanese-American evacuees that were to come later.

I am Kubota's eldest grandchild, and I remember him as a lonely, habitually silent old man who lived with us in our home near Los Angeles for most of my childhood and adolescence. It was the fifties, and my parents had emigrated from Hawaii to the mainland in the hope of a better life away from the old sugar plantation. After some success, they had sent back for my grandparents and taken them in. And it was my grandparents who did the work of the household while my mother and father worked their salaried city jobs. My grandmother cooked and sewed, washed our clothes, and knitted in the front room under the light of a huge lamp with a bright three-way bulb. Kubota raised a flower garden, read up on soils and grasses in gardening books, and planted a zoysia lawn in front and a dichondra one in back. He planted a small patch near the rear block wall with green onions, eggplant, white Japanese radishes, and cucumber. While he hoed and spaded the loamless, clayey earth of Los Angeles, he sang particularly plangent songs in Japanese about plum blossoms and bamboo groves.

Once, in the mid-sixties, after a dinner during which, as always, he had been silent while he worked away at a meal of fish and rice spiced with dabs of Chinese mustard and catsup thinned with soy sauce, Kubota took his own dishes to the kitchen sink and washed them up. He took a clean jelly jar out of the cupboard—the glass was thick and its shape squatty like an old-fashioned. He reached around to the hutch below where he kept his bourbon. He made himself a drink and retired to the living room where I was expected to join him for "talk story," the Hawaiian idiom for chewing the fat.

I was a teenager and, though I was bored listening to stories I'd heard often enough before at holiday dinners, I was dutiful. I took my spot on the couch next to Kubota and heard him out. Usually, he'd tell me about his schooling in Japan where he learned judo along with mathematics and literature. He'd learned the *soroban* there—the abacus, which was the original pocket calculator of the Far East—and that, along with his strong, judo-trained back, got him his first job in Hawaii. This was the moral. "Study *ha-ahd*," he'd say with pidgin emphasis. "Learn read good. Learn speak da kine *good* English." The message is the familiar one taught to any children of immigrants: succeed through education. And imitation. But this time, Kubota reached down into his past and told me a different story. I was thirteen by then, and I suppose he thought me ready for it. He told me about Pearl Harbor, how the planes flew in wing after wing of formations over his old house in La'ie in Hawaii, and how, the next day, after Roosevelt had made his famous "Day of Infamy" speech about the treachery of the Japanese, the FBI agents had come to his door and taken him in, hauled him off to Honolulu for questioning, and held him without charge for several days. I thought he was lying. I thought he was making up a kind of horror story to shock me and give his moral that much more starch. But it was true, I asked around. I brought it up during history class in junior high school, and my teacher, after silencing me and stepping me off to the back of the room, told me that it was indeed so. I asked my mother and she said it was true. I asked my schoolmates, who laughed and ridiculed me for being so ignorant. We lived in a Japanese-American community, and the parents of most of my classmates were the *nisei* who had been interned as teenagers all through the war. But there was a strange silence around all of this. There was a hush, as if one were invoking the ill powers of the dead when one brought it up. No one cared to speak about the evacuation and relocation for very long. It wasn't in our history books, though we were studying World War II at the time. It wasn't in the family albums of the people I knew and whom I'd visit staying over weekends with friends. And it wasn't anything that the family talked about or allowed me to keep bringing up either. I was given the facts, told sternly and pointedly that "it was war" and that "nothing could be done." *"Shikatta ga nai"* is the phrase in Japanese, a kind of resolute and determinist pronouncement on how to deal with inexplicable tragedy. I was to know it but not to dwell on it. Japanese-Americans were busy trying to forget it ever happened and were having a hard enough time building their new lives after "camp." It was as if we had no history for four years and the relocation was something unspeakable.

But Kubota would not let it go. In session after session, for months it seemed, he pounded away at his story. He wanted to tell me the names of the FBI agents. He went over their questions and his responses again and again. He'd tell me how one would try to act friendly toward him, offering him cigarettes while the other, who hounded him with accusations and threats, left the interrogation room. Good cop, bad cop, I thought to myself, already superficially streetwise from stories black classmates told of the Watts riots and from my having watched too many episodes of *Dragnet* and *The Mod Squad.* But Kubota was not interested in my experiences. I was not made yet, and he was determined that his stories be part of my making. He spoke quietly at first, mildly, but once into his narrative and after his drink was down, his voice would rise and quaver with resentment and he'd make his accusations. He gave his testimony to me and I held it at first cautiously in my conscience like it was an heirloom too delicate to expose to strangers and anyone outside of the world Kubota made with his words. "I give you story now," he once said, "and you learn speak good, eh?" It was my job, as the disciple of his preaching I had then become, Ananda to his Buddha, to reassure him with a promise. "You learn speak good like the Dillingham," he'd say another time, referring to the wealthy scion of the grower family who had once run, unsuccessfully, for one of Hawaii's first senatorial seats. Or he'd then invoke a magical name, the name of one of his heroes, a man he thought particularly exemplary and righteous. "Learn speak dah good Ing-rish like *Mistah Inouye,"* Kubota shouted. "He *lick* dah Dillingham even in debate. I saw on *terre-bision* myself." He was remembering the debates before the first senatorial election just before Hawaii was admitted to the Union as its fiftieth state. "You *tell* story," Kubota would end. And I had my injunction.

The town we settled in after the move from Hawaii is called Gardena, the independently incorporated city south of Los Angeles and north of San Pedro harbor. At its northern limit, it borders on Watts and Compton, black towns. To the southwest are Torrance and Redondo Beach, white towns. To the rest of L.A., Gardena is primarily famous for having legalized five-card draw poker after the war. On Vermont Boulevard, its eastern border, there is a dingy little Vegas-like strip of card clubs with huge parking lots and flickering neon signs that spell out "The Rainbow" and "The Horseshoe" in timed sequences of varicolored lights. The town is only secondarily famous as the largest community of Japanese Americans in the United States outside of Honolulu, Hawaii. When I was in high school there, it seemed to me that every *sansei* kid I knew wanted to be a doctor, an engineer, or a pharmacist. Our fathers were gardeners or electricians or nurserymen or ran small businesses catering to other Japanese-Americans. Our mothers worked in civil service for the city or as cashiers for Thrifty Drug. What the kids wanted was a good job, good pay, a fine home, and no troubles. No one wanted to mess with the law—from either side—and no one wanted to mess with language or art. They all talked about getting into the right clubs so that they could go to the right schools. There was a certain kind of sameness, an intensely enforced system of conformity. Style was all. Boys wore moccasin-sewn shoes from Flagg Brothers, black A-1 slacks, and Kensington shirts with high collars. Girls wore their hair up in stiff bouffants solidified in hairspray and knew all the latest dances from the slauson to the funky chicken. We did well in chemistry and in math, no one who was Japanese but me spoke in English class or in history unless called upon, and no one talked about World War II. The day after Robert Kennedy was assassinated, after winning the California Democratic primary, we worked on calculus and elected class coordinators for the prom, featuring the 5th Dimension. We avoided grief. We avoided government. We avoided strong feelings and dangers of any kind. Once punished, we tried to maintain a concerted emotional and social discipline and would not willingly seek to fall out of the narrow margin of protective favor again.

But when I was thirteen, in junior high, I'd not understood why it was so difficult for my classmates, those who were themselves Japanese-American, to talk about the relocation. They had cringed, too, when I tried to bring it up during our discussions of World War II. I was Hawaiian-born. They were mainland-born. Their parents had been in camp, had been the ones to suffer the complicated experience of having to distance themselves from their own history and all things Japanese in order to make their way back and into the American social and economic mainstream. It was out of this sense of shame and a fear of stigma I was only beginning to understand that the *nisei* had silenced themselves. And, for their children, among whom I grew up, they wanted no heritage, no culture, no contact with a defiled history. I recall the silence very well. The Japanese-American children around me were burdened in a way I was not. Their injunction was silence. Mine was to speak.

Away at college, in another protected world in its own way as magical to me as the Hawaii of my childhood, I dreamed about my grandfather. Tired from studying languages, practicing German conjugations or scripting an army's worth of Chinese ideograms on a single sheet of paper, Kubota would come to me as I drifted off into sleep. Or I would walk across the newly mown ball field in back of my dormitory, cutting through a streetside phalanx of ancient eucalyptus trees on my way to visit friends off campus, and I would think of him, his anger, and his sadness.

I don't know myself what makes someone feel that kind of need to have a story they've lived through be deposited somewhere, but I can guess. I think about *The Iliad, The Odyssey, The Peloponnesian Wars* of

Thucydides, and a myriad of the works of literature I've studied. A character, almost a *topoi* he occurs so often, is frequently the witness who gives personal testimony about an event the rest of his community cannot even imagine. The sibyl is such a character. And Procne, the maid whose tongue is cut out so that she will not tell that she has been raped by her own brother-in-law, the king of Thebes. There are the dime novels, the epic blockbusters Hollywood makes into miniseries, and then there are the plain, relentless stories of witnesses who have suffered through horrors major and minor that have marked and changed their lives. I myself haven't talked to Holocaust victims. But I've read their survival stories and their stories of witness and been revolted and moved by them. My father-in-law, Al Thiessen, tells me his war stories again and again and I listen. A Mennonite who set aside the strictures of his own church in order to serve, he was a Marine codeman in the Pacific during World War II, in the Signal Corps on Guadalcanal, Morotai, and Bougainville. He was part of the island-hopping maneuver MacArthur had devised to win the war in the Pacific. He saw friends die from bombs which exploded not ten yards away. When he was with the 298th Signal Corps attached to the Thirteenth Air Force, he saw plane after plane come in and crash, just short of the runway, killing their crews, setting the jungle ablaze with oil and gas fires. Emergency wagons would scramble, bouncing over newly bulldozed land men used just the afternoon before for a football game. Every time we go fishing together, whether it's in a McKenzie boat drifting for salmon in Tillamook Bay or taking a lunch break from wading the rifles of a stream in the Cascades, he tells me about what happened to him and the young men in his unit. One was a Jewish boy from Brooklyn. One was a foul-mouthed kid from Kansas. They died. And he *has* to tell me. And I *have* to listen. It's a ritual payment the young owe their elders who have survived. The evacuation and relocation is something like that.

Kubota, my grandfather, had been ill with Alzheimer's disease for some time before he died. At the house he'd built on Kamehameha Highway in Hau'ula, a seacoast village just down the road from La'ie where he had his store, he'd wander out from the garage or greenhouse where he'd set up a workbench, and trudge down to the beach or up toward the line of pines he'd planted while employed by the Work Projects Administration during the thirties. Kubota thought he was going fishing. Or he thought he was back at work for Roosevelt, planting pines as a windbreak or soilbreak on the windward flank of the Ko'olau Mountains, emerald monoliths rising out of sea and cane fields from Waialua to Kaneohe. When I visited, my grandmother would send me down to the beach to fetch him. Or I'd run down Kam Highway a quarter mile or so and find him hiding in the cane field by the roadside, counting stalks, measuring circumferences in the claw of his thumb and forefinger. The look on his face was confused or concentrated, I didn't know which. But I guessed he was going fishing again. I'd grab him and walk him back to his house on the highway. My grandmother would shut him in a room.

Within a few years, Kubota had a stroke and survived it, then he had another one and was completely debilitated. The family decided to put him in a nursing home in Kahuku, just set back from the highway, within a mile or so of Kahuku Point and the Tanaka Store where he had his first job as a stock boy. He lived there three years, and I visited him once with my aunt. He was like a potato that had been worn down by cooking. Everything on him—his eyes, his teeth, his legs and torso—seemed like it had been sloughed away. What he had been was mostly gone now and I was looking at the nub of a man. In a wheelchair, he grasped my hands and tugged on them—violently. His hands were still thick and, I believed, strong enough to lift me out of my own seat into his lap. He murmured something in Japanese—he'd long ago ceased to speak any English. My aunt and I cried a little, and we left him.

I remember walking out on the black asphalt of the parking lot of the nursing home. It was heat-cracked and eroded already, and grass had veined itself into the interstices. There were coconut trees around, a cane field I could see across the street, and the ocean I knew was pitching a surf just beyond it. The green Ko'olaus came up behind us. Somewhere nearby, alongside the beach, there was an abandoned airfield in the middle of the canes. As a child, I'd come upon it playing one day, and my friends and I kept returning to it, day after day, playing war or sprinting games or coming to fly kites. I recognize it even now when I see it on TV—it's used as a site for action scenes in the detective shows Hollywood always sets in the islands: a helicopter chasing the hero racing away in a Ferrari, or gun dealers making a clandestine rendezvous on the abandoned runway. It was the old airfield strafed by Japanese planes the day the major flight attacked Pearl Harbor. It was the airfield the FBI thought my grandfather had targeted in his night fishing and signaling with the long surf poles he'd stuck in the sandy bays near Kahuku Point.

Kubota died a short while after I visited him, but not, I thought, without giving me a final message. I was on the mainland, in California studying for Ph.D. exams, when my grandmother called me with the news. It was a relief. He'd suffered from his debilitation a long time and I was grateful he'd gone. I went home for the funeral and gave the eulogy. My grandmother and I took his ashes home in a small, heavy metal box wrapped in a black *furoshiki,* a large silk scarf. She showed me the name the priest had given to him on his death, scripted with a calligraphy brush on a long, narrow talent of plain wood. Buddhist commoners, at death, are given priestly names, received symbolically into the clergy. The idea is that, in their next life, one of scholarship and leisure, they might meditate and attain the enlightenment the religion is aimed at. *"Shaku Shūchi,"* the ideograms read. It was Kubota's Buddhist name, incorporating characters from his family and given names. It meant "Shining Wisdom of the Law." He died on Pearl Harbor Day, December 7, 1983.

After years, after I'd finally come back to live in Hawaii again, only once did I dream of Kubota, my grandfather. It was the same night I'd heard HR 442, the redress bill for Japanese Americans, had been signed into law. In my dream that night Kubota was "torching," and he sang a Japanese song, a querulous and wavery folk ballad, as he hung paper lanterns on bamboo poles stuck into the sand in the shallow water of the lagoon behind the reef near Kahuku Point. Then he was at a work table, smoking a hand-rolled cigarette, letting it dangle from his lips Bogart-style as he drew, daintily and skillfully, with a narrow trim brush, ideogram after ideogram on a score of paper lanterns he had hung in a dark shed to dry. He had painted a talismanic mantra onto each lantern, the ideogram for the word "red" in Japanese, a bit of art blended with some superstition, a piece of sympathetic magic appealing to the magenta coloring on the rough skins of the schooling, night-feeding fish he wanted to attract to his baited hooks. He strung them from pole to pole in the dream then, hiking up his khaki worker's pants so his white ankles showed and wading through the shimmering black waters of the sand flats and then the reef. "The moon is leaving, leaving," he sang in Japanese. "Take me deeper in the savage sea." He turned and crouched like an ice racer then, leaning forward so that his unshaven face almost touched the light film of water. I could see the light stubble of beard like a fine, gray ash covering the lower half of his face. I could see his gold-rimmed spectacles. He held a small wooden boat in his cupped hands and placed it lightly on the sea and pushed it away. One of his lanterns was on it and, written in small neat rows like a sutra scroll, it had been decorated with the silvery names of all our dead.

The Conversion of the Jews

Philip Roth

"You're a real one for opening your mouth in the first place," Itzie said. "What do you open your mouth all the time for?"

"I didn't bring it up, Itz, I didn't," Ozzie said.

"What do you care about Jesus Christ for anyway?"

"I didn't bring up Jesus Christ. He did. I didn't even know what he was talking about. Jesus is historical, he kept saying. Jesus is historical." Ozzie mimicked the monumental voice of Rabbi Binder.

"Jesus was a person that lived like you and me," Ozzie continued. "That's what Binder said—"

"Yeah? . . . So what! What do I give two cents whether he lived or not. And what do you gotta open your mouth!" Itzie Lieberman favored closed-mouthedness, especially when it came to Ozzie Freedman's questions. Mrs. Freedman had to see Rabbi Binder twice before about Ozzie's questions and this Wednesday at four-thirty would be the third time. Itzie preferred to keep *his* mother in the kitchen; he settled for behind-the-back subtleties such as gestures, faces, snarls and other less delicate barnyard noises.

"He was a real person, Jesus, but he wasn't like God, and we don't believe he is God." Slowly, Ozzie was explaining Rabbi Binder's position to Itzie, who had been absent from Hebrew School the previous afternoon.

"The Catholics," Itzie said helpfully, "they believe in Jesus Christ, that he's God." Itzie Lieberman used "the Catholics" in its broadest sense—to include the Protestants.

Ozzie received Itzie's remark with a tiny head bob, as though it were a footnote, and went on. "His mother was Mary, and his father probably was Joseph," Ozzie said. "But the New Testament says his real father was God."

"His *real* father?"

"Yeah," Ozzie said, "that's the big thing, his father's supposed to be God."

"Bull."

"That's what Rabbi Binder says, that it's impossible—"

"Sure it's impossible. That stuff's all bull. To have a baby you gotta get laid," Itzie theologized. "Mary hadda get laid."

"That's what Binder says: 'The only way a woman can have a baby is to have intercourse with a man.'"

"He said *that,* Ozz?" For a moment it appeared that Itzie had put the theological question aside. "He said that, intercourse?" A little curled smile shaped itself in the lower half of Itzie's face like a pink mustache. "What you guys do, Ozz, you laugh or something?"

"I raised my hand."

"Yeah? Whatja say?"

"That's when I asked the question."

Itzie's face lit up. "Whatja ask about—intercourse?"

"No, I asked the question about God, how if He could create the heaven and earth in six days, and make all the animals and the fish and the light in six days—the light especially, that's what always gets me, that He could make the light. Making fish and animals, that's pretty good—"

"That's damn good." Itzie's appreciation was honest but unimaginative: it was as though God had just pitched a one-hitter.

"But making light . . . I mean when you think about it, it's really something," Ozzie said. "Anyway, I asked Binder if He could make all that in six days, and He could *pick* the six days he wanted right out of nowhere, why couldn't He let a woman have a baby without having intercourse."

"You said intercourse, Ozz, to Binder?"

"Yeah."

"Right in class?"

"Yeah."

Itzie smacked the side of his head.

"I mean, no kidding around," Ozzie said, "that'd really be nothing. After all that other stuff, that'd practically be nothing."

Itzie considered a moment. "What'd Binder say?"

"He started all over again explaining how Jesus was historical and how he lived like you and me but he wasn't God. So I said I under*stood* that. What I wanted to know was different."

What Ozzie wanted to know was always different. The first time he had wanted to know how Rabbi Binder could call the Jews "The Chosen People" if the Declaration of Independence claimed all men to be created equal. Rabbi Binder tried to distinguish for him between political equality and spiritual legitimacy, but what Ozzie wanted to know, he insisted vehemently, was different. That was the first time his mother had to come.

Then there was the plane crash. Fifty-eight people had been killed in a plane crash at La Guardia. In studying a casualty list in the newspaper his mother had discovered among the list of those dead eight Jewish names (his grandmother had nine but she counted Miller as a Jewish name); because of the eight she said the plane crash was "a tragedy." During free-discussion time on Wednesday Ozzy had brought to Rabbi Binder's attention this matter of "some of his relations" always picking out the Jewish names. Rabbi Binder had begun to explain cultural unity and some other things when Ozzie stood up at his seat and said that what he wanted to know was different. Rabbi Binder insisted that he sit down and it was then that Ozzie shouted that he wished all fifty-eight were Jews. That was the second time his mother came.

"And he kept explaining about Jesus being historical, and so I kept asking him. No kidding Itz, he was trying to make me look stupid."

"So what he finally do?"

"Finally he starts screaming that I was deliberately simple-minded and a wise-guy and that my mother had to come, and this was the last time. And that I'd never get bar-mitzvahed if he could help it. Then, Itz, then he starts talking in that voice like a statue, real slow and deep, and he says that I better think over what I said about the Lord. He told me to go to his office and think it over." Ozzie leaned his body towards Itzie. "Itz, I thought it over for a solid hour, and now I'm convinced God could do it."

Ozzie had planned to confess his latest transgression to his mother as soon as she came home from work. But it was Friday night in November and already dark, and when Mrs. Freedman came through the door she tossed off her coat, kissed Ozzie quickly on the face, and went to the kitchen table to light the three yellow candles, two for the Sabbath and one for Ozzie's father.

When his mother lit the candles she would move her two arms slowly towards her, dragging them through the air, as though persuading people whose minds were half made up. And her eyes would get glassy with tears. Even when his father was alive Ozzie remembered that her eyes had gotten glassy, so it didn't have anything to do with his dying. It had something to do with lighting the candles.

As she touched the flaming match to the unlit wick of a Sabbath candle, the phone rang, and Ozzie, standing only a foot from it, plucked it off the receiver and held it muffled to his chest. When his mother lit candles Ozzie felt there should be no noise; even breathing, if you could manage it, should be softened. Ozzie pressed the phone to his breast and watched his mother dragging whatever she was dragging, and he felt his own eyes get glassy. His mother was a round, tired, gray-haired penguin of a woman whose gray skin had begun to feel the tug of gravity and the weight of her own history. Even when she was dressed up she didn't look like a chosen person. But when she lit candles she looked like something better; like a woman who knew momentarily that God could do anything.

After a few mysterious minutes she was finished. Ozzie hung up the phone and walked to the kitchen table where she was beginning to lay the two places for the four-course Sabbath meal. He told her that she would have to see Rabbi Binder next Wednesday at four-thirty, and then he told her why. For the first time in their life together she hit Ozzie across the face with her hand.

All through the chopped liver and chicken soup part of the dinner Ozzie cried; he didn't have any appetite for the rest.

On Wednesday, in the largest of the three basement classrooms of the synagogue, Rabbi Marvin Binder, a tall, handsome, broad-shouldered many of thirty with thick strong-fibered black hair, removed his watch from his pocket and saw that it was four o'clock. At the rear of the room Yakov Blotnik, the seventy-one-year-old custodian, slowly polished the large window, mumbling to himself, unaware that it was four o'clock or six o'clock, Monday or Wednesday. To most of the students Yakov Blotnik's mumbling, along with his brown curly beard, scythe nose, and two heel-trailing black cats, made him an object of wonder, a foreigner, a relic, towards whom they were alternately fearful and disrespectful. To Ozzie the mumbling had always seemed a monotonous, curious prayer; what made it curious was that old Blotnik had been mumbling so steadily for so many years, Ozzie suspected he had memorized the prayers and forgotten all about God.

"It is now free-discussion time," Rabbi Binder said. "Feel free to talk about any Jewish matter at all—religion, family, politics, sports—"

There was silence. It was a gusty, clouded November afternoon and it did not seem as though there ever was or could be a thing called baseball. So nobody this week said a word about that hero from the past, Hank Greenberg—which limited free discussion considerably.

And the soul-battering Ozzie Freeman had just received from Rabbi Binder had imposed its limitation. When it was Ozzie's turn to read aloud from the Hebrew book the rabbi had asked him petulantly why he didn't read more rapidly. He was showing no progress. Ozzie said he could read faster but that if he did he was sure not to understand what he was reading. Nevertheless, at the rabbi's repeated suggestion Ozzie tried, and showed a great talent, but in the midst of a long passage he stopped short and said he didn't understand a word he was reading, and started in again at a drag-footed pace. Then came the soul battering.

Consequently when free-discussion time rolled around none of the students felt too free. The rabbi's invitation was answered only by the mumbling of feeble old Blotnik.

"Isn't there anything at all you would like to discuss?" Rabbi Binder asked again, looking at his watch. "No questions or comments?"

There was a small grumble from the third row. The rabbi requested that Ozzie rise and give the rest of the class the advantage of his thought.

Ozzie rose. "I forget it now," he said, and sat down in his place.

Rabbi Binder advanced a seat towards Ozzie and poised himself on the edge of the desk. It was Itzie's desk and the rabbi's frame only a dagger's-length away from his face snapped him to sitting attention.

"Stand up again, Oscar," Rabbi Binder said calmly, "and try to assemble your thoughts."

Ozzie stood up. All his classmates turned in their seats and watched as he gave an unconvincing scratch to his forehead.

"I can't assemble any," he announced and plunked himself down.

"Stand up!" Rabbi Binder advanced from Itzie's desk to the one directly in front of Ozzie; when the rabbinical back was turned Itzie gave it five-fingers off the tip of his nose, causing a small titter in the room. Rabbi Binder was too absorbed in squelching Ozzie's nonsense once and for all to bother with titters. "Stand up, Oscar. What's your question about?"

Ozzie pulled a word out of the air. It was the handiest word. "Religion"

"Oh, now you remember?"

"Yes."

"What is it?"

Trapped, Ozzie blurted the first thing that came to him. "Why can't He make anything He wants to make!"

As Rabbi Binder prepared an answer, a final answer, Itzie, ten feet behind him, raised one finger on his left hand, gestured it meaningfully towards the rabbi's back, and brought the house down.

Binder twisted quickly to see what had happened and in the midst of the commotion Ozzie shouted into the rabbi's back what he couldn't have shouted to his face. It was a loud, toneless sound that had the timbre of something stored inside for about six days.

"You don't know! You don't know anything about God!"

The rabbi spun back towards Ozzie. "What?"

"You don't know—you don't—"

"Apologize, Oscar, apologize!" It was a threat.

"You don't—"

Rabbi Binder's hand flicked out at Ozzie's cheek. Perhaps it had only been meant to clamp the boy's mouth shut, but Ozzie ducked and the palm caught him squarely on the nose.

The blood came in a short, red spurt on to Ozzie's shirt front.

The next moment was all confusion. Ozzie screamed, "You bastard, you bastard!" and broke for the classroom door. Rabbi Binder lurched a step backwards, as though his own blood had started flowing violently in the opposite direction, then gave a clumsy lurch forward and bolted out the door after Ozzie. The class followed after the rabbi's huge blue-suited back, and before old Blotnik could turn from his window, the room was empty and everyone was headed full speed up the three flights leading to the roof.

If one should compare the light of day to the life of man: sunrise to birth; sunset—the dropping down over the edge—to death; then as Ozzie Freedman wiggled through the trapdoor of the synagogue roof, his feet kicking backwards bronco-style at Rabbi Binder's outstretched arms—at that moment the day was fifty years old. As a rule, fifty or fifty-five reflects accurately the age of late afternoons in November, for it is in that month, during those hours, that one's awareness of light seems no longer a matter of seeing, but of hearing: light begins clicking away. In fact, as Ozzie locked shut the trapdoor in the rabbi's face, the sharp click of the bolt into the lock might momentarily have been mistaken for the sound of the heavier gray that had just throbbed through the sky.

With all his weight Ozzie kneeled on the locked door; any instant he was certain that Rabbi Binder's should would fling it open, splintering the wood into shrapnel and catapulting his body into the sky. But the door did not move and below him he heard only the rumble of feet, first loud, then dim, like thunder rolling away.

A question shot through his brain. "Can this be *me*?" For a thirteen-year-old who had just labeled his religious leader a bastard, twice, it was not an improper question. Louder and louder the question came to him—"Is it me? Is it me?"—until he discovered himself no longer kneeling, but racing crazily towards the edge of the roof, his eyes crying, his throat screaming, and his arms flying everywhichway as though not his own.

"Is it me? Is it me Me Me Me Me! It has to be me—but is it!"

It is the question a thief must ask himself the night he jimmies open his first window, and it is said to be the question with which bridegrooms quiz themselves before the altar.

In the few wild seconds it took Ozzie's body to propel him to the edge of the roof, his self-examination began to grow fuzzy. Gazing down at the street, he became confused as to the problem beneath the question: was it, is-it-me-who-called-Binder-a-bastard? or, is-it-me-prancing-around-on-the-roof? However, the scene below settled all, for there is an instant in any action when whether it is you or somebody else is academic. The thief crams the money in his pockets and scoots out the window. The bridegroom signs the hotel register for two. And the boy on the roof finds a streetfull of people gaping at him, necks stretched backwards, faces up, as though he were the ceiling of Hayden Planetarium. Suddenly you know it's you.

"Oscar! Oscar Freedman!" A voice rose from the center of the crowd, a voice that, could it have been seen, would have looked like the writing on a scroll. "Oscar Freedman, get down from there. Immediately!" Rabbi Binder was pointing one arm stiffly up at him; and at the end of that arm, one finger aimed menacingly. It was the attitude of a dictator, but one—the eyes confessed all—whose personal valet had spit neatly in his face.

Ozzie didn't answer. Only for a blink's length did he look towards Rabbi Binder. Instead his eyes began to fit together the world beneath him, to sort out people from places, friends from enemies, participants from spectators. In little jagged starlike clusters his friends stood around Rabbi Binder, who was still pointing. The topmost point on a star compounded not of angels but of five adolescent boys was Itzie. What a world it was, with those stars below, Rabbi Binder below . . . Ozzie, who a moment earlier hadn't been able to control his own body, started to feel the meaning of the word control: he felt Peace and he felt Power.

"Oscar Freedman, I'll give you three to come down."

Few dictators give their subjects three to do anything; but, as always, Rabbi Binder only looked dictatorial.

"Are you ready, Oscar?"

Ozzie nodded his head yes, although he had no intention in the world—the lower one or the celestial one he'd just entered—of coming down even if Rabbi Binder should give him a million.

"All right then," said Rabbi Binder. He ran a hand through his black Samson hair as though it were the gesture prescribed for uttering the first digit. Then, with his other hand cutting a circle out of the small piece of sky around him, he spoke. "One!"

There was no thunder. On the contrary, at that moment, as though "one" was the cue for which he had been waiting, the world's least thunderous person appeared on the synagogue steps. He did not so much come out the synagogue door as lean out, onto the darkening air. He clutched at the doorknob with one hand and looked up at the roof.

"Oy!"

Yakov Blotnik's old mind hobbled slowly, as if on crutches and though he couldn't decide precisely what the boy was doing on the roof, he knew it wasn't good—that is, it wasn't-good-for-the-Jews. For Yakov Blotnik life had fractionated itself simply: things were either good-for-the-Jews or no-good-for-the-Jews.

He smacked his free hand to his in-sucked cheek, gently. "Oy, Gut!" And then quickly as he was able, he jacked down his head and surveyed the street. There was Rabbi Binder (like a man at an auction with only three dollars in his pocket, he had just delivered a shaky "Two!"); there were the students, and that was all. So far it-wasn't-so-bad-for-the-Jews. But the boy had to come down immediately before anybody saw. The problem: how to get the boy off the roof?

Anybody who has ever had a cat on the roof knows how to get him down. You call the fire department. Or first you call the operator and you ask her for the fire department. And the next thing there is great jamming of brakes and clanging of bells and shouting of instructions. And then the cat is off the roof. You do the same thing to get a boy off the roof.

That is, you do the same thing if you are Yakov Blotnik and you once had a cat on the roof.

When the engines, all four of them, arrived, Rabbi Binder had four times given Ozzie the count of three. The big hook-and-ladder swung around the corner and one of the firemen leaped from it, plunging headlong towards the yellow fire hydrant in front of the synagogue. With a huge wrench he began to unscrew the top nozzle. Rabbi Binder raced over to him and pulled at his shoulder.

"There's no fire . . ."

The fireman mumbled back over his shoulder, and heatedly, continued working at the nozzle.

"But there's no fire, there's no fire . . ." Binder shouted. When the fireman mumbled again, the rabbi grasped his face with both hands and pointed up at the roof.

To Ozzie it looked as though Rabbi Binder was trying to tug the fireman's head out of his body, like a cork from a bottle. He had to giggle at the picture they made: it was a family portrait—rabbi in black skullcap, fireman in red fire hat, and the little yellow hydrant squatting beside like a kid brother, bareheaded. From the edge of the roof Ozzie waved at the portrait, a one-handed, flapping, mocking wave; in doing it his right foot slipped from under him. Rabbi Binder covered his eyes with his hands.

Firemen work fast. Before Ozzie had even regained his balance, a big round, yellowed net was being held on the synagogue lawn. The firemen who held it looked up at Ozzie with stern, feelingless faces.

One of the firemen turned his head towards Rabbi Binder. "What, is the kid nuts or something?"

Rabbi Binder unpeeled his hands from his eyes, slowly, painfully, as if they were tape. Then he checked: nothing on the sidewalk, no dents in the net.

"Is he gonna jump, or what?" The fireman shouted.

In a voice not at all like a statue, Rabbi Binder finally answered. "Yes. Yes. I think so . . . He's been threatening to . . ."

Threatening to? Why, the reason he was on the roof, Ozzie remembered was to get away; he hadn't even thought about jumping. He had just run to get away, and the truth was that he hadn't really headed for the roof as much as he'd been chased there.

"What's his name, the kid?"

"Freedman," Rabbi Binder answered. "Oscar Freedman."

The fireman looked up at Ozzie. "What is it with you, Oscar? You gonna jump or what?"

Ozzie did not answer. Frankly, the question had just arisen.

"Look, Oscar, if you're gonna jump, jump—and if you're not gonna jump, don't jump. But don't waste our time, willya?"

Ozzie looked at the fireman then at Rabbi Binder. He wanted to see Rabbi Binder cover his eyes one more time.

"I'm going to jump."

And then he scampered around the edge of the roof to the corner, where there was no net below and he flapped his arms at his sides, swishing the air and smacking his palms to his trousers on the downbeat. He began screaming like some kind of engine, "Wheecee . . . wheeeeee," and leaning way out over the edge with the upper half of his body. The firemen whipped around to cover the ground with the net. Rabbi Binder mumbled a few words to Somebody and covered his eyes. Everything happened quickly, jerkily, as in a silent movie. The crowd, which had arrived with the fire engines, gave out a long, Fourth-of-July fireworks oooh-aahhh. In the excitement no one had paid the crowd much heed, except, of course, Yakov Blotnik, who swung from the doorknob counting heads. "Fier und tsvansik . . . finf und tsvantsik . . . Oy, Gut!" It wasn't like this with the cat.

Rabbi Binder peeked through his fingers, checked the sidewalk and net. Empty. But there was Ozzie racing to the other corner. The firemen raced with him but were unable to keep up. Whenever Ozzie wanted to he might jump and splatter himself on the sidewalk, and by the time the firemen scooted to the spot all they could do with their net would be to cover the mess.

"Wheeeeee . . . wheeeeee . . ."

"Hey, Oscar," the winded fireman yelled, "What the hell is this, a game or something?"

"Wheeeee . . . wheeee . . ."

"Hey, Oscar—"

But he was off now to the other corner, flapping his wings fiercely. Rabbi Binder couldn't take it any longer—the fire engines from nowhere, the screaming suicidal boy, the net. He fell to his knees, exhausted, and with his hands curled together in front of his chest like a little dome, he pleaded, "Oscar, stop it, Oscar. Don't jump, Oscar. Please come down . . . Please don't jump."

And further back in the crowd a single voice, a single young voice, shouted a lone word to the boy on the roof.

"Jump!"

It was Itzie. Ozzie momentarily stopped flapping.

"Go ahead, Ozz—jump!" Itzie broke off his point of the star and courageously, with the inspiration not of a wise-guy but of a disciple, stood alone. "Jump, Ozz, jump!"

Still on his knees, his hands still curled, Rabbi Binder twisted his body back. He looked at Itzie, then, agonizingly, back to Ozzie.

"OSCAR, DON'T JUMP! PLEASE, DON'T JUMP . . . please please . . ."

"Jump!" This time it wasn't Itzie but another point of the star. By the time Mrs. Freedman arrived to keep her four-thirty appointment with Rabbi Binder the whole little upside down heaven was shouting and pleading for Ozzie to jump, and Rabbi Binder no longer was pleading with him not to jump, but was crying into the dome of his hands.

Understandably Mrs. Freedman couldn't figure out what her son was doing on the roof. So she asked.

"Ozzie, my Ozzie, what are you doing? My Ozzie, what is it?"

Ozzie stopped wheeeeeing and slowed his arms down to a cruising flap, the kind birds use in soft winds, but he did not answer. He stood against the low, clouded, darkening sky—light clicked down swiftly now, as on a small gear—flapping softly and gazing down at the small bundle of a woman who was his mother.

"What are you doing, Ozzie?" She turned towards the kneeling Rabbi Binder and rushed so close that only a paper-thickness of dusk lay between her stomach and his shoulders.

"What is my baby doing?"

Rabbi Binder gaped at her but he too was mute. All that moved was the dome of his hands; it shook back and forth like a weak pulse.

"Rabbi, get him down! He'll kill himself. Get him down, my only baby . . ."

"I can't," Rabbi Binder said, "I can't . . ." and he turned his handsome head towards the crowd of boys behind him. "It's them. Listen to them."

And for the first time Mrs. Freedman saw the crowd of boys, and she heard what they were yelling.

"He's doing it for them. He won't listen to me. It's them." Rabbi Binder spoke like one in a trance.

"For them?"

"Yes."

"Why for them?"

"They want him to . . ."

Mrs. Freedman raised her two arms upward as though she were conducting the sky. "For them he's doing it!" And thin in a gesture older than pyramids, older than prophets and floods, her arms came slapping down to her sides. "A martyr I have. Look!" She tilted her head to the roof. Ozzie was still flapping softly. "My martyr."

"Oscar, come down, *please,"* Rabbi Binder groaned.

In a startlingly even voice Mrs. Freedman called to the boy on the roof. "Ozzie, come down, Ozzie. Don't be a martyr, my baby."

As though it were a litany, Rabbi Binder repeated her words. "Don't be a martyr, my baby. Don't be a martyr."

"Gawhead, Ozz—*be* a Martin!" It was Itzie. "Be a Martin, be a Martin," and all the voices joined in singing for Martindom, whatever *it* was. "Be a Martin, be a Martin . . ."

Somehow when you're on a roof the darker it gets the less you can hear. All Ozzie knew was that two groups wanted two new things: his friends were spirited and musical about what they wanted; his mother and the rabbi were even-toned, chanting, about what they didn't want. The rabbi's voice was without tears now and so was his mother's.

The big net stared up at Ozzie like a sightless eye. The big, clouded sky pushed down. From beneath it looked like a gray corrugated board. Suddenly, looking up into that unsympathetic sky, Ozzie realized all the strangeness of what these people, his friends, were asking: they wanted him to jump, to kill himself; they were singing about it now—it made them that happy. And there was an even greater strangeness: Rabbi Binder was on his knees, trembling. If there was a question to be asked now it was not "Is it me?" but rather "Is it us? . . . Is it us?"

Being on the roof, it turned out, was a serious thing. If he jumped would the singing become dancing? Would it? What would jumping stop? Yearningly, Ozzie wished he could rip open the sky, plunge his hands through, and pull out the sun; and on the sun, like a coin, would be stamped JUMP or DON'T JUMP.

Ozzie's knees rocked and sagged a little under him as though they were setting him for a dive. His arms tightened, stiffened, froze, from shoulders to fingernails. He felt as if each part of his body were going to vote as to whether he should kill himself or not—and each part as though it were independent of *him.*

The light took an unexpected click down and the new darkness, like a gag, hushed the friends singing for this and the mother and rabbi chanting for that.

Ozzie stopped counting votes, and in a curiously high voice, like one who wasn't prepared for speech, he spoke.

"Mamma?"

"Yes. Oscar."

"Mamma, get down on your knees, like Rabbi Binder."

"Oscar—"

"Get down on your knees," he said, "or I'll jump."

Ozzie heard a whimper, then a quick rustling, and when he looked way down where his mother had stood he saw the top of a head and beneath that a circle of dress. She was kneeling beside Rabbi Binder.

He spoke again. "Everybody kneel." There was the sound of everybody kneeling.

Ozzie looked around. With one had he pointed towards the synagogue entrance. "Make *him* kneel."

There was a noise, not of kneeling, but of body-and-cloth stretching. Ozzie could hear Rabbi Binder saying in a gruff whipser, ". . . or he'll *kill* himself," and when next he looked there was Yakov Blotnik off the doorknob and for the first time in his life upon his knees in the Gentile posture of prayer.

As for the firemen—it is not as difficult as one might imagine to hold a net taut while you are kneeling.

Ozzie look around again; and then he called to Rabbi Binder.

"Rabbi?"

"Yes, Oscar."

"Rabbi Binder, do you believe in God?"

"Yes."

"Do you believe God can do Anything?" Ozzie leaned his head out into the darkness. "Anything?"

"Oscar, I think—"

"Tell me you believe God can do Anything."

There was a second's hesitation. Then: "God can do Anything."

"Tell me you believe God can make a child without intercourse."

"He can."

"Tell me!"

"God," Rabbi Binder admitted, "can make a child without intercourse."

"Mamma, you tell me."

"God can make a child without intercourse," his mother said.

"Make *him* tell me." There was no doubt about who *him* was.

In a few moments Ozzie heard an old comical voice say something to the increasing darkness about God.

Next, Ozzie made everybody say it. And then he made them all say they believed in Jesus Christ—first one at a time, then all together.

When the catechizing was through it was the beginning of evening. From the street it sounded as if the boy on the roof might have sighed.

"Ozzie?" A woman's voice dared to speak. "You'll come down now?"

There was no answer, but the woman waited, and when a voice finally did speak it was thin and crying, and exhausted as that of an old man who had just finished pulling the bells.

"Mamma, don't you see—you shouldn't hit me. He shouldn't hit me. You shouldn't hit me about God, Mamma. You should never hit anybody about God—"

"Ozzie, please come down now."

"Promise me, promise me you'll never hit anybody about God."

He had asked only his mother, but for some reason everyone kneeling in the street promised he would never hit anybody about God.

Once again there was silence.

"I can come down now, Mamma," the boy on the roof finally said. He turned his head both ways as though checking traffic lights. "Now I can come down . . ."

And he did, right into the center of the yellow net that glowed in the evening's edge like an overgrown halo.

Two Kinds

Amy Tan

In 1993, the movie based on Amy Tan's novel The Joy Luck Club *helped make her one of the country's best-known Asian American writers. Born in Oakland, California, she writes about California's immigrant Chinese—their journey between the values and traumas of the Chinese past and the challenges of the American future. Her novel* The Kitchen God's Wife *(1991) continued to explore the role that ties to the mainland culture and the interlocking relationships of the large traditional family play in the lives of Chinese Americans. Her novel* The Hundred Secret Senses *was published in 1995.*

My mother believed you could be anything you wanted to be in America. You could open a restaurant. You could work for the government and get good retirement. You could buy a house with almost no money down. You could become rich. You could become instantly famous.

"Of course you can be prodigy, too," my mother told me when I was nine. "You can be best anything. What does Auntie Lindo know? Her daughter, she is only best tricky."

America was where all my mother's hopes lay. She had come here in 1949 after losing everything in China: her mother and father, her family home, her first husband, and two daughters, twin baby girls. But she never looked back with regret. There were so many ways for things to get better.

We didn't immediately pick the right kind of prodigy. At first my mother thought I could be a Chinese Shirley Temple. We'd watch Shirley's old movies on TV as though they were training films. My mother would poke my arm and say, "*Ni kan,*"—You watch. And I would see Shirley tapping her feet, or singing a sailor song, or pursing her lips into a very round O while saying, "Oh my goodness."

"*Ni kan,*" said my mother as Shirley's eyes flooded with tears. "You already know how. Don't need talent for crying!"

Soon after my mother got this idea about Shirley Temple, she took me to a beauty training school in the Mission district and put me in the hands of a student who could barely hold the scissors without shaking. Instead of getting big fat curls, I emerged with an uneven mass of crinkly black fuzz. My mother dragged me off to the bathroom and tried to wet down my hair.

"You look like Negro Chinese," she lamented, as if I had done this on purpose.

The instructor of the beauty training school had to lop off these soggy clumps to make my hair even again. "Peter Pan is very popular these days," the instructor assured my mother. I now had hair the length of a boy's, with straight-across bangs that hung at a slant two inches above my eyebrows. I liked the haircut and it made me actually look forward to my future fame.

In fact, in the beginning, I was just as excited as my mother, maybe even more so. I pictured this prodigy part of me as many different images, trying each one on for size. I was a dainty ballerina girl standing by the curtains waiting to hear the right music that would send me floating on my tiptoes. I was like the Christ child lifted out of the straw manger, crying with holy indignity. I was Cinderella stepping from her pumpkin carriage with sparkly cartoon music filling the air.

In all of my imaginings, I was filled with a sense that I would soon become *perfect*. My mother and father would adore me. I would be beyond reproach. I would never feel the need to sulk for anything.

But sometimes the prodigy in me became impatient. "If you don't hurry up and get me out of here, I'm disappearing for good," it warned. "And then you'll always be nothing."

Every night after dinner; my mother and I would sit at the Formica kitchen table. She would present new tests, taking her examples from stories of amazing children she had read in *Ripley's Believe It or Not*, or *Good Housekeeping*, *Reader's Digest*, and a dozen other magazines she kept in a pile in our bathroom. My mother got these magazines from people whose houses she cleaned. And since she cleaned many houses each week, we had a great assortment. She would look through them all, searching for stories about remarkable children.

The first night she brought out a story about a three-year-old boy who knew the capitals of all the states and even most of the European countries. A teacher was quoted as saying the little boy could also pronounce the names of the foreign cities correctly.

"What's the capital of Finland?" my mother asked me, looking at the magazine story.

All I knew was the capital of California, because Sacramento was the name of the street we lived on in Chinatown. "Nairobi!" I guessed, saying the most foreign word I could think of. She checked to see if that was possibly one way to pronounce "Helsinki" before showing me the answer.

The tests got harder—multiplying numbers in my head, finding the queen of hearts in a deck of cards, trying to stand on my head without using my hands, predicting the daily temperatures in Los Angeles, New York, and London.

One night I had to look at a page from the Bible for three minutes and then report everything I could remember. "Now Jehoshaphat had riches and honor in abundance and . . . that's all I remember, Ma," I said.

And after seeing my mother's disappointed face once again, something inside of me began to die. I hated the tests, the raised hopes and failed expectations. Before going to bed that night, I looked in the mirror above the bathroom sink and when I saw only my face staring back—and that it would always be this ordinary face—I began to cry. Such a sad, ugly girl! I made high-pitched noises like a crazed animal, trying to scratch out the face in the mirror.

And then I saw what seemed to be the prodigy side of me—because I had never seen that face before. I looked at my reflection, blinking so I could see more clearly. The girl staring back at me was angry, powerful. This girl and I were the same. I had new thoughts, willful thoughts, or rather thoughts filled with lots of won'ts. I won't let her change me, I promised myself I won't be what I'm not.

So now on nights when my mother presented her tests, I performed listlessly, my head propped on one arm. I pretended to be bored. And I was. I got so bored I started counting the bellows of the foghorns out on the bay while my mother drilled me in other areas. The sound was comforting and reminded me of the cow jumping over the moon. And the next day, I played a game with myself, seeing if my mother would give up on me before eight bellows. After a while I usually counted only one, maybe two bellows at most. At last she was beginning to give up hope.

Two or three months had gone by without any mention of my being a prodigy again. And then one day my mother was watching *The Ed Sullivan Show* on TV. The TV was old and the sound kept shorting out. Every time my mother got halfway up from the sofa to adjust the set, the sound would go back on and Ed would be talking. As soon as she sat down, Ed would go silent again. She got up, the TV broke into loud piano music. She sat down. Silence. Up and down, back and forth, quiet and loud. It was like a stiff embraceless dance between her and the TV set. Finally she stood by the set with her hand on the sound dial.

She seemed entranced by the music, a little frenzied piano piece with this mesmerizing quality, sort of quick passages and then teasing lilting ones before it returned to the quick playful parts.

"*Ni kan*," my mother said, calling me over with hurried hand gestures. "Look here."

I could see why my mother was fascinated by the music. It was being pounded out by a little Chinese girl, about nine years old, with a Peter Pan haircut. The girl had the sauciness of a Shirley Temple. She was proudly modest like a proper Chinese child. And she also did this fancy sweep of a curtsy, so that the fluffy skirt of her white dress cascaded slowly to the floor like the petals of a large carnation.

In spite of these warning signs, I wasn't worried. Our family had no piano and we couldn't afford to buy one, let alone reams of sheet music and piano lessons. So I could be generous in my comments when my mother bad-mouthed the little girl on TV.

"Play note right, but doesn't sound good! No singing sound," complained my mother.

"What are you picking on her for?" I said carelessly.

"She's pretty good. Maybe she's not the best, but she's trying hard." I knew almost immediately I would be sorry I said that.

"Just like you," she said. "Not the best. Because you not trying." She gave a little huff as she let go of the sound dial and sat down on the sofa.

The little Chinese girl sat down also to play an encore of "Anitra's Dance" by Grieg. I remember the song, because later on I had to learn how to play it.

Three days after watching *The Ed Sullivan Show* my mother told me what my schedule would be for piano lessons and piano practice. She had talked to Mr. Chong, who lived on the first floor of our apartment building. Mr. Chong was a retired piano teacher and my mother had traded housecleaning services for weekly lessons and a piano for me to practice on every day, two hours a day, from four until six.

When my mother told me this, I felt as though I had been sent to hell. I whined and then kicked my foot a little when I couldn't stand it anymore.

"Why don't you like me the way I am? I'm *not* a genius! I can't play the piano. And even if I could, I wouldn't go on TV if you paid me a million dollars!" I cried.

My mother slapped me. "Who ask you be genius?" she shouted. "Only ask you be your best. For you sake. You think I want you be genius? Hnnh! What for! Who ask you!"

"So ungrateful," I heard her mutter in Chinese. "If she had as much talent as she has temper, she would be famous now."

Mr. Chong, whom I secretly nicknamed Old Chong, was very strange, always tapping his fingers to the silent music of an invisible orchestra. He looked ancient in my eyes. He had lost most of the hair on top of his head and he wore thick glasses and had eyes that always looked tired and sleepy. But he must have been younger than I thought, since he lived with his mother and was not yet married.

I met Old Lady Chong once and that was enough. She had this peculiar smell like a baby that had done something in its pants. And her fingers felt like a dead person's, like an old peach I once found in the back of the refrigerator; the skin just slid off the meat when I picked it up.

I soon found out why Old Chong had retired from teaching piano. He was deaf. "Like Beethoven!" he shouted to me. "We're both listening only in our head!" And he would start to conduct his frantic silent sonatas.

Our lessons went like this. He would open the book and point to different things, explaining their purpose: "Key! Treble! Bass! No sharps or flats! So this is C major! Listen now and play after me!"

And then he would play the C scale a few times, a simple chord, and then, as if inspired by an old, unreachable itch, he gradually added more notes and running trills and a pounding bass until the music was really something quite grand.

I would play after him, the simple scale, the simple chord, and then I just played some nonsense that sounded like a cat running up and down on top of garbage cans. Old Chong smiled and applauded and then said, "Very good! But now you must learn to keep time!"

So that's how I discovered that Old Chong's eyes were too slow to keep up with the wrong notes I was playing. He went through the motions in half-time. To help me keep rhythm, he stood behind me, pushing down on my right shoulder for every beat. He balanced pennies on top of my wrists so I would keep them still as I slowly played scales and arpeggios. He had me curve my hand around an apple and keep that shape when playing chords. He marched stiffly to show me how to make each finger dance up and down, staccato like an obedient little soldier.

He taught me all these things, and that was how I also learned I could be lazy and get away with mistakes, lots of mistakes. If I hit the wrong notes because I hadn't practiced enough, I never corrected myself. I just kept playing in rhythm. And Old Chong kept conducting his own private reverie.

So maybe I never really gave myself a fair chance. I did pick up the basics pretty quickly, and I might have become a good pianist at that young age. But I was so determined not to try, not to be anybody different that I learned to play only the most earsplitting preludes, the most discordant hymns.

Over the next year, I practiced like this, dutifully in my own way. And then one day I heard my mother and her friend Lindo Jong both talking in a loud bragging tone of voice so others could hear. It was after church, and I was leaning against the brick wall wearing a dress with stiff white petticoats. Auntie Lindo's daughter, Waverly, who was about my age, was standing farther down the wall about five feet away. We had grown up together and shared all the closeness of two sisters squabbling over crayons and dolls. In other words, for the most part, we hated each other. I thought she was snotty. Waverly Jong had gained a certain amount of fame as "Chinatown's Littlest Chinese Chess Champion."

"She bring home too many trophy," lamented Auntie Lindo that Sunday. All day she play chess. All day I have no time do nothing but dust off her winnings." She threw a scolding look at Waverly, who pretended not to see her.

"You lucky you don't have this problem," said Auntie Lindo with a sigh to my mother.

And my mother squared her shoulders and bragged: "Our problem worser than yours. If we ask Jing-mei wash dish, she hear nothing but music. It's like you can't stop this natural talent."

And right then, I was determined to put a stop to her foolish pride.

* * *

A few weeks later, Old Chong and my mother conspired to have me play in a talent show which would be held in the church hall. By then, my parents had saved up enough to buy me a second-hand piano, a black Wurlitzer spinet with a scarred bench. It was the showpiece of our living room.

For the talent show, I was to play a piece called "Pleading Child" from Schumann's *Scenes from Childhood.* It was a simple, moody piece that sounded more difficult than it was. I was supposed to memorize the whole thing, playing the repeat parts twice to make the piece sound longer. But I dawdled over it, playing a few bars and then cheating, looking up to see what notes followed, I never really listened to what I was playing. I daydreamed about being somewhere else, about being someone else.

The part I liked to practice best was the fancy curtsy: right foot out, touch the rose on the carpet with a pointed foot, sweep to the side, left leg bends, look up and smile.

My parents invited all the couples from the Joy Luck Club to witness my debut. Auntie Lindo and Uncle Tin were there. Waverly and her two older brothers had also come. The first two rows were filled with children both younger and older than I was. The littlest ones got to go first. They recited simple nursery rhymes, squawked out tunes on miniature violins, twirled Hula Hoops, pranced in pink ballet tutus, and when they bowed or curtsied, the audience would sigh in unison, "Awww," and then clap enthusiastically.

When my turn came, I was very confident. I remember my childish excitement. It was as if I knew, without a doubt, that the prodigy side of me really did exist. I had no fear whatsoever, no nervousness. I remember thinking to myself, This is it! This is it! I looked out over the audience, at my mother's blank face, my father's yawn, Auntie Lindo's stiff-lipped smile, Waverly's sulky expression. I had on a white dress layered with sheets of lace, and a pink bow in my Peter Pan haircut. As I sat down I envisioned people jumping to their feet and Ed Sullivan rushing up to introduce me to everyone on TV.

And I started to play. It was so beautiful. I was so caught up in how lovely I looked that at first I didn't worry how I would sound: So it was a surprise to me when I hit the first wrong note and I realized something didn't sound quite right. And then I hit another and another followed that. A chill started at the top of my head and began to trickle down. Yet I couldn't stop playing, as though my hands were bewitched. I kept thinking my fingers would adjust themselves back, like a train switching to the right track. I played this strange jumble through two repeats, the sour notes staying with me all the way to the end.

When I stood up, I discovered my legs were shaking. Maybe I had just been nervous and the audience, like Old Chong, had seen me go through the right motions and had not heard anything wrong at all. I swept my right foot out, went down on my knee, looked up and smiled. The room was quiet, except for Old Chong, who was beaming and shouting, "Bravo! Bravo! Well done!" But then I saw my mother's face, her stricken face. The audience clapped weakly, and as I walked back to my chair, with my whole face quivering as I tried not to cry, I heard a little boy whisper loudly to his mother, "That was awful," and the mother whispered back, "Well, she certainly tried."

And now I realized how many people were in the audience, the whole world it seemed. I was aware of eyes burning into my back. I felt the shame of my mother and father as they sat stiffly throughout the rest of the show.

We could have escaped during intermission. Pride and some strange sense of honor must have anchored my parents to their chairs. And so we watched it all: the eighteen-year-old boy with a fake mustache who did a magic show and juggled flaming hoops while riding a unicycle. The breasted girl with white makeup who sang from *Madama Butterfly* and got honorable mention. And the eleven-year-old boy who won first prize playing a tricky violin song that sounded like a busy bee.

After the show, the Hsus, the Jongs, and the St. Clairs from the Joy Luck came up to my mother and father.

"Lots of talented kids," Auntie Lindo said vaguely, smiling broadly.

"That was somethin' else," said my father, and I wondered if he was referring to me in a humorous way, or whether he even remembered what I had done.

Waverly looked at me and shrugged her shoulders. "You aren't a genius like me," she said matter-of-factly. And if I hadn't felt so bad, I would have pulled her braids and punched her stomach.

But my mother's expression was what devastated me: a quiet, blank look that said she had lost everything. I felt the same way, and it seemed as if everybody were now coming up, like gawkers at the scene of an accident, to see what parts were actually missing. When we got on the bus to go home, my father was humming the busy-bee tune and my mother was silent. I kept thinking she wanted to wait until we got home before shouting at me. But when my father unlocked the door to our apartment, my mother walked in and then went to the back, into the bedroom. No accusations. No blame. And in a way, I felt disappointed. I had been waiting for her to start shouting, so I could shout back and cry and blame her for all my misery.

I assumed my talent-show fiasco meant I never had to play the piano again. But two days later, after school, my mother came out of the kitchen and saw me watching TV.

"Four clock," she reminded me as if it were any other day. I was stunned, as though she were asking me to go through the talent-show torture again. I wedged myself more tightly in front of the TV.

"Turn off TV," she called from the kitchen five minutes later.

I didn't budge. And then I decided. I didn't have to do what my mother said anymore. I wasn't her slave. This wasn't China. I had listened to her before and look what happened. She was the stupid one.

She came out from the kitchen and stood in the arched entryway of the living-room. "Four clock," she said once again, louder.

"I'm not going to play anymore," I said nonchalantly. "Why should I? I'm not a genius."

She walked over and stood in front of the TV. I saw her chest was heaving up and down in an angry way.

"No!" I said, and I now felt stronger, as if my true self had finally emerged. So this was what had been inside me all along.

"No! I won't!" I screamed.

She yanked me by the arm, pulled me off the floor, snapped off the TV. She was frighteningly strong, half pulling, half carrying me toward the piano as I kicked the throw rugs under my feet. She lifted me up and onto the hard bench. I was sobbing by now, looking at her bitterly. Her chest was heaving even more and her mouth was open, smiling crazily as if she were pleased I was crying.

"You want me to be someone that I'm not!" I sobbed. "I'll never be the kind of daughter you want me to be!"

"Only two kinds of daughters," she shouted in Chinese. "Those who are obedient and those who follow their own mind! Only one kind of daughter can live in this house. Obedient daughter!"

"Then I wish I wasn't your daughter. I wish you weren't my mother," I shouted. As I said these things I got scared. I felt like worms and toads and slimy things were crawling out of my chest, but it also felt good, as if this awful side of me had surfaced, at last.

"Too late change this," said my mother shrilly.

And I could sense her anger rising to its breaking point. I wanted to see it spill over. And that's when I remembered the babies she had lost in China, the ones we never talked about. "Then I wish I'd never been born!" I shouted. "I wish I were dead! Like them."

It was as if I had said the magic words. Alakazam!—and her face went blank, her mouth closed, her arms went slack, and she backed out of the room, stunned, as if she were blowing away like a small brown leaf, thin, brittle, lifeless.

It was not the only disappointment my mother felt in me. In the years that followed, I failed her so many times, each time asserting my own will, my right to fall short of expectations. I didn't get straight As. I didn't become class president. I didn't get into Stanford. I dropped out of college.

For unlike my mother, I did not believe I could be anything I wanted to be. I could only be me.

And for all those years, we never talked about the disaster at the recital or my terrible accusations afterward at the piano bench. All that remained unchecked, like a betrayal that was now unspeakable. So I never found a way to ask her why she had hoped for something so large that failure was inevitable.

And even worse, I never asked her what frightened me the most: Why had she given up hope?

For after our struggle at the piano, she never mentioned my playing again. The lessons stopped. The lid to the piano was closed, shutting out the dust, my misery, and her dreams.

So she surprised me. A few years ago, she offered to give me the piano for my thirtieth birthday. I had not played in all those years. I saw the offer as a sign of forgiveness, a tremendous burden removed.

"Are you sure?" I asked shyly. "I mean, won't you and Dad miss it?"

"No, this is your piano," she said firmly. "Always your piano. You only one can play."

"Well, I probably can't play anymore," I said. "It's been years."

"You pick up fast," said my mother, as if she knew this was certain. "You have natural talent. You could been genius if you want to."

"No I couldn't."

"You just not trying," said my mother. And she was neither angry nor sad. She said it as if to announce a fact that could never be disproved. "Take it," she said.

But I didn't at first. It was enough that she had offered it to me. And after that, every time I saw it in my parents' living room, standing in front of the bay windows, it made me feel proud, as if it were a shiny trophy I had won back.

Last week I sent a tuner over to my parents' apartment and had the piano reconditioned, for purely sentimental reasons. My mother had died a few months before and I had been getting things in order for my father, a little bit at a time. I put the jewelry in special silk pouches. The sweaters she had knitted in yellow, pink, bright orange—all the colors I hated—I put those in mothproof boxes. I found some old Chinese silk dresses, the kind with little slits up the sides. I rubbed the old silk against my skin, then wrapped them in tissue and decided to take them home with me.

After I had the piano tuned, I opened the lid and touched the keys. It sounded even richer than I remembered. Really, it was a very good piano. Inside the bench were the same exercise notes with handwritten scales, the same second-hand music books with their covers held together with yellow tape.

I opened up the Schumann book to the dark little piece I had played at the recital. It was on the left-hand side of the page, "Pleading Child." It looked more difficult than I remembered. I played a few bars, surprised at how easily the notes came back to me.

And for the first time, or so it seemed, I noticed the piece on the right-hand side. It was called "Perfectly Contented." I tried to play this one as well. It had a lighter melody but the same flowing rhythm and turned out to be quite easy. "Pleading Child" was shorter but slower; "Perfectly Contented" was longer but faster. And after I played them both a few times, I realized they were two halves of the same song.

1920

Toni Morrison

It had to be as far away from the Sundown House as possible. And her grandmother's middle-aged nephew who lived in a Northern town called Medallion was the one chance she had to make sure it would be. The red shutters had haunted both Helene Sabat and her grandmother for sixteen years. Helene was born behind those shutters, daughter of a Creole whore who worked there. The grandmother took Helene away from the soft lights and flowered carpets of the Sundown House and raised her under the dolesome eyes of a multicolored Virgin Mary, counseling her to be constantly on guard for any sign of her mother's wild blood.

So when Wiley Wright came to visit his Great Aunt Cecile in New Orleans, his enchantment with the pretty Helene became a marriage proposal—under the pressure of both women. He was a seaman (or rather a lakeman, for he was a ship's cook on one of the Great Lakes lines), in port only three days out of every sixteen.

He took his bride to his home in Medallion and put her in a lovely house with a brick porch and real lace curtains at the window. His long absences were quite bearable for Helene Wright, especially when, after some nine years of marriage, her daughter was born.

Her daughter was more comfort and purpose than she had ever hoped to find in this life. She rose grandly to the occasion of motherhood—grateful, deep down in her heart, that the child had not inherited the great beauty that was hers: that her skin had dusk in it, that her lashes were substantial but not undignified in their length, that she had taken the broad flat nose of Wiley (although Helene expected to improve it somewhat) and his generous lips.

Under Helene's hand the girl became obedient and polite. Any enthusiasms that little Nel showed were calmed by the mother until she drove her daughter's imagination underground.

Helene Wright was an impressive woman, at least in Medallion she was. Heavy hair in a bun, dark eyes arched in a perpetual query about other people's manners. A woman who won all social battles with presence and a conviction of the legitimacy of her authority. Since there was no Catholic church in Medallion then, she joined the most conservative black church. And held sway. It was Helene who never turned her head in church when latecomers arrived; Helene who establishd the practice of seasonal altar flowers; Helene who introduced the giving of banquets of welcome to returning Negro veterans. She lost only one battle—the pronunciation of her name. The people in the Bottom refused to say Helene. They called her Helen Wright and left it at that.

All in all her life was a satisfactory one. She loved her house and enjoyed manipulating her daughter and her husband. She would sigh sometimes just before falling asleep, thinking that she had indeed come far enough away from the Sundown House.

So it was with extremely mixed emotions that she read a letter from Mr. Henri Martin describing the illness of her grandmother, and suggesting she come down right away. She didn't want to go, but could not bring herself to ignore the silent plea of the woman who had rescued her.

It was November. November, 1920. Even in Medallion there was a victorious swagger in the legs of white men and a dull-eyed excitement in the eyes of colored veterans.

Helene thought about the trip South with heavy misgiving but decided that she had the best protection: her manner and her bearing, to which she would add a beautiful dress. She bought some deep-brown wool and three-fourths of a yard of matching velvet. Out of this she made herself a heavy but elegant dress with velvet collar and pockets.

Nel watched her mother cutting the pattern from newspapers and moving her eyes rapidly from a magazine model to her own hands. She watched her turn up the kerosene lamp at sunset to sew far into the night.

The day they were ready, Helene cooked a smoked ham, left a note for her lake-bound husband, in case he docked early, and walked head high and arms stiff with luggage ahead of her daughter to the train depot.

It was a longer walk than she remembered, and they saw the train steaming up just as they turned the corner. They ran along the track looking for the coach pointed out to them by the colored porter. Even at that they made a mistake. Helene and her daughter entered a coach peopled by some twenty white men and women. Rather than go back and down the three wooden steps again, Helene decided to spare herself some embarrassment and walked on through to the colored car. She carried two pieces of luggage and a string purse; her daughter carried a covered basket of food.

As they opened the door marked Colored Only, they saw a white conductor coming toward them. It was a chilly day but a light skim of sweat glistened on the woman's face as she and the little girl struggled to hold the door open, hang on to their luggage and enter all at once. The conductor let his eyes travel over the pale yellow woman and then stuck his little finger into his ear, jiggling it free of wax. "What you think you doin', gal?"

Helene looked up at him.

So soon. So soon. She hadn't even begun the trip back. Back to her grandmother's house in the city where the red shutters glowed, and already she had been called "gal." All the old vulnerabilities, all the old fears of being somehow flawed gathered in her stomach and made her hands tremble. She had heard only that one word; it dangled above her wide-brimmed hat, which had slipped, in her exertion, from its carefully leveled placement and was now tilted in a bit of a jaunt over her eye.

Thinking he wanted her tickets, she quickly dropped both the cowhide suitcase and the straw one in order to search for them in her purse. An eagerness to please and an apology for living met in her voice. "I have them. Right here somewhere, sir. . . ."

The conductor looked at the bit of wax his fingernail had retrieved.

"What was you doin' back in there? What was you doin' in that coach yonder?"

Helene licked her lips. "Oh . . . I . . ." Her glance moved beyond the white man's face to the passengers seated behind him. Four or five black faces were watching, two belonging to soldiers still in their shit-colored uniforms and peaked caps. She saw their closed faces, their locked eyes, and turned for compassion to the gray eyes of the conductor.

"We made a mistake, sir. You see, there wasn't no sign. We just got in the wrong car, that's all. Sir."

"We don't 'low no mistakes on this train. Now git your butt on in there."

He stood there staring at her until she realized that he wanted her to move aside. Pulling Nel by the arm, she pressed herself and her daughter into the foot space in front of a wooden seat. Then, for no earthly reason, at least no reason that anybody could understand, certainly no reason that Nel understood then or later, she smiled. Like a street pup that wags its tail at the very doorjamb of the butcher shop he has been kicked away from only moments before, Helene smiled. Smiled dazzlingly and coquettishly at the salmon-colored face of the conductor.

Nel looked away from the flash of pretty teeth to the other passengers. The two black soldiers, who had been watching the scene with what appeared to be indifference, now looked stricken. Behind Nel was the bright and blazing light of her mother's smile; before her the midnight eyes of the soldiers. She saw the muscles of their faces tighten, a movement under the skin from blood to marble. No change in the expression of the eyes, but a hard wetness that veiled them as they looked at the stretch of her mother's foolish smile.

As the door slammed on the conductor's exit, Helene walked down the aisle to a seat. She looked about for a second to see whether any of the men would help her put the suitcases in the overhead rack. Not a man moved. Helene sat down, fussily, her back toward the men. Nel sat opposite, facing both her mother and the soldiers, neither of whom she could look at. She felt both pleased and ashamed to sense that these men, unlike her father, who worshiped his graceful, beautiful wife, were bubbling with a hatred for her mother that had not been there in the beginning but had been born with the dazzling smile. In the silence that preceded the train's heave, she looked deeply at the folds of her mother's dress. There in the fall of the heavy brown wool she held her eyes. She could not risk letting them travel upward for fear of seeing that the hooks and eyes in the placket of the dress had come undone and exposed the custard-colored skin underneath. She stared at the hem, wanting to believe in its weight but knowing that custard was all that it hid. If this tall, proud woman, this woman who was very particular about her friends, who slipped into church with unequaled elegance, who could quell a roustabout with a look, if *she* were really custard, then there was a chance that Nel was too.

It was on that train, shuffling toward Cincinnati, that she resolved to be on guard—always. She wanted to make certain that no man ever looked at her that way. That no midnight eyes or marbled flesh would ever accost her and turn her into jelly.

For two days they rode; two days of watching sleet turn to rain, turn to purple sunsets, and one night knotted on the wooden seats (their heads on folded coats), trying not to hear the snoring soldiers. When they changed trains in Birmingham for the last leg of the trip, they discovered what luxury they had been in through Kentucky and Tennessee, where the rest stops had all had colored toilets. After Birmingham there were none. Helene's face was drawn with the need to relieve herself, and so intense was her distress she finally brought herself to speak about her problem to a black woman with four children who had got on in Tuscaloosa.

"Is there somewhere we can go to use the restroom?"

The woman looked up at her and seemed not to understand. "Ma'am?" Her eyes fastened on the thick velvet collar, the fair skin, the high-tone voice.

"The restroom," Helene repeated. Then, in a whisper, "The toilet."

The woman pointed out the window and said, "Yes ma'am. Yonder." Helene looked out of the window halfway expecting to see a comfort station in the distance; instead she saw gray-green trees leaning over tangled grass. "Where?"

"Yonder," the woman said. "Meridian. We be pullin' in direc'lin." Then she smiled sympathetically and asked, "Kin you make it?"

Helene nodded and went back to her seat trying to think of other things—for the surest way to have an accident would be to remember her full bladder.

At Meridian the women got out with their children. While Helene looked about the tiny stationhouse for a door that said COLORED WOMEN, the other woman stalked off to a field of high grass on the far side of the track. Some white men were leaning on the railing in front of the stationhouse. It was not only their tongues curling around toothpicks that kept Helene from asking information of them. She looked around for the other woman and, seeing just the top of her head rag in the grass, slowly realized where "yonder" was. All of them, the fat woman and her four children, three boys and a girl, Helene and her daughter, squatted there in the four o'clock Meridian sun. They did it again in Ellisville, again in Hattiesburg, and by the time they reached Slidell, not too far from Lake Pontchartrain, Helene could not only fold leaves as well as the fat woman, she never felt a stir as she passed the muddy eyes of the men who stood like wrecked Dorics under the station roofs of those towns.

The lift in spirit that such an accomplishment produced in her quickly disappeared when the train finally pulled into New Orleans.

Cecile Sabat's house leaned between two others just like it on Elysian Fields. A Frenchified shotgun house, it sported a magnificent garden in the back and a tiny wrought-iron fence in the front. On the door hung a black crepe wreath with purple ribbon. They were too late. Helene reached up to touch the ribbon, hesitated, and knocked. A man in a collarless shirt opened the door. Helene identified herself and he said he was Henri Martin and that he was there for the settin'-up. They stepped into the house. The Virgin Mary clasped her hands in front of her neck three times in the front room and once in the bedroom where Cecile's body lay. The old woman had died without seeing or blessing her granddaughter.

No one other than Mr. Martin seemed to be in the house, but a sweet odor as of gardenias told them that someone else had been. Blotting her lashes with a white handkerchief, Helene walked through the kitchen to the back bedroom where she had slept for sixteen years. Nel trotted along behind, enchanted with the smell, the candles and the strangeness. When Helene bent to loosen the ribbons of Nel's hat, a woman in a yellow dress came out of the garden and onto the back porch that opened into the bedroom. The two women looked at each other. There was no recognition in the eyes of either. Then Helene said, "This is your . . . grandmother, Nel." Nel looked at her mother and then quickly back at the door they had just come out of.

"No. That was your great-grandmother. This is your grandmother. My mother . . ."

Before the child could think, her words were hanging in the gardenia air. "But she looks so young."

The woman in the canary-yellow dress laughed and said she was forty-eight, "an old forty-eight."

Then it was she who carried the gardenia smell. This tiny woman with the softness and glare of a canary. In that somber house that held four Virgin Marys, where death sighed in every corner and candles sputtered, the gardenia smell and canary-yellow dress emphasized the funeral atmosphere surrounding them.

The woman smiled, glanced in the mirror and said, throwing her voice toward Helene, "That your only one?"

"Yes," said Helene.

"Pretty. A lot like you."

"Yes. Well. She's ten now."

"Ten? Vrai? Small for her age, no?"

Helene shrugged and looked at her daughter's questioning eyes. The woman in the yellow dress leaned forward. "Come. Come, chère."

Helene interrupted. "We have to get cleaned up. We been three days on the train with no chance to wash or . . ."

"Comment t'appelle?"

"She doesn't talk Creole."

"Then you ask her."

"She wants to know your name, honey."

With her head pressed into her mother's heavy brown dress, Nel told her and then asked, "What's yours?"

"Mine's Rochelle. Well. I must be going on." She moved closer to the mirror and stood there sweeping hair up from her neck back into its halo-like roll, and wetting with spit the ringlets that fell over her ears. "I been here, you know, most of the day. She pass on yesterday. The funeral tomorrow. Henri takin' care." She struck a match, blew it out and darkened her eyebrows with the burnt head. All the while Helene and Nel watched her. The one in a rage at the folded leaves she had endured, the wooden benches she had slept on, all to miss seeing her grandmother and seeing instead that painted canary who never said a word of greeting or affection or . . .

Rochelle continued, "I don't know what happened to de house. Long time paid for. You be thinkin' on it? Oui?" Her newly darkened eyebrows queried Helene.

"Oui." Helene's voice was chilly. "I be thinkin' on it."

"Oh, well. Not for me to say . . ."

Suddenly she swept around and hugged Nel—a quick embrace tighter and harder than one would have imagined her thin soft arms capable of.

"'Voir! 'Voir!" and she was gone.

In the kitchen, being soaped head to toe by her mother, Nel ventured an observation. "She smelled so nice. And her skin was so soft."

Helene rinsed the cloth. "Much handled things are always soft."

"What does 'vwah' mean?"

"I don't know," her mother said. "I don't talk Creole." She gazed at her daughter's wet buttocks. "And neither do you."

When they got back to Medallion and into the quiet house they saw the note exactly where they had left it and the ham dried out in the icebox.

"Lord, I've never been so glad to see this place. But look at the dust. Get the rags, Nel. Oh, never mind. Let's breathe awhile first. Lord, I never thought I'd get back here safe and sound. Whoo. Well it's over. Good and over. Praise His name. Look at that. I told that old fool not to deliver any milk and there's the can curdled to beat all. What gets into people? I told him not to. Well, I got other things to worry 'bout. Got to get a fire started. I left it ready so I wouldn't have to do nothin' but light it. Lord, it's cold. Don't just sit there, honey. You could be pulling your nose . . ."

Nel sat on the red-velvet sofa listening to her mother but remembering the smell and the tight, tight hug of the woman in yellow who rubbed burned matches over her eyes.

Late that night after the fire was made, the cold supper eaten, the surface dust removed, Nel lay in bed thinking of her trip. She remembered clearly the urine running down and into her stockings until she learned how to squat properly; the disgust on the face of the dead woman and the sound of the funeral drums. It had been an exhilarating trip but a fearful one. She had been frightened of the soldiers' eyes on the train, the black wreath on the door, the custard pudding she believed lurked under her mother's heavy dress, the feel of unknown streets and unknown people. But she had gone on a real trip, and now she was different. She got out of bed and lit the lamp to look in the mirror. There was her face, plain brown eyes, three braids and the nose her mother hated. She looked for a long time and suddenly a shiver ran through her.

"I'm me," she whispered. "Me."

Nel didn't know quite what she meant, but on the other hand she knew exactly what she meant.

"I'm me. I'm not their daughter. I'm not Nel. I'm me. Me."

Each time she said the word *me* there was a gathering in her like power, like joy, like fear. Back in bed with her discovery, she stared out the window at the dark leaves of the horse chestnut.

"Me," she murmured. And then, sinking deeper into the quilts, "I want . . . I want to be . . . wonderful. Oh, Jesus, make me wonderful."

The many experiences of her trip crowded in on her. She slept. It was the last as well as the first time she was ever to leave Medallion.

For days afterward she imagined other trips she would take, alone though, to faraway places. Contemplating them was delicious. Leaving Medallion would be her goal. But that was before she met Sula, the girl she had seen for five years at Garfield Primary but never played with, never knew, because her mother said Sula's mother was sooty. The trip, perhaps, or her new found me-ness, gave her the strength to cultivate a friend in spite of her mother.

When Sula first visited the Wright house, Helene's curdled scorn turned to butter. Her daughter's friend seemed to have none of the mother's slackness. Nel, who regarded the oppressive neatness of her home with dread, felt comfortable in it with Sula, who loved it and would sit on the red-velvet sofa for ten to twenty minutes at a time—still as dawn. As for Nel, she preferred Sula's woolly house, where a pot of something was always cooking on the stove; where the mother, Hannah, never scolded or gave directions; where all sorts of people dropped in; where newspapers were stacked in the hallway, and dirty dishes left for hours at a time in the sink, and where a one-legged grandmother named Eva handed you goobers from deep inside her pockets or read you a dream.

Sister Monroe

Maya Angelou

In the Christian Methodist Episcopal church the children's section was on the right, cater-cornered from the pew that held those ominous women called the Mothers of the Church. In the young people's section the benches were placed close together, and when a child's legs no longer comfortably fitted in the narrow space, it was an indication to the elders that that person could now move into the intermediate area (center church). Bailey and I were allowed to sit with the other children only when there were informal meetings, church socials or the like. But on the Sundays when Reverend Thomas preached, it was ordained that we occupy the first row, called the mourners' bench. I thought we were placed in front because Momma was proud of us, but Bailey assured me that she just wanted to keep her grandchildren under her thumb and eye.

Reverend Thomas took his text from Deuteronomy. And I was stretched between loathing his voice and wanting to listen to the sermon. Deuteronomy was my favorite book in the Bible. The laws were so absolute, so clearly set down, that I knew if a person truly wanted to avoid hell and brimstone, and being roasted forever in the devil's fire, all she had to do was memorize Deuteronomy and follow its teaching, word for word. I also liked the way the word rolled off the tongue.

Bailey and I sat alone on the front bench, the wooden slats pressing hard on our behinds and the backs of our thighs. I would have wriggled just a bit, but each time I looked over at Momma, she seemed to threaten, "Move and I'll tear you up," so, obedient to the unvoiced command, I sat still. The church ladies were warming up behind me with a a few hallelujahs and praise the Lords and Amens, and the preacher hadn't really moved into the meat of the sermon.

It was going to be a hot service.

On my way into church, I saw Sister Monroe, her open-faced gold crown glinting when she opened her mouth to return a neighborly greeting. She lived in the country and couldn't get to church every Sunday, so she made up for her absences by shouting so hard when she did make it that she shook the whole church. As soon as she took her seat, all the ushers would move to her side of the church because it took three women and sometimes a man or two to hold her.

Once she hadn't been to church for a few months (she had taken off to have a child), she got the spirit and started shouting, throwing her arms around and jerking her body, so that the ushers went over to hold her down, but she tore herself away from them and ran up to the pulpit. She stood in front of the altar, shaking like a freshly caught trout. She screamed at Reverend Taylor. "Preach it. I say, preach it." naturally he kept on preaching as if she wasn't standing there telling him what to do. Then she screamed an extremely fierce "I said, preach it" and stepped up on the altar. The Reverend kept on throwing out phrases like home-run balls and Sister Monroe made a quick break and grasped for him. For just a second, everything and everyone in the church except Reverend Taylor and Sister Monroe hung loose like stockings on a washline. Then she caught the minister by the sleeve of his jacket and his coattail, then she rocked him from side to side.

I have to say this for our minister, he never stopped giving us the lesson. The usher board made its way to the pulpit, going up both aisles with a little more haste than is customarily seen in church. Truth to tell, they fairly ran to the minister's aid. Then two of the deacons, in their shiny Sunday suits, joined the ladies in white on the pulpit, and each time they pried Sister Monroe loose from the preacher he took another deep breath and kept on preaching, and Sister Monroe grabbed him in another place, and more firmly. Reverend Taylor was helping his rescuers as much as possible by jumping around when he got a chance. His voice at one point got so low it sounded like a roll of thunder, then Sister Monroe's "Preach it" cut through the roar, and we all wondered (I did, in any case) if it would ever end. Would they go on forever, or get tired out at last like a game of blindman's bluff that lasted too long, with nobody caring who was "it"?

I'll never know what might have happened, because magically the pandemonium spread. The spirit infused Deacon Jackson and Sister Willson, the chairman of the usher board, at the same time. Deacon Jackson, a tall, thin, quiet man, who was also a part-time Sunday school teacher, gave a scream like a falling tree, leaned back on thin air and punched Reverend Taylor on the arm. It must have hurt as much as it caught the Reverend unawares. There was a moment's break in the rolling sounds and Reverend Taylor jerked around surprised, and hauled off and punched Deacon Jackson. In the same second Sister Willson caught his tie, looped it over her fist a few times, and pressed down on him. There wasn't time to laugh or cry before all three of them were down on the floor behind the altar. Their legs spiked out like kindling wood.

Sister Monroe, who had been the cause of the all excitement, walked off the dais, cool and spent, and raised her flinty voice in the hymn, "I came to Jesus, as I was worried, wounded, and sad, I found in Him a resting place and He has made me glad."

The minister took advantage of already being on the floor and asked in the choky little voice if the church would kneel with him to offer a prayer of thanksgiving. He said we had been visited with a mighty spirit, and let the whole church say Amen.

On the next Sunday, he took his text from the eighteenth chapter of the Gospel according to St. Luke, and talked quietly but seriously about the Pharisees, who prayed in the streets so that the public would be impressed with their religious devotion. I doubt that anyone got the message—certainly not those to whom it was directed. The deacon board, however, did appropriate funds for him to buy a new suit. The other was a total loss.

Swaddling Clothes

Translated by Ivan Morris

Yukio Mishima

He was always busy, Toshiko's husband. Even tonight he had to dash off to an appointment, leaving her to go home alone by taxi. But what else could a woman expect when she married an actor—an attractive one? No doubt she had been foolish to hope that he would spend the evening with her. And yet he must have known how she dreaded going back to their house, unhomely with its Western-style furniture and with the bloodstains still showing on the floor.

Toshiko had been oversensitive since girlhood: that was her nature. As the result of constant worrying she never put on weight, and now, an adult woman, she looked more like a transparent picture than a creature of flesh and blood. Her delicacy of spirit was evident to her most casual acquaintance.

Earlier that evening, when she had joined her husband at a night club, she had been shocked to find him entertaining friends with an account of "the incident." Sitting there in his American-style suit, puffing at a cigarette, he had seemed to her almost a stranger.

"It's a fantastic story," he was saying, gesturing flamboyantly as if in an attempt to outweigh the attractions of the dance band. "Here this new nurse for our baby arrives from the employment agency, and the very first thing I notice about her is her stomach. It's enormous—as if she had a pillow stuck under her kimono! No wonder, I thought, for I soon saw that she could eat more than the rest of us put together. She polished off the contents of our rice bin like that. . . ." He snapped his fingers. "'Gastric dilation'—that's how she explained her girth and her appetite. Well, the day before yesterday we heard groans and moans coming from the nursery. We rushed in and found her squatting on the floor, holding her stomach in her two hands, and moaning like a cow. Next to her our baby lay in his cot, scared out of his wits and crying at the top of his lungs. A pretty scene, I can tell you!"

"So the cat was out of the bag?" suggested one of their friends, a film actor like Toshiko's husband.

"Indeed it was! And it gave me the shock of my life. You see, I'd completely swallowed that story about 'gastric dilation.' Well, I didn't waste any time. I rescued our good rug from the floor and spread a blanket for her to lie on. The whole time the girl was yelling like a stuck pig. By the time the doctor from the maternity clinic arrived, the baby had already been born. But our sitting room was a pretty shambles!"

"Oh, that I'm sure of!" said another of their friends, and the whole company burst into laughter.

Toshiko was dumbfounded to hear her husband discussing the horrifying happening as though it were no more than an amusing incident which they chanced to have witnessed. She shut her eyes for a moment and all at once she saw the newborn baby lying before her: on the parquet floor the infant lay, and his frail body was wrapped in bloodstained newspapers.

Toshiko was sure that the doctor had done the whole thing out of spite. As if to emphasize his scorn for this mother who had given birth to a bastard under such sordid conditions, he had told his assistant to wrap the baby in some loose newspapers, rather than proper swaddling. This callous treatment of the newborn child had offended Toshiko. Overcoming her disgust at the entire scene, she had fetched a brand-new piece of flannel from her cupboard and, having swaddled the baby in it, had lain him carefully in an armchair.

This all had taken place in the evening after her husband had left the house. Toshiko had told him nothing of it, fearing that he would think her oversoft, oversentimental; yet the scene had engraved itself deeply in her mind. Tonight she sat silently thinking back on it, while the jazz orchestra brayed and her husband chatted cheerfully with his friends. She knew that she would never forget the sight of the baby, wrapped in stained newspapers and lying on the floor—it was a scene fit for a butchershop. Toshiko, whose own life had been spent in solid comfort, poignantly felt the wretchedness of the illegitimate baby.

I am the only person to have witnessed its shame, the thought occurred to her. The mother never saw her child lying there in its newspaper wrappings, and the baby itself of course didn't know. I alone shall have to preserve that terrible scene in my memory. When the baby grows up and wants to find out about his birth, there will be no one to tell him, so long as I preserve silence. How strange that I should have this feeling of guilt! After all, it was I who took him up from the floor, swathed him properly in flannel, and laid him down to sleep in the armchair.

They left the night club and Toshiko stepped into the taxi that her husband had called for her. "Take this lady to Ushigomé," he told the driver and shut the door from the outside. Toshiko gazed through the window at her husband's smiling face and noticed his strong, white teeth. Then she leaned back in the seat, oppressed by the knowledge that their life together was in some way too easy, too painless. It would have been difficult for her to put her thoughts into words. Through the rear window of the taxi she took a last look at her husband. He was striding along the street toward his Nash car, and soon the back of his rather garish tweed coat had blended with the figures of the passers-by.

The taxi drove off, passed down a street dotted with bars and then by a theater, in front of which the throngs of people jostled each other on the pavement. Although the performance had only just ended, the lights had already been turned out and in the half dark outside it was depressingly obvious that the cherry blossoms decorating the front of the theater were merely scraps of white paper.

Even if that baby should grow up in ignorance of the secret of his birth, he can never become a respectable citizen, reflected Toshiko, pursuing the same train of thoughts. Those soiled newspaper swaddling clothes will be the symbol of his entire life. But why should I keep worrying about him so much? Is it because I feel uneasy about the future of my own child? Say twenty years from now, when our boy will have grown up into a fine, carefully educated young man, one day by a quirk of fate he meets the other boy, who then will also have turned twenty. And say that the other boy, who has been sinned against, savagely stabs him with a knife. . . .

It was a warm, overcast April night, but thoughts of the future made Toshiko feel cold and miserable. She shivered on the back seat of the car.

No, when the time comes I shall take my son's place, she told herself suddenly. Twenty years from now I shall be forty-three. I shall go to that young man and tell him straight out about everything—about his newspaper swaddling clothes, and about how I went and wrapped him in flannel.

The taxi ran along the dark wide road that was bordered by the park and by the Imperial Palace moat. In the distance Toshiko noticed the pinpricks of light which came from the blocks of tall office buildings.

Twenty years from now that wretched child will be in utter misery. He will be living a desolate, hopeless, poverty-stricken existences—a lonely rat. What else could happen to a baby who has had such a birth? He'll be wandering through the streets by himself, cursing his father, loathing his mother.

No doubt Toshiko derived a certain satisfaction from her somber thoughts: she tortured herself with them without cease. The taxi approached Hanzomon and drove past the compound of the British Embassy. At that point the famous rows of cherry trees were spread out before Toshiko in all their purity. On the spur of the moment she decided to go and view the blossoms by herself in the dark night. It was a strange decision for a timid and unadventurous young woman, but then she was in a strange state of mind and she dreaded the return home. That evening all sorts of unsettling fancies had burst open in her mind.

She crossed the wide street—a slim, solitary figure in the darkness. As a rule when she walked in the traffic Toshiko used to cling fearfully to her companion, but tonight she darted alone between the cars and a moment later had reached the long narrow park that borders the Palace moat. Chidorigafuchi, it is called—the Abyss of the Thousand Birds.

Tonight the whole park had become a grove of blossoming cherry trees. Under the calm cloudy sky the blossoms formed a mass of solid whiteness. The paper lanterns that hung from wires between the trees had been put out; in their place electric light bulbs, red, yellow, and green, shone dully beneath the blossoms. It was well past ten o'clock and most of the flower-viewers had gone home. As the occasional passers-by strolled through the park, they would automatically kick aside the empty bottles or crush the waste paper beneath their feet.

Newspapers, thought Toshiko, her mind going back once again to those happenings. Bloodstained newspapers. If a man were ever to hear of that piteous birth and know that it was he who had lain there, it would ruin his entire life. To think that I, a perfect stranger, should from now on have to keep such a secret—the secret of a man's whole existence. . . .

Lost in these thoughts, Toshiko walked on through the park. Most of the people still remaining there were quiet couples; no one paid her any attention. She noticed two people sitting on a stone bench beside the moat, not looking at the blossoms, but gazing silently at the water. Pitch black it was, and swathed in heavy shadows. Beyond the moat the somber forest of the Imperial Palace blocked her view. The trees reached up, to form a solid dark mass against the night sky. Toshiko walked slowly along the path beneath the blossoms hanging heavily overhead.

On a stone bench, slightly apart from the others, she noticed a pale object—not, as she had at first imagined, a pile of cherry blossoms, nor a garment forgotten by one of the visitors to the park. Only when she came closer did she see that it was a human form lying on the bench Was it, she wondered, one of those miserable drunks often to be seen sleeping in public places? Obviously not, for the body had been systematically covered with newspapers, and it was the whiteness of those papers that had attracted Toshiko's attention. Standing by the bench, she gazed down at the sleeping figure.

It was a man in a brown jersey who lay there, curled up on layers of newspapers, other newspapers covering him. No doubt this had become his normal night residence now that spring had arrived. Toshiko gazed down at the man's dirty, unkempt hair, which in places had become hopelessly matted. As she observed the sleeping figure wrapped in its newspapers, she was inevitably reminded of the baby who had lain on the floor in its wretched swaddling clothes. The shoulder of the man's jersey rose and fell in the darkness in time with his heavy breathing.

It seemed to Toshiko that all her fears and premonitions had suddenly taken concrete form. In the darkness the man's pale forehead stood out, and it was a young forehead, though carved with the wrinkles of long poverty and hardship. His khaki trousers had been slightly pulled up; on his sockless feet he wore a pair of battered gym shoes. She could not see his face and suddenly had an overmastering desire to get one glimpse of it.

She walked to the head of the bench and looked down. The man's head was half buried in his arms, but Toshiko could see that he was surprisingly young. She noticed the thick eyebrows and the fine bridge of his nose. His slightly open mouth was alive with youth.

But Toshiko had approached too close. In the silent night the newspaper bedding rustled, and abruptly the man opened his eyes. Seeing the young woman standing directly beside him, he raised himself with a jerk, and his eyes lit up. A second later a powerful hand reached out and seized Toshiko by her slender wrist.

She did not feel in the least afraid and made no effort to free herself. In a flash the thought had struck her. Ah, so the twenty years have already gone by! The forest of the Imperial Palace was pitch dark and utterly silent.